Digital Design with
Standard MSI and LSI

Digital Design with Standard MSI and LSI

DESIGN TECHNIQUES
FOR THE MICROCOMPUTER AGE

SECOND EDITION

THOMAS R. BLAKESLEE

Vice President-Engineering
Logisticon Inc, Sunnyvale Calif

A WILEY-INTERSCIENCE PUBLICATION

JOHN WILEY & SONS

New York • Chichester • Brisbane • Toronto

Library of Congress Cataloging in Publication Data:

Blakeslee, Thomas R 1937–
    Digital design with standard MSI and LSI.

    "A Wiley-Interscience publication."
    Includes bibliographies and index.
    1. Logic circuits. 2. Digital integrated
circuits. I. Title.

TK7868.L6B55   1979   621.3819'58'35   78-24201
ISBN 0-471-05222-1

Printed in the United States of America

10 9 8 7 6 5 4 3

Preface

The integrated circuit revolution has rendered many of the traditional goals of digital design obsolete. Since a single inexpensive IC can now contain thousands of components, minimization of gates and flip-flops is no longer the primary goal of design.

In the first edition of this book I attempted to develop a complete design approach which would reflect these new realities. Developments in the past few years have confirmed the validity of the strategies I presented. In this edition I have updated and expanded the book to include some of the dramatic new developments in the area of microcomputers.

As important as microcomputers are, they are only one of the many tools that must be understood by the working digital designer. I have tried to bring together in this book a balanced collection of design tools which will continue to be useful as the IC revolution develops.

Since the specific details of microcomputers will continue to change rapidly for many years, I have concentrated on *basic concepts* and philosophies as much as possible. Many present-day books on microcomputers attempt to survey the operating *details* of a large number of specific microprocessors. Since these details tend to become obsolete every two years, they are more economically available from the IC manufacturers. I would therefore recommend that this book be supplemented with at least one up-to-date microcomputer users' manual or IC data catalog.

This book emphasizes system design, because the LSI and MSI building blocks are really system components, that can be traded off with other mechanical and electrical components for an optimum overall result. Since the cost of the logic itself is becoming a negligible part of total system cost, it is increasingly important to attack the system problem as a whole. Though the book concentrates on the use of *standard* LSI and MSI products, most of the principles are equally valid where product volume makes the use of custom LSI circuits practical.

Briefly, the design technique centers around using standardardized "bargain components" to handle most system requirements, then filling in, as needed, with small-scale ICs. The primary aim is *minimizing IC package count,* rather than the traditional minimization of the number of flip-flops and gate inputs. Many practical considerations normally ignored by logic texts are also included, because ignorance of these problems causes consequences far more disastrous than just wasting a few gates.

I have tried to make the book useful to working professionals and students alike by covering basic concepts, terminology, and abbreviations in a glossary rather than in the text. In this way the professional and the advanced student do not have to be bored with concepts they already know, yet the beginner can learn the terms, as needed, by looking them up in the glossary.

The first two chapters establish and justify by example the design philosophy used throughout the remainder of the book. While the emphasis in Chapters 3 to 5 is on using MSI and LSI circuits where possible, the basic techniques of traditional logic design are also presented since they are still needed to "fill in the cracks" between MSI and LSI circuits and for very small subsystems. An intuitive state assignment method is presented that allows minimizing the total logic without obscuring system operation with mathematical abstractions.

Chapters 7 through 10 cover the programming concept, programming languages, microcomputer development systems, and microcomputer hardware design techniques.

Chapter 11 discusses the time dimension as a method of making multiple use of circuits for bit serial or time-shared operation. Since circuit speeds are increasing and LSI circuits are interconnection limited, this is becoming an increasingly important technique.

Chapters 6 and 12 consider the nasty realities of the real world such as race conditions, hangup states, noise, reflections, and cross talk. Chapter 13 covers some of the techniques, such as negative feedback and incremental operation, that are used to simplify modern input/output devices. Several successful real computer peripheral devices are described as examples. Chapter 14 discusses, from a strictly practical point of view, the use of statistics in reliability and buffer overflow calculations. Finally, Chapter 15 discusses the social consequences of engineering and the possibilities of using the fantastic capabilities of ICs to improve the quality of life.

Thomas R. Blakeslee

Woodside, California
February 1979

Contents

Digital Design with
Standard MSI and LSI

CHAPTER ONE

PHILOSOPHY
Adapting the Job to the Bargain Components

When we consider the incredible range of seemingly unrelated tasks being done by **digital*** **integrated circuits** (ICs), it seems that digital logic is, indeed, about to take over the world. Since most of these jobs are not basically digital or even electrical, why are they done so much more efficiently with digital logic? By exploring the answers to this question, we can gain an under-standing of the real reasons for the tremendous savings offered by the digital approach and thereby learn how to fully exploit its potential.

1.1 STANDARDIZATION

Standardization is the key to the fantastic material wealth we have today. When a large number of identical (or nearly identical) items are manufactured on a production line, tremendous savings are possible. A large production volume makes it feasible to invest heavily in special production equipment and set up an efficient production line. The gulf between the bargains available in high-volume standard products and the cost of things that cannot be produced

---

* Boldface type indicates the first use of a term in the Glossary.

in volume has grown steadily since Henry Ford started the trend. Today you can buy a complete typewriter, with all its intricate parts, for less than it would cost to have one of the key levers made in a machine shop. Think of what it would cost to have the hundreds of intricate parts in a $10 wristwatch custom made by a machinist!

The "batch" process used to manufacture ICs is perhaps the ultimate example of this kind of efficient production. Thousands of repetitions of the same circuit are produced on a single **wafer** of silicon only 3 in. in diameter (see Fig. 1-1). Each time a step in the manufacturing process is done, it is done for thousands of circuits in a single operation. This means that ICs can be made *very* economically, *but only if we can use a large volume of identical circuits.* Except for the final packaging, the labor required to make one circuit is the same as that to make a thousand! Of course, the processing of the wafers themselves is most efficient if done on a high-volume production line basis. If every wafer produced contains 1000 circuits, just 100 wafers a day gives us 100,000 integrated circuits! It is obvious, then, that we must use an IC type in tremendous volume to reap the full benefits of this production technique. If we are making a portable radio, a television set, or a pocket calculator, this is no problem—we can just design a custom circuit and the volume will be high simply because the volume of our product is high.

1.2 STANDARD DIGITAL CIRCUITS

Most digital systems, unfortunately, are produced in only moderate quantities. The only way we can really reap the benefits of high-volume IC production is if the same basic integrated circuits can be used in many *different* systems throughout the industry. This is a reality today only because of a very special quality possessed by **binary** logic.

The binary number system is the simplest possible because digits can have only one of two values: 1 or 0. Because of this, the number of possible combinations is quite restricted. For example, the multiplication tables, which we spent years memorizing in the decimal (10 valued) number system, are trivial in the binary system.* The whole system of binary (Boolean) algebra can be developed, with proofs, in a few pages (see Refs. 1 and 2). It is this simplicity that has made standardization of digital ICs possible. It is actually possible to build *any* logic system, including a large computer, entirely from a single **gate** circuit type and a single **flip-flop** circuit type.† In the early days of ICs this is exactly what was done. By building the entire computer out of NAND gates

---

* The complete multiplication tables are $1 \times 1 = 1$, $1 \times 0 = 0$, and $0 \times 0 = 0$. Addition is equally simple: $1 + 1 = 0$ (and carry 1), $1 + 0 = 1$, and $0 + 0 = 0$.

† Actually, it *could* be done with a single type of gated flip-flop, but this is somewhat wasteful.

(a)

(b)

Fig. 1-1. The key to IC economy: 30 wafers, each capable of producing hundreds of ICs, can be processed simultaneously (Applied Materials AMG-500 Reactor System).

and *J-K* flip-flops, quite a sizable volume of these two circuit types could be consumed by a single company making only 100 or so computers.

This ultimate in standardization is practical only with digital logic. High-volume mechanical parts can be made very economically, as the $5 alarm clock proves, but too many variations are possible in mechanical components to achieve the kind of standardization we now have in digital ICs. Clock gears, for example, can have any number of teeth and be any of an infinite number of sizes. Catalogs of standard gears do exist, but they have pages and pages of tables of different gear sizes and numbers of teeth. With such a large selection, it is impossible to produce standard gear components for general use in anywhere near the required volume.

With digital ICs standardization is easy. The logic equivalent of the speed-reducing gear train in a clock is a chain of identical, standard, flip-flops—each of which reduces the speed by a factor of 2. These flip-flops are identical to the ones used in a computer, a tape unit, or any other logic device. For this reason, it is quite practical to build just one clock, out of standard digital components (in fact many hobbyists have done it), but it would take thousands of dollars worth of custom-machines parts to make just one mechanical clock.

We thus have the key to the digital logic takeover of the world: *standardized bargain components*. The advantages of standardization are great even when the job does not fit the components. The example of the clock illustrates this. Since we ultimately want a mechanical representation of the time, it certainly seems logical to make the clock mechanical in the first place. Although the standard flip-flops are nice, they produce nothing but electrical 1 and 0 outputs, which then have to be decoded and converted into something we can see. This is a lot of extra trouble, but the result is much less expensive.

The tremendous savings from using the standardized components more than offset the inefficiency of adapting the components to the application. This is not an exceptional case. As a matter of fact, the gulf between the capability per dollar of ICs and any other approach makes it almost always true that, if we can somehow adapt the job to these components, the result will be more economical.

Now that we know why digital logic is taking over so many nondigital jobs, we can generalize this principle into a general design technique and slogan: *"Fit the job to the bargain components."* All that is required for this technique to work is that the inefficiency in adapting to the bargain component is less than the advantage offered by its use. In the case of integrated circuits versus mechanical components the gulf is very wide indeed. A 100:1 cost advantage is not unusual where the quantities are small. This means that we would be ahead if our efficiency is greater than 1%! Since the technology gap between mechanical parts and ICs is growing all the time, this design technique, which is quite effective now, should prove even more effective in the future.

This technique is by no means limited to the **systems design** level. There are also 100:1 cost differences between different ICs doing the same job (e.g., between discrete flip-flops and random access memory ICs or between **gates** and **read-only memories**). The same principle of trying to use the bargain components also works in **logic design**.

These 100:1 ratios between the cost of mechanical components and the cost of ICs performing analogous functions may seem a little hard to accept. The IC revolution has happened so quickly that appreciation of its true impact, and the art of making full use of its capabilities, has lagged far behind.

If we consider the rate of development of integrated circuit technology between 1959 and 1977, it is easy to see why ICs have surpassed all other technologies so quickly. Although the technologies for making mechanical components have evolved only gradually, IC technology has had an explosive, exponential growth. As Figure 1-2 shows, the number of components that can be squeezed on an IC chip has doubled every year during the 18-year period from 1959 to 1977. More important, this trend seems certain to continue for many more years! If we correct for inflation we find that the cost of a state-of-the-art IC chip has remained constant since 1959. The net result is that *the cost per component in IC's has decreased by a factor of a million in 20 years!*

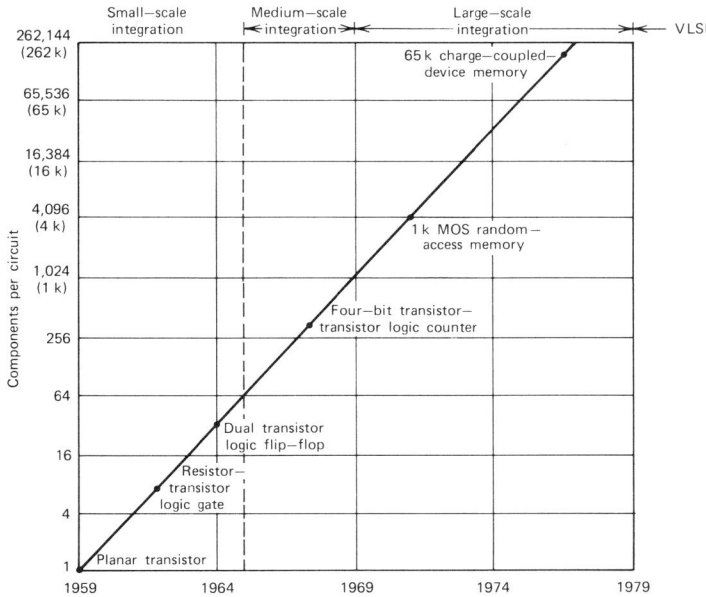

Fig. 1-2. Number of components per chip in state-of-the-art ICs (from Noyce, 1977).

1.3 SOME EXAMPLES OF THE METHOD

Since the relationship between the bargain component and the item it replaces is often very abstract, let us illustrate the principle with a few examples.

THE DESK CALCULATOR.   The desk calculator has a mechanical input (keyboard) and mechanical output (visible numbers). For years the job was done by an incredible collection of gears, levers, and springs. In the late 1960s somebody started making them with digital ICs. Though it was necessary to convert from mechanical to binary coded electrical signals, then back to visual, with **nixie tubes,** the result was better in every way.

Although the **small-scale integrated (SSI)** version represented an adaptation to a bargain component on the system design level, an evolution continued at the logic design level with the introduction of still larger scale ICs. In 1971 the ultimate version was introduced—a single IC that does everything.

PULSE-CODE-MODULATION TELEPHONY.   In this system 24 voice signals are sent over a single telephone line. The signals are sampled and converted to digital codes, which are sent serially over the line. At the other end they are converted back to 24 **analog** signals and reconstructed. In spite of all this conversion back and forth, the digital system is actually more economical than equivalent analog systems. In a sense, we are trading off a mechanical component (the 23 lines eliminated) for digital logic. This system is described in more detail in Section 13.11.

A STATUS BOARD.   A status board of 1200 lamps arranged in a 40×30 matrix indicates the status of 1200 points in a missile system. Although 1200 light bulbs driven by 1200 flip-flops and 1200 lamp drivers could be used, the job is perfectly suited to a combination of two bargain components: a TV monitor and 1200 **bits** of MOS **shift register.** As the beam scans across the face of the TV set, the MOS shift register produces a signal that unblanks the beam if a "1" status is stored for the point corresponding to the beam position. We thus obtain a 40×30 array of bright or dark spots on the TV screen.

As the preceding examples show, the relationship between the bargain component and the item it replaces is not always obvious. For instance, in the second example, 23 telephone lines are traded for some digital logic circuits. Although it is certainly impossible to generalize about how many ICs equal one telephone line, it can be said that logic built with ICs is so economical that if it can be used to replace an expensive "component" like a telephone line, it will probably save money. In other words, ICs are generally called "bargain components" and telephone lines are generally *not* a bargain.

1.4 OTHER ADVANTAGES

Up to this point, we have mainly discussed cost, but standardization has many other advantages that are often more important. A standard component is made in great volume, used by many people, and manufactured by several sources. As a result, *design flaws, reliability problems, and manufacturing difficulties are worked out* by the initial users of the component. Although we must solve these problems for ourselves on any custom design, with established standard components they have been worked out already. The design cycle is greatly shortened with standard components because the components can be designed in immediately from information on the specification sheet. Since the parts are usually in stock at distributors, we can often have them in hand before the design is finished if they are ordered early in the design phase. Since several distributors stock the part, it is unlikely that production will stop due to parts shortages. Finally, servicing the final product in the field is greatly simplified because the standard parts are stocked by distributors in cities throughout the world. Replacement parts can simply be purchased locally by the service technician.

While all of the above benefits are available with hundreds of "bargain component" ICs, very few mechanical components are so standardized. We can, therefore, rate almost all mechanical components as "definitely not bargains." In the initial design of a system, then, we try to reduce the mechanical components to a minimum—even if this means greatly increasing the electronic complexity.

Though almost all ICs are bargains compared to other types of components, some stand out above the others as real bargains. *We can thus apply our rule of "adapting the job to the bargain components" at the logic design level also.* The idea is to try to use bargain **large-scale integrated (LSI)** or **medium-scale integrated (MSI)** circuits for the bulk of the logic *job to be done*, then use SSI circuits as necessary to "hold everything together." Because the bargain LSI components offer so much more capability for the money, they can be used quite inefficiently and still produce a net savings. As before, we often have to adapt a lot to fit in the bargain component. A large part of this book is devoted to showing just how this is done.

As a general rule, the more complex an IC is, the bigger bargain it is. The reason for this is obvious if we remember the batch fabrication process shown in Figure 1-1. Most of the labor cost in an IC is in the final packaging and testing steps that must be done on the individual IC chips. Because 1000 or so ICs are processed together on one 4-in. wafer, the cost per IC on the wafer is almost negligible. A complex LSI circuits may require a larger area chip, but *even with the largest chips, one wafer produces hundreds of ICs*. The cost of producing LSI and MSI circuits is therefore not much greater than that of SSI circuits. Equally exciting is the fact, that, since most IC failures are package or interconnection failures, the failure rate of LSI circuits is almost the same as that of small-scale circuits!

The selling price of LSI circuits is currently much higher than that of small-scale circuits. Part of the reason for this is temporary. New processes are often used in LSI circuits so the development costs of that process must be paid off, and many more parts must be scrapped because the manufacturing process is so new. But there is another fundamental reason for the higher cost: the more complex a component is the less general it is, so *standardization is difficult* and it is impossible to reach truly high volume production. It is easy to standardize nuts and bolts, but not so easy, for example, to standardize pistons.

This problem of standardization for volume production is what makes the concept of using bargain LSI and MSI products so important. Certain products catch on and become "industry standards." The circuits are always made by many manufacturers so competition keeps the price at rock bottom. The low price, however, causes more and more designers to use them in their systems, causing the volume to increase further. As they become better and better "bargains," it becomes more and more desirable to adapt them into systems, even where they do not quite fit, so the volume continues to increase and the price continues to fall.

We can thus see that the principle of "adapting the job to the bargain components" is a natural force that tends to cause "de facto standardization" of certain components without any central control. This is a very healthy situation because, of all the countless new ICs announced, this stampede usually occurs on only the best one of a type—the rest tend to stay expensive and gradually disappear. The standard part, once established, stays on top until something significantly better comes along. Two excellent examples of this de facto standardization are Texas Instruments' 7400 series **TTL** family and Mostek's 4116, 16 k bit **random access memory (RAM)**.

1.5 CHOOSING COMPONENTS FOR FUTURE COST

The IC field changes so fast that designs are often obsolete within 2 years. Since it often takes a year or two to really build up production of a product using ICs, a logic system designer must learn to anticipate the future. In this way he can minimize the *total* parts cost over the life of the product rather than their cost at the time of design.

Figure 1-3 shows the price trends in two different families of digital logic. In 1969 one might have been tempted to choose **RTL** logic for economy, even though TTL performed much better. It is true that there is no need to pay for

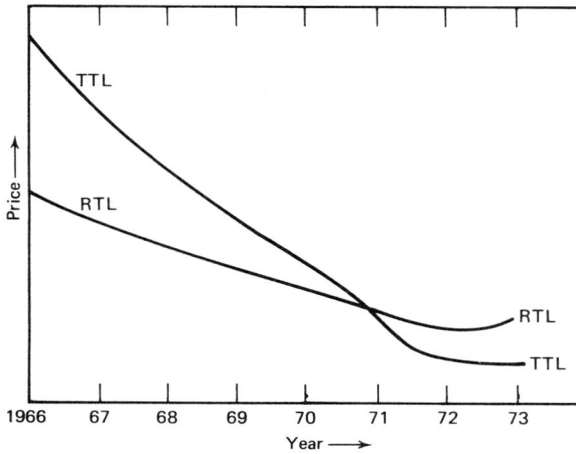

Fig. 1-3.   Prices of RTL and TTL gates, 1966–1973.

performance you do not need, but a little knowledge of the normal life cycle of components and what is inside the ICs would have made a wise designer choose TTL. As Figure 1-3 shows, a choice of RTL would be a mistake on all counts. Not shown is the fact that none of the MSI and LSI products available today will work directly with RTL, so a changeover would have to be made to take advantage of MSI on new products. This would cause a very messy double standard in the company, making it necessary to stock both RTL and TTL in manufacturing and field service spares as long as the RTL product was still around.

If we look at the normal life cycle of an IC (Fig. 1-4), we see that when the

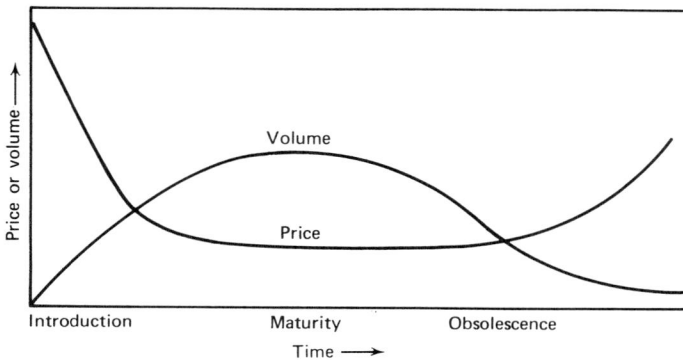

Fig. 1-4.   IC product life cycle.

product is first introduced, the price is quite high because, of course, the volume is zero. The price at this point has little relation to the cost to produce the part but is more a matter of strategy. If the component performs better than anything else available, it is not unusual to ask a *very* high price for it until competition develops; there are always designers who need higher performance and will pay almost any price for it. A large-quantity order at this point may go for one-fifth of the catalog price.

If the circuit is unsuccessful, no volume develops, no competition develops, and it dies. A successful product, however, soon is "second sourced" by other manufacturers. It thus enters a phase where competition forces the price down to something related to the actual manufacturing cost. As volume increases, price decreases. This is partly due to increased volume and partly due to the "learning curve."

The learning curve (Fig. 1-5) is simply a decrease in manufacturing cost as more experience is accumulated from actually shipping the product. There is a learning curve for each circuit and another one for the process in general. Thus a process (like bipolar, P-channel MOS or N-channel MOS) also has a life cycle like the one shown in Figure 1-4. The speed with which the price of an individual circuit falls is related to the degree of development of the process used to manufacture it. The rate of decrease in price is always greater for a new part or process than for an older, more mature one.

After about 2 years of steadily increasing volume (Fig. 1-4) and steadily decreasing prices, a point is reached (maturity) where another, clearly superior product becomes available. From this point on, the price stops falling, but the volume continues to increase for a couple more years. The volume grows

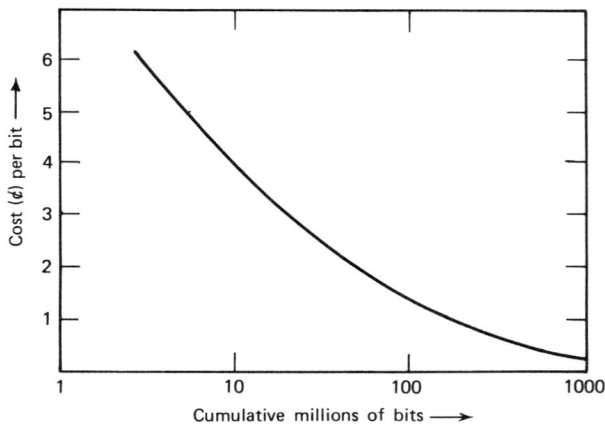

Fig. 1-5. Typical price learning curve—long shift register.

mainly because of continuing shipments of products using the circuit. There is no need to lower the price further because nobody will use the part in new designs anyway, and those who have it designed in already are stuck with it. At this point the part is "past its prime." Finally, the part becomes obsolete; the volume starts to fall, and eventually prices actually rise a little as the volume falls and manufacturers drop out of the market.

The case of the 1969 TTL versus RTL decision of Figure 1-3 now becomes quite clear. At that time RTL had almost reached "maturity" while TTL was just starting out. Since the actual chip size for TTL was smaller than for RTL and the process was virtually the same, the manufacturing cost of TTL was potentially lower. Moreover, it was well known in 1968 that MSI was on its way and would certainly be compatible with the new logic family when it came. It would therefore have been much cheaper, *over the product life of the system* in question, if it had been designed with TTL—even though TTL was more expensive at the time!

Figure 1-6 shows a similar example. Here the contest is between designing-in a core memory and using MOS IC memory. If the decision was made in 1971, it might seem that core was the more economical choice. However, since core was a "mature" product and MOS was just beginning, it is clear that MOS memory was the better choice. Here again, looking inside the component is useful. A core memory is quite complicated and involves many parts and technologies. IC memory, on the other hand, is made by a process that is actually simpler than the one used to make TTL ICs (which were

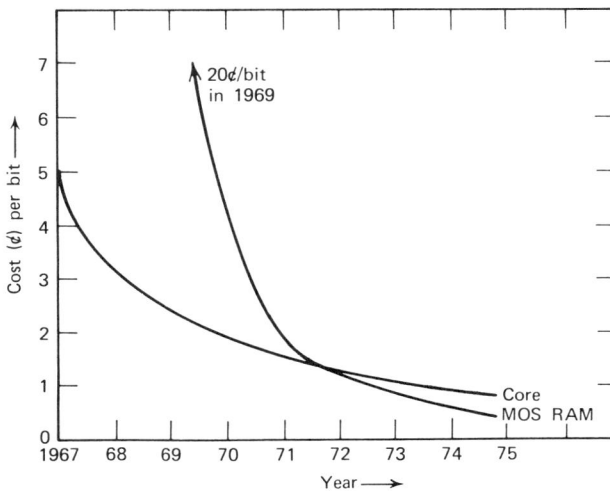

Fig. 1-6. Memory cost (high volume), MOS versus core.

selling for 15 to 20¢ each at that time). The only difference was that the MOS process was in an earlier stage of the learning curve development. Core memory was already a fairly mature product by 1971. In fact, most of the decrease in cost per bit between 1967 and 1970 was actually due to integrated circuits! IC core drivers and **sense amplifiers** reduced the cost of core memories to only slightly more than the cost of the core planes themselves.

1.6   FINDING BARGAIN COMPONENTS

Designers of truly high-volume products like pocket calculators, TV sets, and computer memories can specify components that fit their requirements exactly. The great majority of actual design work, however, is for systems made in only moderate volume. The designer of this type of system must learn to find existing bargain components and adapt them to his needs. *One of the best sources of real bargains is components actually designed for high volume products.*

A pocket calculator, for example, has a very powerful LSI circuit and numeric display that can often be adapted to other applications. By just simulating key operations with other logic, we can perform quite a bit of logic and storage and display the result for a very low price. The keyboard in a calculator, because it is made in such volume, is also a bargain. If we can adapt our design so we can use exactly the same key layout as a calculator uses, we can often buy the keyboard for half of what a custom-designed keyboard would cost. We effectively get a "free-ride" on the volume generated by the application for which it was designed. Besides saving money, we also get the reliability and availability advantages of a standard, volume component.

REFERENCES

1.   M. Phister, *Logical Design of Digital Computers,* Wiley, New York, 1958, Chapter 3.
2.   George Boole, *An Investigation of the Laws of Thought,* Dover, New York, 1954.

BIBLIOGRAPHY

The Manufacture of ICs

Carr, W. N., and Mize, J. P., *MOS/LSI Design and Application,* McGraw-Hill, New York, 1972 Chapters 1, 2, and 10.

Hibbard, Robert, *Integrated Circuits—A Basic Course for Engineers and Technicians,* McGraw-Hill, New York, 1969.

Development of IC Technology

Graham, Robert F., "Semiconductor Memories: Evolution or Revolution," *Datamation,* June 1969.

Martino, J., "Tools for Looking Ahead," *IEEE Spectrum,* October 1972, pp. 32–40.

Moore, Gordon E., "What Level of LSI is Best for You?" *Electronics,* February 16, 1970, pp. 126–129.

Mudge, J., and K. Taft, "V-ATE Memory Scores a New High in Combining Speed and Bit Density," *Electronics,* July 17, 1972, pp. 65–69.

Noyce, Robert, "Microelectronics," *Scientific American,* September 1977, pp. 62–69.

Special Issue: "The Great Takeover" (of IC Technology), *Electronics,* October 25, 1973, pp. 68–191.

Pulse-Code-Modulation Telephony

Davis, C. G., "An Experimental Time Multiplex Terminal," *Bell System Technical Journal,* January 1962.

CHAPTER TWO

THE GOALS
Of Digital System Design

One of the most surprising exercises we can do in the analysis of costs in a logic system is to compare total system cost to total IC cost in an actual system. The cost of ICs has fallen so low that *the integrated circuit cost is now less than 10% of the total system cost!* This percentage varies somewhat depending on the production volume of the system (it is much lower for one-of-a-kind systems). But the unmistakable trend is for IC cost to become increasingly *a negligible part* of total system cost. The same trend is found in failure reports—more and more, ICs are becoming a negligible source of system failures.

The reason for this trend is simple. Although ICs have improved their cost and reliability at a fantastic rate, the mechanical and packaging aspects of the system have improved only slightly. There has been no revolution in the power supplies, cooling and packaging concepts—only a relatively slow evolution. The situation is quite similar to the case of the supersonic airliner. We can now fly between London and New York in about the same time as it takes to drive to the airport, park, check in, and wait for baggage on the ground. The reason is the same—spectacular development of the aircraft without comparable improvements in ground facilities.

This similarity between IC and airliner development is a similarity in kind only. The rate at which ICs have developed is unique. The cost per function of ICs has improved by a factor of 200,000 in the 18-year period between 1959 and 1977. If aviation had developed that quickly, the New York–London airfare would now be a fraction of a cent! If the speed of airliners had kept pace with IC speed improvements, we would be able to fly from New York to London in less than a minute.

With this unprecedented rate of development, it is easy to see why IC technology has so far surpassed our ability to use it. Already we find ourselves in the ridiculous position where the cost of the ICs in a system is becoming a negligible part of the total cost. Furthermore, this trend of *halving the cost per function every year* seems to be continuing unabated.

To the modern logic system designer, this idea of negligible IC cost is very important because *it invalidates the traditional goals of logic design*. Until now, logic design has been a problem of mathematically simplifying the logic to minimize the number of flip-flops and gate inputs. If the cost and reliability impact of these circuits is negligible, then we must develop more valid design goals based on minimizing the real sources of cost and failure in modern logic systems.

Before proceeding with a cost analysis of a typical system let us first describe the typical modular packaging system on which the analysis will be based.

2.1 MODULAR PACKAGING

Since the days of discrete transistor logic, the dominant system for packaging large logic systems has been the "card file" system shown in Figure 2-1. This system is still the most economical today, though LSI is making more and more systems fit on a single **printed circuit (PC) board**. The reason for the superiority of this system is **standardization**. Any size system can be built up from the same basic mechanical parts.

The basic unit is a PC board with a connector on one edge. These PC boards hold the ICs and other components and interconnect them with a special, photographically produced pattern of copper interconnections. Although the circuit pattern is special for each board, the board size and shape, connector mounting, and so on, are standardized. This standardization makes possible many economies in the manufacture and testing of the board. The board is an optimum size subsystem for handling and testing. System servicing can often be done by replacing one of these boards. In more and more systems today a single PC board is all that is required. To build larger systems we work our way up the hierarchy of Figure 2-1 as far as required by the system size.

The next system level is variously called a **subrack**, card file, or card cage. This is a sheet metal assembly with connector sockets and mechanical guides for many PC boards. Each subrack holds up to 32 PC boards. The interconnection of the PC sockets is usually done with **wire wrap**, except for the

IC Package   Basic unit

PC Board   Up to 60 ICs

Subrack
(card file)   Up to 30 PC boards

Subrack group   Up to three subracks
                                              (bolted together)

Rack (cabinet)   Power, cooling, and up to two subrack groups
                                              (various size options)

Group of racks   Up to four racks

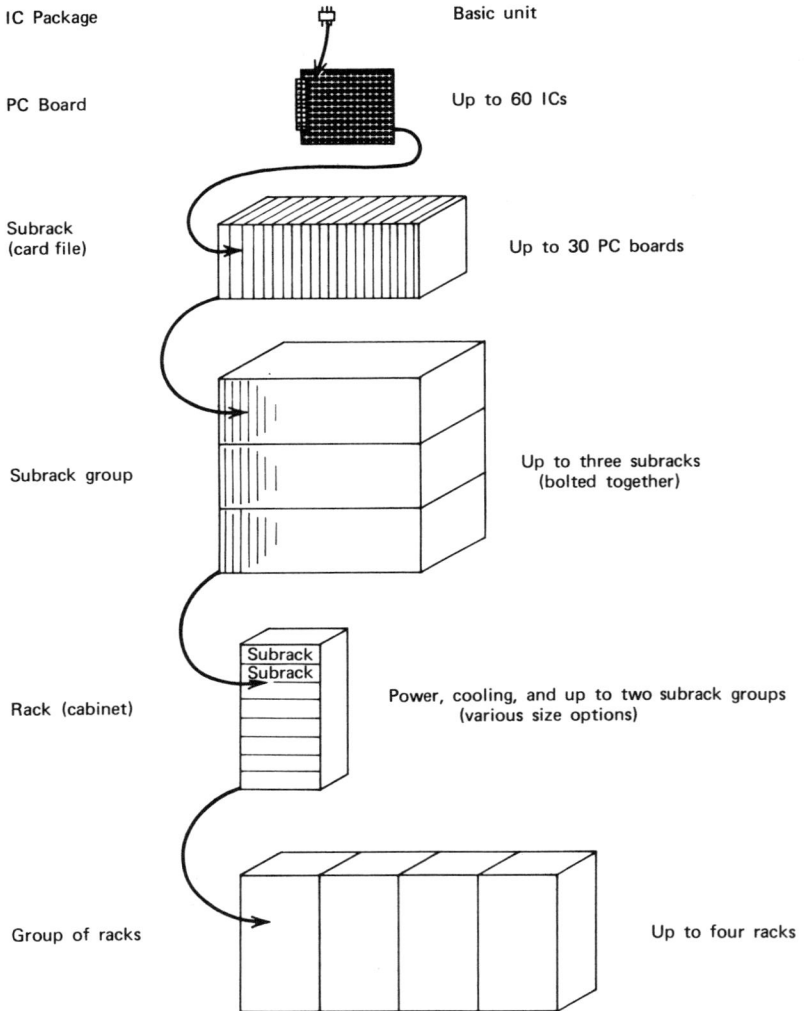

Fig. 2-1.   Modular card file packaging.

ground and DC power distribution, which is done by a PC ground plane
"mother board." Sometimes all or part of the signal interconnections are done
by a **multilayer** mother board. For even larger systems two to four subracks
can be bolted together and share a larger ground plane mother board. This
allows them to be handled and wire wrapped as a unit.

The next packaging level is the **rack**. This is usually a standard 19-in. relay rack of the same type that was standard before the first electronic computer was built. These racks usually come with cooling fans, power supplies, and protective covers. A large rack can often take two assemblies of three or four subracks. For even larger systems racks can be bolted together in groups of four or more. Although many variations of this system are in use, they differ mainly in details. This system will serve as a model in our analysis of where all the money goes in a digital system.

2.2   WHERE DOES ALL THE MONEY GO?

Accounting for the costs in a digital system is very much like accounting for where your paycheck goes. If you simply add up the major, visible costs, you conclude that you can put half of your paycheck in the bank every month! The only way to get a correct result is to add up every nickel and dime expense. In Table 2-1 we have an example of where the costs seem to be going in late 1977. Only costs that are proportional to system size are shown. The actual dollar amounts shown are average amounts for moderate volume production of a medium-sized computer at this time. The actual numbers will change with time, but the general principle of being "nickeled and dimed to death" is universal.

2.3   CONCLUSION: MINIMIZE PACKAGE COUNT

As Table 2-1 shows, there is no single item that we can concentrate on to minimize system cost. One very important fact, however, is that *all the costs are very nearly proportional to the number of ICs in the system.* In other words, a system with half as many IC packages requires approximately half as many PC boards, half as many subracks, half the power, half the design effort, half the servicing cost, and so on. It therefore appears that *the real way to minimize system cost is to minimize the IC package count.* Fortunately, with LSI and MSI ICs this is quite a practical goal. In Chapters 3 and 4 we see that it is often possible to trade one, more expensive, LSI or MSI package for many small-scale ICs.

The total of all these costs per IC is shown at the bottom of Table 2-1. Since this cost is many times the cost of most ICs, it may come as quite a shock. It is easy to check the validity of this number, however, by simply working backward from total cost figures:

$$\text{overhead cost per IC} = \frac{\text{total logic system cost} - \text{total IC cost}}{\text{number of ICs in system}}$$

TABLE 2-1 DIGITAL SYSTEM COSTS THAT ARE PROPORTIONAL TO
THE NUMBER OF ICs

Item	Cost/Unit	ICs/Unit[a]	Cost/IC[b]
PC board	$20/board	50/board	$0.40
PC connector	$5/board	50/board	0.10
Subrack (metalwork, guides, labor)	$200/subrack	1000/subrack	0.20
Backplane (subrack ground-power plane)	$100/subrack	1000/subrack	0.10
Wire wrap (automatic)	6¢/wire	1/wire	0.06
Power supplies ($1/W, 70% utilized)	$1.50/W	10/W	0.15
Rack (including doors, fans, power distributor)	$500/rack	4000/rack	0.13
IC ordering, receiving, and inventory	2¢/IC	1	0.02
IC testing	8¢/IC	1	0.08
PC board testing	$5/board	50/board	0.10
System checkout	$800/rack	4000/rack	0.20
Bypass capacitors (including insertion labor)	$1/board	50/board	0.02
IC insertion and soldering	6¢/IC	1	0.06
Interconnecting cables	$80/rack	4000/rack	0.02
Maintenance panel	$300/rack	4000/rack	0.07
System assembly	$200/rack	4000/rack	0.05
System packing and shipment	$200/rack	4000/rack	0.05
System design, drafting and prototype checkout (÷ 100 systems)	$2000/rack	4000/rack	0.50
Service cost for 5 years	$4000/rack	4000/rack	1.00
Total overhead cost per IC			= $3.31

[a] Since it is impractical to fill PC boards completely, an average (rather than maximum) number of ICs per board, subrack, and so on, is assumed.

[b] $\text{Cost/IC} = \dfrac{(\text{cost/unit})}{(\text{ICs/unit})}$

The total logic system cost should include the amortized design cost and servicing cost and any other costs that are proportional to system size. It should not include the cost of peripheral devices not affected by system size. Once we know the overhead cost per IC we can design for minimum system cost much more intelligently. For example, we can decide on the "breakeven point" for trading two SSI packages for one MSI package that does the same job, as follows. If

$$O = \text{overhead cost per IC}$$
$$S = \text{cost of an SSI package}$$
$$M = \text{cost of an MSI package}$$

then the breakeven point is when the system cost of one MSI package equals the system cost of the two SSI packages replaced.

$$M + O = 2(S + O)$$

Solving for M:

$$M = 2(S + O) - O = 2S + 2O - O = 2S + O$$

For example, if $O = \$3.31$ and $S = \$0.17$, + then $M = 2 \times 0.17 + 3.31 = \$3.65$ is the breakeven point. In other words, if it is possible to replace two 17¢ SSI ICs with one MSI package, we should do it—as long as the MSI package costs less than \$3.65. Since the cost per gate or flip-flop in an MSI circuit is always less than the cost of the same gates or flip-flops in SSI, an MSI circuit that costs this much must have at least $3.65/0.34 = 11$ times as many gates or flip-flops than the two SSI packages replaced.

We thus have a more economical, more reliable mechanization of the same function—even though it uses 12 times as many gates or flip-flops! It is obvious, then, that our goal in logic design should *not* be to minimize the number of gates or flip-flops. A much more valid goal today is to minimize the number of IC packages. In a sense, this important rule is just a variation of our rule of Chapter 1, because the way we minimize is to use MSI, a "bargain component" even though it does not exactly fit the job. In the above example MSI can be used with only 9% efficiency, yet we still come out ahead.

2.4 QUANTIZED PRICE INCREASES

The modular packaging concept described earlier (Fig. 2-1) provides a certain degree of standardization in the manufacture of a logic system. As with all standardization, the benefits are offset, to some degree, by inefficiencies due to the standard not quite fitting the application. The cabinet, for example, may be capable of holding some 250 PC boards—but what if the system uses only 125? Of course, if the production volume is adequate, we can design and build a cabinet half as large, but then the production volume will be divided between the large and the small cabinet, making both more expensive.

We have the same problem with all the other parts of the system; half-full

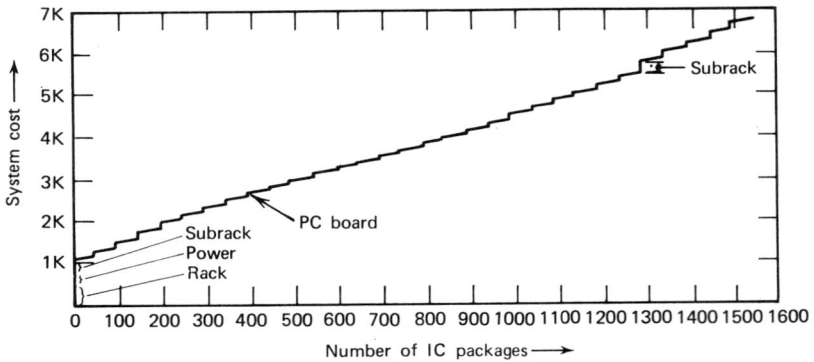

Fig. 2-2.   System cost as a function of number of ICs.

subracks, half-full PC boards, and half-loaded power supplies. This, indeed, is one reason a truly high-volume device like a TV set can be made so economically—everything can be custom designed to fit the application. When the product volume lowers, we must choose between ill-fitting standard components and expensive special components.

In the computer, peripheral equipment, and instruments field, system quantities for a given model are usually in the hundreds. At this level, we are forced to standardize on a modular packaging system like that shown in Figure 2-1. As a result, the system cost tends to vary in quantum jumps, as shown in Figure 2-2. We start out with the basic cost of the rack, fans, power supplies, and one subrack. As we increase the number of ICs in the system, we have little jumps in the cost each time we fill up another PC board. These jumps are equivalent to the cost of many ICs, yet when a PC board is full, it only takes one more IC to effect this jump in cost. When the subrack is full, we have another, larger jump where one additional IC package means we have to add a whole additional subrack.

These crucial points must be given serious consideration early in the design. It is very desirable, for ease of servicing and testing, to subdivide the logic into well defined functional modules. It is almost impossible to do this and, at the same time, completely fill each PC board and subrack. This problem will be discussed further later, but it is the reason that we have assumed incomplete utilization of the packaging modules in the costs in Table 2-1.

2.5   THE *REAL* GOALS OF DESIGN

Almost all of our dicussion so far has been concerned with minimizing cost. Yet the goals of digital system design, in order of importance, are as follows:

1. The system *works* (many never do)
2. Reliability
3. Simplicity of concept (easy manufacturing and service)
4. Economy

Though economy is the last item, we will continue to talk chiefly in terms of cost, mainly because it is a real number that we can analyze with some certainty. In addition, *all* of these goals are optimized in the same way. Systems that "never get off the ground" are usually overly complex. Reliability and economy are optimum when the number of components is minimized. If we just follow our two rules—"fit the job to the bargain component" and "minimize IC package count"—we can optimize all four factors.

The only time these goals may be in conflict is in choosing the quality of the components. Certainly it is false economy to buy a cheaper part if reliability is sacrificed. Fortunately, one seldom has such a choice with ICs, for the cheapest way to make ICs is to maximize the "yield" by processing them perfectly. Although there are certainly variations in the quality of ICs shipped by different manufacturers, these differences do not usually correlate with their prices.

2.6 THE GENERAL DIGITAL SYSTEM

Now that we have established some very general design principles, we can describe a general digital system and how to design it.

Figure 2-3 shows a general digital system. It consists of (a) one or more "input devices," which convert mechanical movement, position, light, and so on, to binary logic signals, (b) the "digital logic subsystem," which produces binary output signals that are functions of the inputs received, and (c) one or more "output devices," which convert the binary output signals into the desired mechanical* outputs.

The input device consists of an electromechanical* device and interface cir-

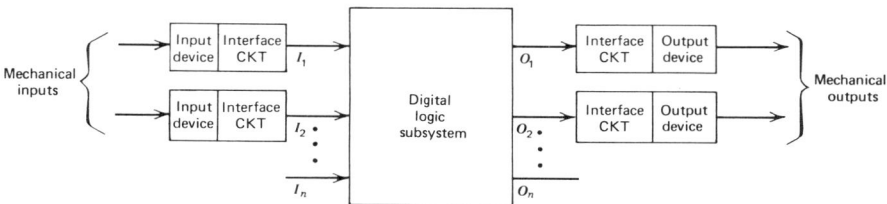

Fig. 2-3. A general digit system.

* Mechanical in the broadest sense—including light, temperature, sound, and so on.

cuits to convert the electrical signals (which are usually weak, analog signals) into binary logic signals. The output device is just the reverse: the interface circuits convert the binary logic signals to the proper form (usually powerful analog or digital signals) to operate an electromechanical device.

In Chapter 13 several modern input and output devices are described in detail. The following list shows come common examples:

Input Devices	Output Devices
Keyboard	CRT display
Tape reader	Printer
Card reader	Loudspeaker
Badge reader	Card punch
Optical character reader	Alarm bell
Microphone	Paper tape punch
Thermocouple	Valve
Pressure transducer	Digital display
Shaft angle encoder	Magnetic tape
Photocell	Motor
Tachometer	Solenoid
Switch	Lamp

A truly complete, useful system always has a mechanical input and output. Although many useful electronic products have purely electrical inputs and outputs, they are really subsystems—not useful by themselves. For example, a computer is a subsystem that is useless by itself. It is only when mechanical input and output devices are connected to it that we have a computer *system*.

All designs should start at this total system level, because that is where the greatest opportunity for true creativity exists. Once the overall system concept is defined, our choices become increasingly limited as we proceed from general concepts to actual design details. At the basic system concept level, a change of approach can often actually reduce the size of the job by a factor of 100 or more. It is obvious, then, that we should not rush through this phase of the design as simplifications at this stage of the design can greatly reduce the magnitude of the remaining design effort. Most design failures are a result of rushing through this initial phase. With a faulty or overcomplicated system concept, the machinery of design, checkout, and production is started on the wrong course. The more time invested in the faulty concept, the harder it is to change—even if it becomes obvious that the initial concept was wrong. The machinery just grinds on until the project is canceled or the company fails.

2.7   THE DESIGN SPECIFICATION

The best way to avoid a fiasco is to try to remain in the initial system-concept phase as long as possible. When the correct system approach becomes clear, it should be described in a detailed, written specification. In this way, many of the design details are worked out. By solving these detail questions at this stage, the design concept becomes much clearer and flaws in the concept may become apparent. The important thing is that a first pass be made at the design without any real commitment to the concept. Since it exists only in the form of a specification rather than hardware or a detailed design, it can be easily modified.

Another advantage of a detailed design specification is that, by having management and marketing approve it, misunderstandings and unpleasant surprises can be avoided. Management can, in effect, review the product before it is too late to change it.

2.8   AVOIDING "BAD" COMPONENTS

The first step in the basic system design process is to decide on the electromechanical components to be used for input and output. Certain types of mechanical components are inherently "bad" from a reliability point of view and should be avoided. Because digital logic is relatively inexpensive, we can design the logic portion of the system to fit essentially any interface to the electromechanical part. This gives us great flexibility in choosing the mechanical components and often allows a tradeoff of more logic for less mechanical hardware.

The table below lists some components that have been sources of trouble; their use should be avoided. The criteria on this table are reliability first and cost second. Fortunately, the two often go together since they both result from simplicity and standardization.

Preferred	"Bad" (Avoid using)
Light-emitting diodes (LEDs)	
LED digital readouts	Interconnecting cables
Television monitors	Motors with brushes
Stepper motors	Switches
Solenoids	Relays
Phototransistors	Trimpot adjustments
Photographic film	Lamps
Magnetic pickups	
Keyboards	

Although the preferred components are bargains, as mechanical components go, we should generally try to use a minimum of mechanical components in the system.

Possibly the best way to show how to trade off logic for the "bad" components is to give examples for each item:

INTERCONNECTING CABLES. Although cables cannot be eliminated, the number of cables can be greatly reduced by sending data bits serially (one at a time in some fixed order) over a single line. This often results in a *decrease* in the amount of IC logic required. Section 11.8 covers this technique in detail.

MOTORS WITH BRUSHES. These can be replaced by induction motors or by stepper motors. The digital logic can produce the required AC (square-wave) signal directly.

SWITCHES. Although it is difficult to eliminate switches completely, each switch need have only one contact. This contact should simply generate a logic signal. Digital logic can then perform the rest of the switching. If the system has a keyboard and display, switch positions can often be changed be pressing a sequence of number keys. The present state of the "switch" can be presented on the display.

Many of the old, mechanical calculators had a large array of keys with 10 keys for each digit position. On an electronic calculator it is much more economical and reliable to have only 10 keys and to let the logic shift the digits into a register in the order they are entered.

RELAYS. Solid-state equivalents are available for almost every relay function. Control signals should be switched over by normal logic gates.

TRIMPOT ADJUSTMENTS. The digital approach is inherently free of adjustments and therefore requires no trimpots. Instead of having a trimpot to adjust the period of an analog monostable circuit, for example, a digital counter can do the same job. Chapter 13 describes many other examples of doing analog jobs digitally.

LAMPS. **Light-emitting diodes (LEDs)** can replace lamps directly. Since LEDs actually work better at a low duty cycle, wiring, logic, and drive electronics can often be time shared. By driving a row at a time cyclically (faster than the eye can see), the number of drivers can be reduced to one per column and one per row.

2.9   THE BLACK-BOX CONCEPT

One of the most useful concepts in system and logic design is the "black box." Because of the tremendous complexities of most digital systems, our minds are simply incapable of grasping the entire system, in all its detail, all at once. The black-box concept allows us to subdivide the system into bite-size chunks that are more easily digested. We do this by ignoring certain details that are unimportant at the system level we are considering.

When we are thinking at the highest, complete system level, we need not be concerned with the details of the actual logic gates and flip-flops in the system. For this reason, that entire part of the system is simply represented by a black box in the system diagram (see Fig. 2-3). In this way we can ignore the details of how the logic will be performed and concentrate on the question at hand—the choice of electromechanical components.

If we are discussing system operation on an even higher system level, we do not even care about the details of how the electromechanical devices are connected into the system. At this level the entire system is one big black box with *mechanical* inputs and outputs. Figure 2-4 shows a specific example: a pocket calculator with keyboard input and visual display output. We can write a detailed specification of exactly what this black box will do without ever considering the details of what is inside the box. For example, for the desk calculator we can describe exactly what we should see on the output as a result of any sequence of keys being pressed. Each time we press a key, a definite and well defined change should occur on the output. This change depends not only on which key we pressed but also on all other keys that have been pressed since the last time the CLEAR key was pressed. We thus have a complete definition of what the black box of Figure 2-4 must do; we have defined a design problem.

To proceed with the design, we must now begin a process of subdividing this design problem into a series of smaller, more easily handled problems. Figure 2-5 illustrates this process of subdivision for the example of the pocket calculator. At each successive level of the design process we are interested in a greater degree of detail, so the number of subdivisions necessary to give us understandable units increases. Fortunately, as we pass from the general system level to the detailed design of subsystems, we can increasingly ignore the overall picture. Once the subsystems are decided upon and their functions specified, we can essentially ignore the rest of the system and attack each black box as an independent design problem, which may or may not require further subdivision for solution.

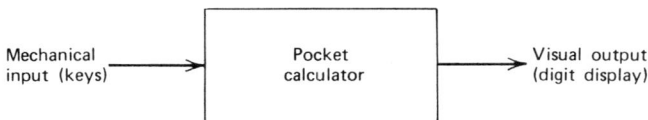

Fig. 2-4. A pocket calculator system.

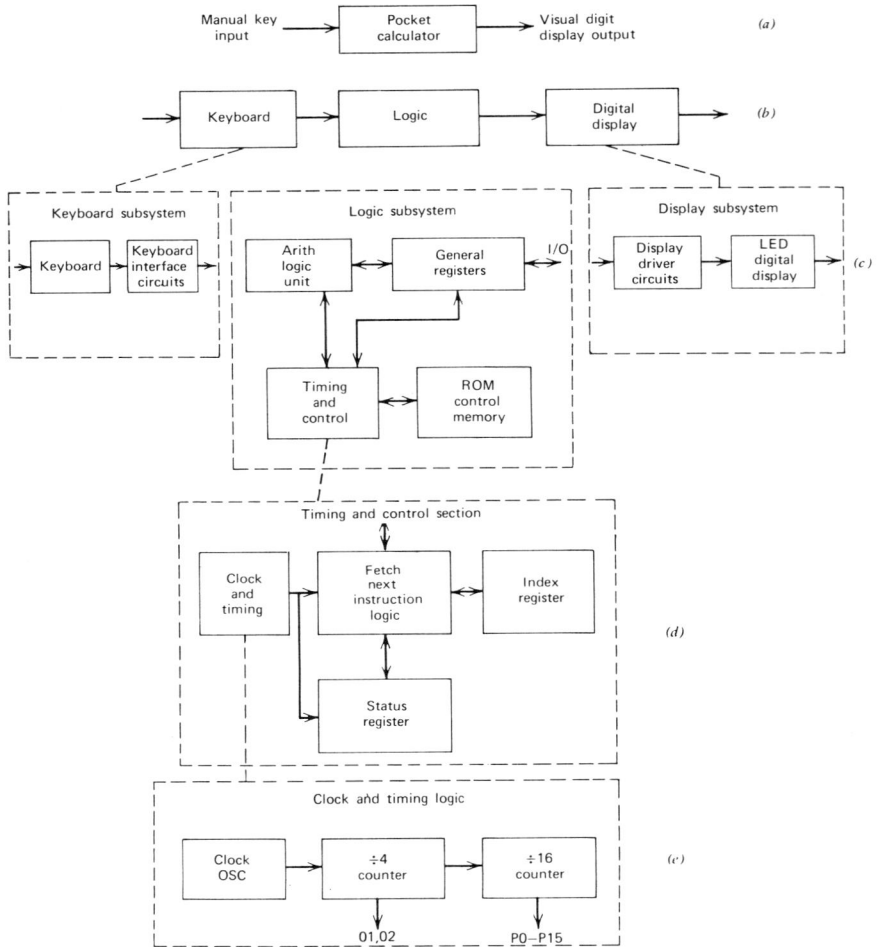

Fig. 2-5. "Black-box" subdivision of a calculator system. (d) One of eight boxes shown. (e) One of many boxes shown.

We thus progress from ignoring circuit details to ignoring system details, always focusing only on the area pertinent to the current stage of design. In this way we always keep the magnitude of the problem at a size that is optimum for our mental abilities. If we subdivide the problem in too gentle steps, we are wasting time attacking the problem in tiny nibbles. If we try to make overly large leaps, we are "biting off more than we can chew." An example would be trying to do the detailed logic design of the whole calculator in Figure 2.5*a* without subdividing it at least to the level of 2-5*c*. Just how many levels of design are needed depends on the complexity of the system and the individual designer's mental capacity.

This process of subdivision is useful in understanding as well as designing systems. We can handle systems of any complexity with this technique. In explaining how a system works, we first explain the system in general terms, then we explain the major subsystems and their building blocks until we work our way down to the small details. In designing very large systems it is often necessary to divide the design effort among several engineers. By giving each one a well-specified black-box subsystem to design, a minimum of confusion would result.

2.10 SYSTEM PARTITIONING

There is a distinct parallel between the conceptual subdivision of a system into various subsystem levels and the modular packaging concept described earlier. Both are attempts to divide the job into pieces that are more easily handled, and both have a branching, treelike structure, with the total system as the trunk of the tree. It is therefore obviously desirable to try to unify these two systems of subdivision. When this is done, each subrack, PC board, and so on, will actually represent a specific subsystem.

Although there are certain conflicts in making these two types of subdivisions coincide, there are many more requirements in common. For example, it is desirable to minimize the number of input and output signal connections both in a conceptual subsystem and in a packaging module. It is convenient that the optimum size for PC board modules is also a good-size conceptual black-box size for making the final design step to actual IC package interconnections.

This partitioning into subsystems is normally done in the written design specification stage. It is a very important design decision that should be made with quite a thorough knowledge of the hardware that will be involved in each subsystem. Since the subsystems are designed with the narrower viewpoint of simply optimizing the subsystem design, chances for real "breakthrough" reductions in system complexity, once this level is reached, are quite small. On really large systems it is often useful to write a detailed design specification on each subsystem, similar in purpose to the design specification for the overall system. Both the hardware and the paperwork for the system are thus subdivided into the same "tree" structure.

EXERCISES

1. Make a table similar to Table 2-1 with actual costs obtained from records on a real system.
    (a) Use this table to calculate overhead cost per IC.
    (b) Check the result by comparing the actual manufacturing cost of the system with the cost calculated by adding the actual IC cost to the overhead cost per IC multiplied by the number of ICs.
2. Calculate the overhead cost per IC for a system with a manufacturing cost of $16,000. The system contains 5000 ICs costing a total of $1800. The engineering cost of the system was $60,000. A total of 40 systems are made, with an average lifetime of 5 years. Servicing cost for each system is $2000/year.
3. In a system with packaging as shown on Figure 2-1 and costs as shown on Table 2-1:
    (a) What is the additional system cost if one additional IC is added to the system?
    (b) What is the additional cost of adding that IC if, as a result, we must add another PC board?
    (c) What if all the subracks are full so that an additional subrack must be added as a result of adding the one IC?
4. (a) Draw a system (like Fig. 2-4) of a TV system showing all inputs and outputs.
    (b) Subdivide the system into subsystems to the level where a TV receiver is one block.
    (c) Subdivide the TV receiver into four subsystems, showing all inputs and outputs of the receiver (including controls).

BIBLIOGRAPHY

Packaging

Fordiani, Wm O., "New Vistas for CAM," *1971 Wescon Technical Papers,* IEEE, 1971.

Harper, C., *Handbook of Electronic Packaging,* McGraw-Hill, New York, 1969.

"Packaging With Integrated Circuits Course," *The Electronic Engineer,* March–August 1972 (a complete survey by various authors).

Costs

Samaras, Thomas T., "Estimating Project Costs—What the Textbooks Don't Tell," *Electronics,* February 28, 1972, pp. 90–92.

Black Boxes

Phister, M., *Logical Design of Digital Computers,* Wiley, New York 1958, pp. 144–148.

CHAPTER THREE

COMBINATIONAL LOGIC I
Traditional Logic Design

3.1 INTRODUCTION

In the next three chapters we get down to the specifics of how to design the digital logic subsystem of Figure 2-3. This design job is traditionally called "logic design" and has been the subject of many textbooks. We treat it here as simply a more detailed level of system design where the components are IC packages. Our purpose is to minimize the number of IC packages. Since we use abstract equations and diagrams at this phase of the design, our chances of finding ways to adapt the problem to standard ICs are greatly reduced. For this reason, we should postpone starting this phase of the design as long as possible.

Once we start on the wrong approach, we can spend weeks carefully applying techniques for minimizing logic gate inputs and still end up using 100 logic gates for a job that could have been done with a single LSI circuit.

The digital logic subsystem is purely digital in that all inputs are in the form of binary electrical signals consisting of wires with one of two voltage levels on them representing 1 or 0. Outputs are likewise binary voltage levels. These levels are converted to and from the required mechanical inputs and outputs by the interface circuits and input/output devices shown in Figure 2-3.

3.2 COMBINATION VERSUS SEQUENTIAL

For any given set of input signals a specific set of output signals is required by the application. Usually the relationship between the inputs and the required outputs is affected by what has gone before. For example, in a pocket calculator, the result displayed when you press the "=" key depends on the number and operation keys you pressed while entering the problem. A logic subsystem of this type is called "sequential" logic and must include some form of memory. Although most practical logic systems require some memory, it is

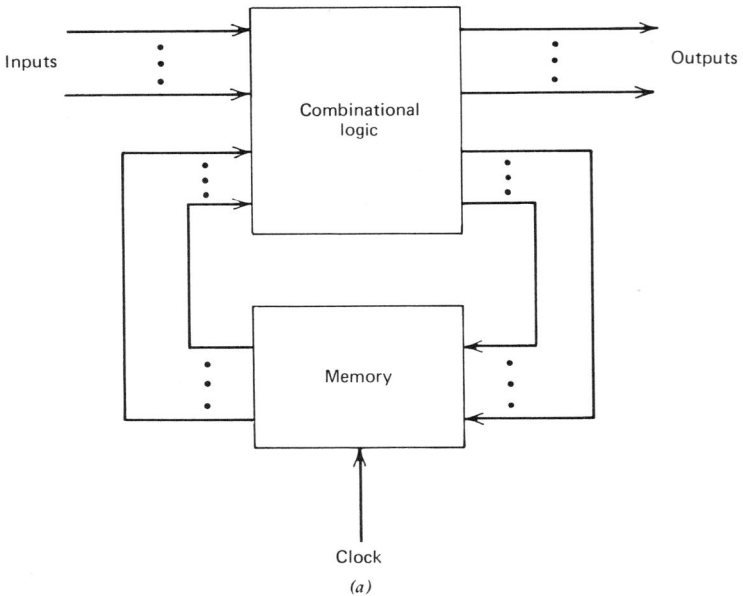

(a)

Level of Integration	Combinational Logic	Memory
Small (SSI)	Gates	Flip—flops
Medium (MSI)	Decoders, multiplexers, adders	Registers, counters
Large (LSI)	Read—only memory (ROM) Programmable logic array (PLA)	Random—access memory (RAM) Large shift registers

(b)

Fig. 3-1. (a) Subdividing sequential logic. (b) Circuit types for various levels of integration.

useful to subdivide the subject of logic design into two types: sequential and combinational. *A purely combinational logic subsystem has no memory, and therefore its outputs are completely defined by its present inputs.* The analysis and design of combinational logic is therefore much easier.

Any sequential logic subsystem can be broken up, as shown in Figure 3-1a, into a memory portion and a purely combinational logic portion. This subdivision of the problem makes it much easier to work with. Once the required combinational logic is defined, the much simpler problem of designing purely combinational logic can be dealt with. Figure 3-1*b* shows typical circuit types for the various levels of integration. Since the larger levels of integration represent better "bargains," we will always try to use the largest level of integration possible. Chapter 5 describes how we define the combinational logic required for a sequential logic system. But let us assume for now that we have defined a combinational logic function and have to mechanize it in the best possible way.

In the next two chapters we develop an assortment of logic mechanization techniques using small-, medium-, and large-scale ICs. First, however, let us consider how logic functions are expressed.

3.3 DESCRIBING LOGIC FUNCTIONS I: THE TRUTH TABLE

The first step in any design problem is to define the problem concisely. Usually the problem is initially defined in a general way, so we must translate the general description into either *logic equation* or *truth table* form. As an example, Figure 3-2*a* shows a type of digital display that consists of seven illuminated segments. Let us examine the problem of designing a logic network that will turn on the correct segments (as shown in Fig. 3-2*b*) in response to a **binary coded decimal (BCD)** input. The desired logic network is shown as a black box in Figure 3-2*c*. The four input wires indicate the number to be displayed by the binary pattern of 1's and 0's (high and low voltage levels) on them. The seven outputs of the logic network must turn on in the proper pattern to display the desired digit.

From the word description and the pictured displays in Figures 3-2*a* and *b* we can tabulate the desired outputs for each valid combination of inputs as shown in Figure 3-2*d*. The left-hand column marked "decimal displayed" is not part of the truth table as such but is often useful in making the table easier to read and understand. In this case each row on the table represents a different number displayed, and each column represents an input or an output signal. Since the input codes are BCD representations of the digit displayed, the four input columns are generated by simply writing in the BCD codes for each of the decimal digits. The output columns are then filled in by inspecting the

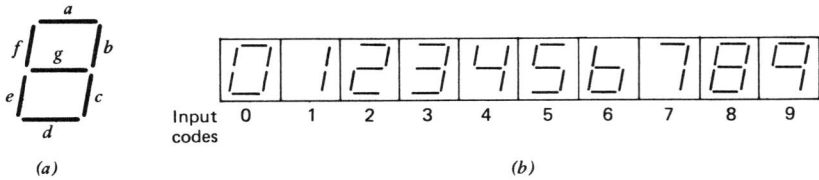

(a)

(b)

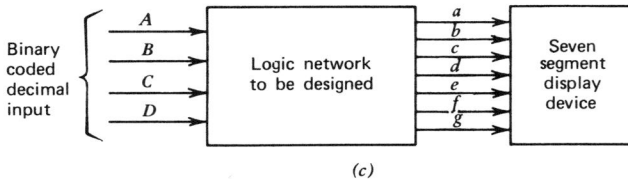

(c)

Decimal displayed	Inputs				Outputs						
	A	B	C	D	a	b	c	d	e	f	g
0	0	0	0	0	1	1	1	1	1	1	0
1	0	0	0	1	0	1	1	0	0	0	0
2	0	0	1	0	1	1	0	1	1	0	1
3	0	0	1	1	1	1	1	1	0	0	1
4	0	1	0	0	0	1	1	0	0	1	1
5	0	1	0	1	1	0	1	1	0	1	1
6	0	1	1	0	0	0	1	1	1	1	1
7	0	1	1	1	1	1	1	0	0	0	0
8	1	0	0	0	1	1	1	1	1	1	1
9	1	0	0	1	1	1	1	0	0	1	1

(d)

Fig. 3-2. (a) Segment identification. (b) Desired displays. (c) Logic design problem. (d) Truth table.

displays pictured in Figure 3-2b and filling in a 1 if the segment is shown or a 0 if it is not. For example, the 1 requires only segments b and c, so we write 0110000 in that row. We can thus fill out the entire table from the word and picture description of the application.

We have thus translated the problem statement into the concise, standardized language of the truth table from the generalized requirement of driving a seven-segment display. We are now one step closer to choosing a final design. Actually, many of the most important design decisions (such as the decision to use a BCD code to indicate the digit and to use a seven-segment display) have already been made. If we had chosen a display that requires one driver for each digit, the table would have 10 output columns with a single 1 in each row corresponding to the digit to be displayed.

Actually the truth table of Figure 3-2d is incomplete. Since there are 16 possible combinations of four inputs, we could make a more complete truth table by making all 16 rows and writing X's in the output columns in the six extra rows to indicate that we "don't care" what the outputs are for these codes. Their omission from the table, however, indicates the same thing.

3.4 DESCRIBING LOGIC FUNCTION II: THE BOOLEAN EQUATION

The truth table is easy to generate and very effective for showing several logic functions at once. It also shows their relationship to the real world, since each column and row has a definite meaning in the real world.

Another very useful way of expressing logic functions is the logic equation. A logic equation expresses only one function, so that, for example, a separate logic equation could be written for each output function (a–g) in the truth table of Figure 3-2d. Since logic equations are very similar in form to discrete gating structures, they are very useful when the logic is to be mechanized with logic gates. Many problems are stated in such a way that logic equations can be written directly from the word description of the problem. Though logic equations could be written directly from the pictures in Figure 3-2b, their relationship to the real problem would be very abstract.

Boolean algebra is a greatly simplified algebra where variables have only one of two values. These values can be called 0 or 1, true or false, high or low, or whatever. The only mathematical operations in Boolean algebra are "not," "or," and "and." These are mathematically complementation, addition, and multiplication, but their meanings correspond exactly to common English usage. For this reason an English language statement of many functions can be written directly as Boolean logic equations. The mathematical symbols ', +, and ·, respectively, are used to indiciate "not," "or," and "and." Looking at Figure 3-2b we see that segment e of our seven-segment display must be on when displaying the digits 0, or 2, or 6, or 8. Using the + symbol in place of "or," we can write a logic equation for the variable e as follows:

$$e = 0 + 2 + 6 + 8 \qquad (3\text{-}1)$$

This is an equation in terms of the displayed digit, but we really want an equation in terms of the input variables, A, B, C, and D. The input code for 0 is "not-A and not-B and not-C and not-D" (see Fig. 3-2d); we can therefore write an equation:

$$\text{digit } 0 = A' \cdot B' \cdot C' \cdot D' \tag{3-2}*$$

Likewise

$$\text{digit } 2 = A' \cdot B' \cdot C \cdot D' \tag{3-3}$$

$$\text{digit } 6 = A' \cdot B \cdot C \cdot D' \tag{3-4}$$

and

$$\text{digit } 8 = A \cdot B' \cdot C' \cdot D' \tag{3-5}$$

We can now substitute these expressions for the digits in terms of inputs into the original equation for e as follows:

$$e = A' \cdot B' \cdot C' \cdot D' + A' \cdot B' \cdot C \cdot D' + A' \cdot B \cdot C \cdot D' + A \cdot B' \cdot C' \cdot D' \tag{3-6}$$

We thus have a logic equation for output e as a function of the inputs A, B, C, and D. We can write similar equations for each of the other outputs as follows:

$$a = A' \cdot B' \cdot C' \cdot D' + A' \cdot B' \cdot C \cdot D' + A' \cdot B' \cdot C \cdot D + A' \cdot B \cdot C' \cdot D$$
$$+ A' \cdot B \cdot C \cdot D + A \cdot B' \cdot C' \cdot D' + A \cdot B' \cdot C' \cdot D \tag{3-7}$$

$$b = A' \cdot B' \cdot C' \cdot D' + A' \cdot B' \cdot C' \cdot D + A' \cdot B' \cdot C \cdot D' + A' \cdot B' \cdot C \cdot D$$
$$+ A' \cdot B \cdot C' \cdot D' + A' \cdot B \cdot C \cdot D + A \cdot B' \cdot C' \cdot D' + A' \cdot B' \cdot C' \cdot D \tag{3-8}$$

$$c = A' \cdot B' \cdot C' \cdot D' + A' \cdot B' \cdot C' \cdot D + A' \cdot B' \cdot C \cdot D + A' \cdot B \cdot C' \cdot D'$$
$$+ A' \cdot B \cdot C' \cdot D + A' \cdot B \cdot C \cdot D' + A' \cdot B \cdot C \cdot D + A \cdot B' \cdot C' \cdot D'$$
$$+ A \cdot B' \cdot C' \cdot D \tag{3-9}$$

$$d = A' \cdot B' \cdot C' \cdot D' + A' \cdot B' \cdot C \cdot D' + A' \cdot B' \cdot C \cdot D + A' \cdot B \cdot C' \cdot D$$
$$+ A' \cdot B \cdot C \cdot D' + A \cdot B' \cdot C' \cdot D' \tag{3-10}$$

$$f = A' \cdot B' \cdot C' \cdot D' + A' \cdot B \cdot C' \cdot D' + A' \cdot B \cdot C' \cdot D + A' \cdot B \cdot C \cdot D'$$
$$+ A \cdot B' \cdot C' \cdot D' + A \cdot B' \cdot C' \cdot D \tag{3-11}$$

$$g = \text{A}' \cdot B' \cdot C' \cdot D' + A' \cdot B' \cdot C \cdot D + A' \cdot B \cdot C' \cdot D' + A' \cdot B \cdot C' \cdot D$$
$$+ A' \cdot B \cdot C \cdot D' + A \cdot B' \cdot C' \cdot D \tag{3-12}$$

It is obvious that the truth table of Figure 3-2d is better suited to expressing this type of logic function compactly.

---

* This would be spoken as "digit zero equals A prime and B prime and C prime and D prime." A bar over the negated variable is also sometimes used, for example, $\overline{A}\ \overline{B}\ \overline{C}\ \overline{D} = A'\ B'\ C'\ D'$ or $\overline{A\ B\ C} = (A\ B\ C)'$. The prime notation has the advantage of being typeable and computer enterable for automated design.

Other types of problem, however, are better expressed directly as logic equations. For example, a seat-belt interlock system for an automobile is supposed to prevent starting unless all passengers have their seat belts fastened. The starter motor should turn on when the start switch is turned *and* the driver's seat belt is fastened *and* each of the other seat belts is fastened *or* there is no weight on the seat. We can write a logic equation directly from this statement. First, we give shorter names to each sensor signal in the system:

$$SS = \text{start switch}$$

$$SB1 = \text{drivers seat belt fastened}$$

$$\left.\begin{array}{l} SB2 = \\ SB3 = \\ SB4 = \end{array}\right\} \text{other seat belts}$$

$$\left.\begin{array}{l} W2 = \\ W3 = \\ W4 = \end{array}\right\} \text{weight sensed on seats 2, 3, and 4}$$

We then write a logic equation directly from the statement:

$$\text{START} = SS \cdot SB1 \cdot (SB2 + W2') \cdot (SB3 + W3') \cdot (SB4 + W4') \quad (3\text{-}13)$$

Since there are eight input variables, a truth table would have to have $2^8 = 256$ rows to cover all possible combinations of input variables. Clearly, this is a case where a logic equation is better suited for defining the function.

3.5 THE CANONICAL MINTERM FORM

As with normal algebra, Boolean expressions can take many forms. The equations just given for the seven-segment display are in a cumbersome form called the *canonical form*. There are two types of canonical forms: minterm and maxterm. The equations shown are in minterm form. A *minterm* is simply a logic product containing every variable once and only once (either plain or primed). There are $2^4 = 16$ minterms of four variable each shown below:

$$m_0 = A' \cdot B' \cdot C' \cdot D'$$

$$m_1 = A' \cdot B' \cdot C' \cdot D$$

$$m_2 = A' \cdot B' \cdot C \cdot D'$$

$$m_3 = A' \cdot B' \cdot C \cdot D$$

$$m_4 = A' \cdot B \cdot C' \cdot D'$$

$$m_5 = A' \cdot B \cdot C' \cdot D$$

$$\vdots$$

$$m_{14} = A \cdot B \cdot C \cdot D'$$

$$m_{15} = A \cdot B \cdot C \cdot D \qquad (3\text{-}14)$$

These numbered minterms (note that the numbers correspond to the binary weighted value) are simply a useful shorthand for writing functions in canonic form. For example, equation 3-7 could simply be written as

$$a = m_0 + m_2 + m_3 + m_5 + m_7 + m_8 + m_9 \qquad (3\text{-}15)$$

Since the minterm numbers directly correspond to rows on the truth table and displayed digits, this form of writing the equation is not only shorter but also easier to write directly from Figure 3-2b.

The other canonic form, the *maxterm* is less often used. Maxterms are simply sums of all variables and are indicated by M as shown below:

$$M_0 = A' + B' + C' + D'$$

$$M_1 = A' + B' + C' + D$$

$$M_2 = A' + B' + C + D'$$

$$\vdots$$

$$M_{15} = A + B + C + D \qquad (3\text{-}16)$$

Note that each maxterm is true for all but one combination of the variables. For example, M_0 is true as long as at least one of the variables is false and thus true for all combinations except 1111. Any function can be written as a product of maxterms. For example, the maxterm form of equations 3-7 and 3-15 would be

$$a = M_{14} \cdot M_{11} \cdot M_9 \qquad (3\text{-}17)$$

The maxterm form is essentially equivalent to the grammatical "double negative." Since we are unaccustomed to thinking in these terms, it is less useful for initially defining a function from a word description.

3.6 MECHANIZING LOGIC FUNCTIONS WITH GATES

Though MSI and LSI have made it possible to mechanize many logic functions much more economically than the traditional method of connecting discrete gates, much logic must still be done at the gate level—even in the most sophisticated designs. Although it is often possible to do the bulk of the logic in a large system using MSI and LSI circuits, there are always many little details that must be done with discrete gates. The ratio of gate IC packages to MSI and LSI packages is still about 2 : 1 in very small systems and about 1 : 1 in large systems. Though the discrete gate part of the system may perform the minority of the logic "work" in the system, it represents a large part of the logic designer's detail design and checkout work. It is therefore important that the designer fully master the techniques of gate minimization.

Logic gates perform functions that can be described both by logic equations and by truth tables. Figures 3-3a through 3-3c show two input gates for

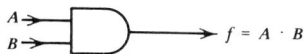

(a) AND gate: A, B inputs → $f = A \cdot B$

Input		Output
A	B	f
0	0	0
0	1	0
1	0	0
1	1	1

(Output true if all inputs true)

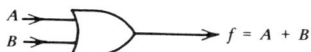

(a)

(b) OR gate: A, B inputs → $f = A + B$

Input		Output
A	B	f
0	0	0
0	1	1
1	0	1
1	1	1

(Output true if any input true)

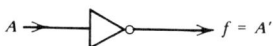

(b)

(c) Inverter: A input → $f = A'$

Input	Output
A	f
0	1
1	0

(Output true if input false)

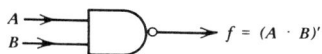

(c)

(d) NAND gate: A, B inputs → $f = (A \cdot B)'$

Input		Output
A	B	f
0	0	1
0	1	1
1	0	1
1	1	0

(Output false if all inputs true)

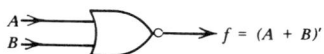

(d)

(e) NOR gate: A, B inputs → $f = (A + B)'$

Input		Output
A	B	f
0	0	1
0	1	0
1	0	0
1	1	0

(Output false if any input true)

(e)

Fig. 3-3. (a) AND gate. (b) OR gate. (c) Inverter (not). (d) NAND gate. (e) NOR gate.

performing the three basic logic functions. Figures 3-3d shows the NAND gate, which is a popular combination of the AND function and the NOT function. NAND can be thought of as a contraction of NOT AND. Figure 3-3e shows a NOR gate, which is a contraction of NOT OR. Note that the NAND and NOR gate symbols are composites of the AND and OR gate symbols with a small "bubble" added to indicate inversion of the output. Though all the gates are shown with two inputs, they can have as many inputs as desired. Package pin limitations make it fairly standard to package together *four two-input gates, three three-input gates, two four-input gates, or one eight-input gate per package.*

Figure 3-4a shows how AND and OR gates can be connected in a network to mechanize a logic equation. Notice the one-to-one correspondence between the four-input AND gates and the four-variable product terms of the equations. Once the product terms are generated, an OR gate is used to form a logical sum, producing the desired function. If the primed input terms required at the AND gate inputs are unavailable elsewhere, they can be generated with four **inverters**. This one-to-one relationship between the gate structure and the logic equation makes it easy to see why logic equations are so useful in gating design.

3.7   SIMPLIFYING LOGIC FUNCTIONS

Figure 3-4b shows another mechanization that actually does the same job as Figure 3-4a! The reason is simply that the circuit of Figure 3-4a is a mechanization of the raw equation, written from the problem definition. This is really similar to the problem-solving process in normal algebra. The first equation that we write from the word statement of the problem requires mathematical simplification before it can be solved. Likewise, Boolean equations should be put in their simplest form before they are mechanized into logic gates. Though logic equations can be simplified by manipulations at the equation level, an uncomplicated diagram technique makes simplification possible by just following a straightforward procedure.

Before describing the actual procedure, let us see why the simplified circuit in Figure 3-4b does so much with so few gates. The difference is that each of the two input gates ignores two of the variables. Their outputs will therefore go true for *four different minterms*. The circuit in Figure 3-4a uses four different four-input AND gates to define four minterms, with each gate defining one minterm. The trick, then, is to use gates with n fewer inputs to define groups of 2^n minterms that fit the requirements of the function. No harm is done if some of the desired minterms fit into both groups, nor if we define extra minterms that will never appear on the inputs. In this problem we will

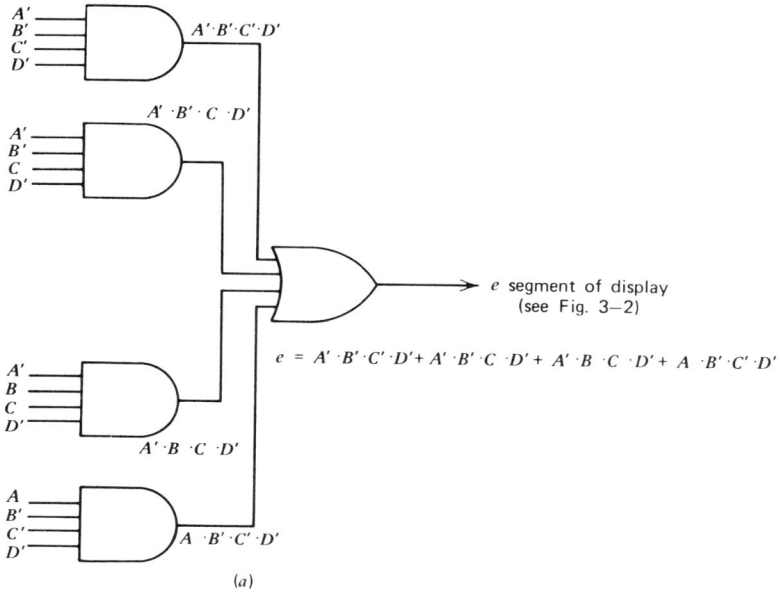

$$e = A'\cdot B'\cdot C'\cdot D' + A'\cdot B'\cdot C \cdot D' + A'\cdot B \cdot C \cdot D' + A \cdot B'\cdot C'\cdot D'$$

(a)

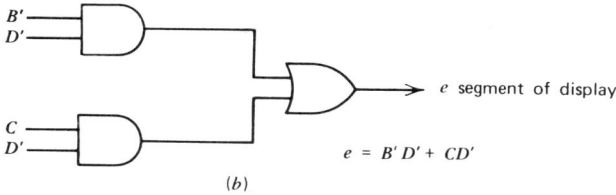

$$e = B'D' + CD'$$

(b)

Fig. 3-4. (a) AND/OR mechanization of equation 3-6. (b) Simplified mechanization of the same function.

have only BCD input codes 0–9, so we *"don't care"* about minterms 10 through 15.

The function $B'\cdot D'$ will produce the four minterms containing $B'\cdot D'$ with all four possible combinations of the variables A and C. In other words,

$$B'\cdot D' = B'\cdot D'\cdot(A'\cdot C' + A'\cdot C + A\cdot C' + A\cdot C)$$

$$= A'\cdot B'\cdot C'\cdot D' + A'\cdot B'\cdot C\cdot D' + A\cdot B'\cdot C'\cdot D' + A\cdot B'\cdot C\cdot D'$$

$$= m_0 \qquad\qquad + m_2 \qquad\qquad + m_8 \qquad\qquad + m_{10} \qquad\qquad (3\text{-}18)$$

Equation 3-6 in canonical minterm form is

$$e = m_0 + m_2 + m_6 + m_8 \qquad (3\text{-}19)$$

So we see that $B' \cdot D'$ gives us three of our desired minterms plus m_{10}, which we "don't care" about because the BCD 10 code is not allowed. The only minterm still needed is m_6.

The other function, $C \cdot D'$, produces four more minterms:

$$C \cdot D' = C \cdot D'(A' \cdot B' + A' \cdot B + A \cdot B' + A \cdot B)$$

$$= A' \cdot B' \cdot C \cdot D' + A' \cdot B \cdot C \cdot D' + A \cdot B' \cdot C \cdot D' + A \cdot B \cdot C \cdot D'$$

$$= m_2 + m_6 + m_{10} + m_{14} \qquad (3\text{-}20)$$

We thus get the m_6 we needed, another m_2, and two more "don't care" minterms, m_{10} and m_{14}.

In functions where there are no "don't care" minterms we can simplify by combining pairs of terms that are the same in all but one variable. For example,

$$A \cdot B \cdot C \cdot D + A \cdot B \cdot C \cdot D' = A \cdot B \cdot C \cdot (D + D') = A \cdot B \cdot C \qquad (3\text{-}21)$$

Mathematically we essentially factor out the common term $A \cdot B \cdot C$ and, since $D + D'$ is always true (because if D is false, D' is always true), we can write the function simply as $A \cdot B \cdot C$. The difficulty in finding such terms, and the problem of including "don't care" minterms usually make it easier to use a diagram technique to simplify logic functions.

3.8   DESCRIBING LOGIC FUNCTIONS III: VEITCH DIAGRAMS

The Veitch diagram (or map) is a special form of truth table, for one function, that is arranged so that *minterms that can be combined are adjacent*. Figure 3-5a shows a Veitch diagram for functions of four variables. There are 16 squares, one for each minterm. The numbers inside the squares indicate the minterm number represented by that square. The brackets labeled A, B, C, and D indicate the *regions where the indicated variables are true*; therefore $B = 1$ in all the squares in the top half and $B = 0$ in the squares in the bottom half.

Any function of four variables can be represented by simply filling in 1's and 0's to indicate the function, as with a truth table. For example, Figure 3-5b represents the function $f = A' \cdot B \cdot C' \cdot D$. Note that the 1 is in the A' region (since it is outside the region labeled A) *and* in the B region *and* the C' region *and* the D region. The diagram is a representation of the function, just as a logic equation or a truth table is.

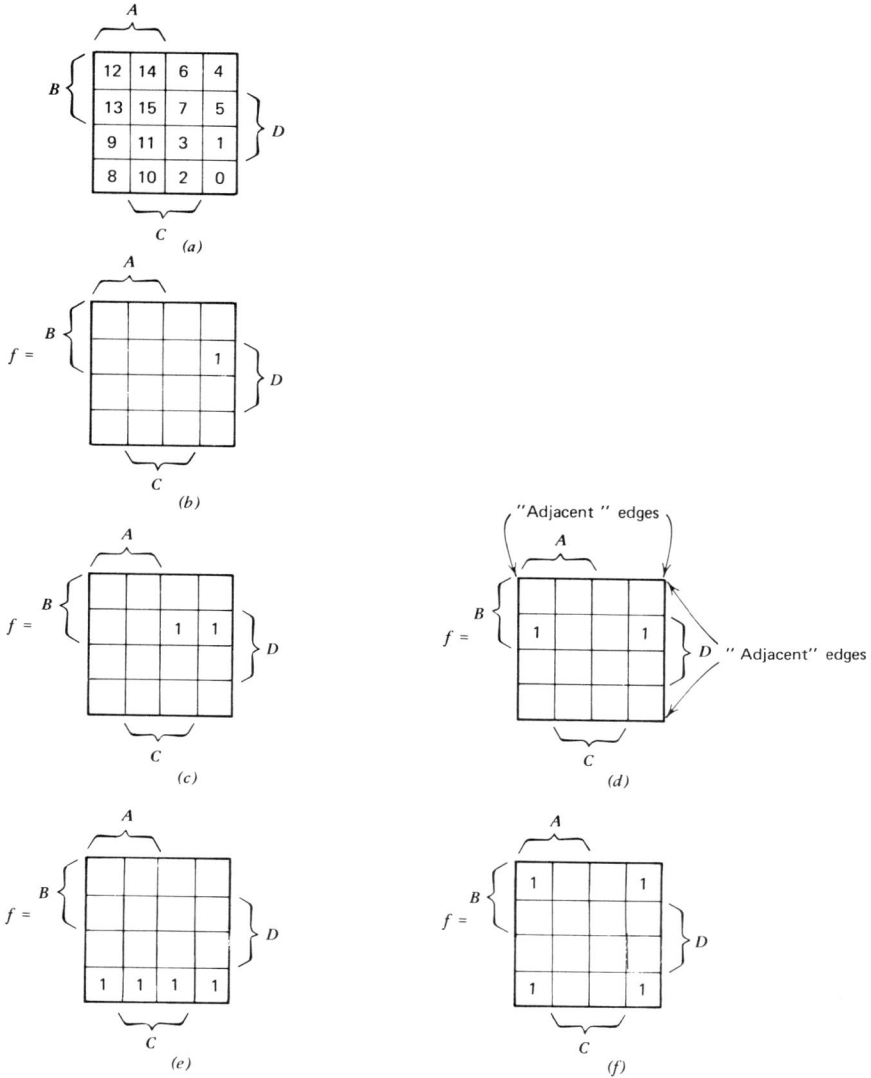

Fig. 3-5. Mapping a four-variable function: (a) minterm numbers; (b) function: $f = A' \cdot B \cdot C' \cdot D = M_5$; (c) function: $f = A' \cdot B \cdot D$; (d) function: $f = B \cdot C' \cdot D$; (e) function: $f = B'D'$; (f) function: $f = C' \cdot D'$.

The real value of the diagram representation of a function can be seen from Figure 3-5c. The product of only three variables gives a pair of *adjacent* ones. In fact, *any* product of three variables (three input gates) gives a pair of adjacent minterms on the diagram. Figure 3-5d illustrates one catch to this rule: *We must stretch the definition of "adjacent" to include opposite edges of the diagram.* If we visualize the diagram wrapped around a cylinder so that the edges meet, this definition is not too farfetched (however, it must be simultaneously wrapped in a vertical *and* a horizontal cylinder so that the four corners meet).

Figures 3-5e and f illustrate that products of only two of the variables (two input gates) give groups of four adjacent minterms. Again we must visualize the edges of the diagram as touching the opposite edge to call the four 1's in Figure 3-5f adjacent. Single variables, of course, give adjacent groups of eight minterms (the labeled regions of the diagram).

To demonstrate the usefulness of Veitch diagrams as a simplification tool, let us make a diagram of the function:

$$f = A \cdot B' \cdot C' \cdot D' + A \cdot B' \cdot C \cdot D' + A' \cdot B' \cdot D' \qquad (3\text{-}22)$$

We can enter 1's in the proper boxes for each term by simply finding the correct box by elimination. For example, on the first term the *a* indicates the left half of the map; then the B' narrows it down to the four boxes in the lower left corner; then C' restrict it to two boxes one above the other in the lower left corner; finally the D' limits it to the lower left corner box (m_8). The next term leads us to write another 1, and finally the third term gives us two 1's. The resulting diagram looks just like Figure 3-5e, so we can tell at a glance that the function of equation 3-22 can be written simply as $f = B'D'$.

Simplifying by Veitch diagram, then, consists of simply writing the function on the diagram term by term, then reading it off in groups that are as large as possible. When the simplified function consists of several product terms, it is useful to note which minterms have been eliminated by encircling the area defined by the product term each time a new term is written.

Figure 3-6 shows some examples. Notice that it is permissible to include a minterm in several terms if it helps make the terms shorter (larger groups of minterms). Notice also that the "circles" along the edge (Figures 3-6b, e, and f) are circles only if we visualize opposite edges of the map as touching. We get contiguous groups of two, four, or eight minterms by ignoring one, two, or three of the input variables.

3.8.1 "Don't Care" Conditions

Quite often some of the possible combinations of input values never occur. In this case we "don't care" what the function does if these input combinations

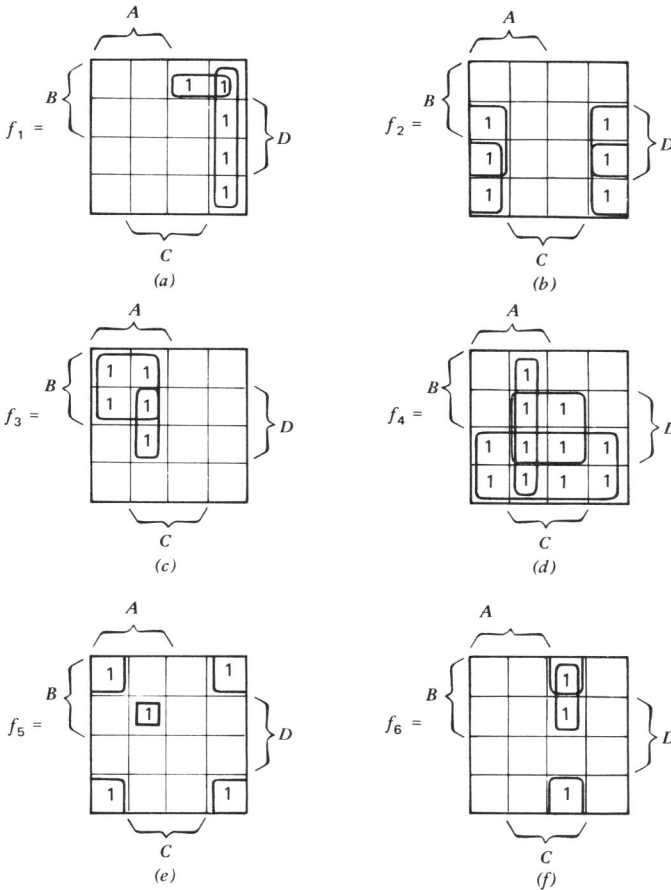

Fig. 3-6. Mapping examples: (a) $f_1 = A' \cdot C' + A' \cdot B \cdot D'$; (b) $f_2 = C' \cdot D + B' \cdot C'$; (c) $f_3 = A \cdot B + A \cdot C \cdot D$; (d) $f_4 = B' + C \cdot D + A \cdot C$; (e) $f_5 = C' \cdot D' + A \cdot B \cdot C \cdot D$; (f) $f_6 = A' \cdot C \cdot D' + A' \cdot B \cdot C$.

appear. Diagramming makes it easy to take advantage of these "don't care" conditions by letting the "don't care" minterms be 1 or 0, depending on which value results in a simpler expression.

Figure 3-7a shows an example of the use of "don't cares" (redundancies) to simplify the seven-segment display function for segment e (from Fig. 3-2) previously referred to. Since minterms 10 through 15 will never occur, we put X's on the diagram in those positions. We then put 1's on the diagram for minterms m_0, m_2, m_6, and m_8 (since segment e must be on when displaying 0, 2, 6, and 8). We now mechanize all four 1's on the map with two two-variable

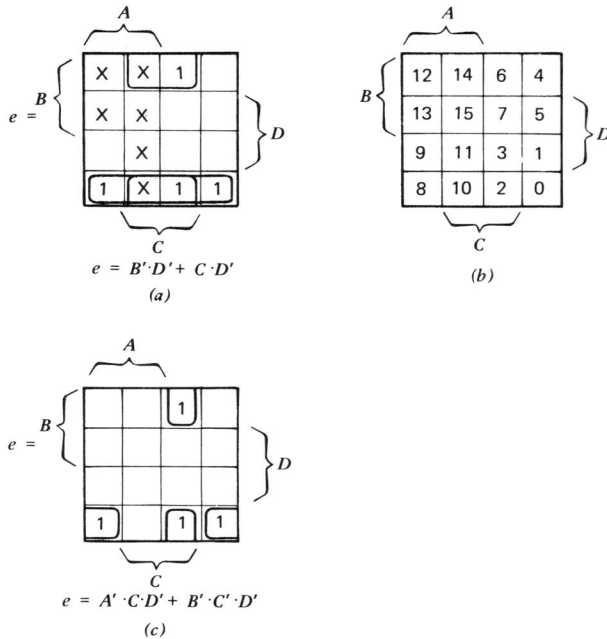

$$e = B' \cdot D' + C \cdot D'$$
(a)

(b)

$$e = A' \cdot C \cdot D' + B' \cdot C' \cdot D'$$
(c)

Fig. 3-7. Mapping examples with "don't care" minterms: (a) segment e with "don't cares"; (b) minterm numbers (for reference); (c) segment e without "don't cares."

terms. This is made possible by letting minterms 10 and 14 equal 1 and the rest of the X's equal 0. The resulting function is considerably simpler than the two three-variable terms required if the "don't cares" are not used, as shown in Figure 3-7c.

3.8.2   A Design Example

The actual design procedure for the seven-segment display driver of Figure 3-2 can be performed in a few minutes on a single sheet of quadrille paper as follows (see Fig. 3-8):

1.   A Veitch diagram with the minterm numbers written in each box is drawn on the top of the paper for reference (a), and seven blank map outlines are drawn on the same paper—one for each segment.

2.   The "don't care" X's are then drawn in for minterms m_{10} through m_{15} in all seven maps.

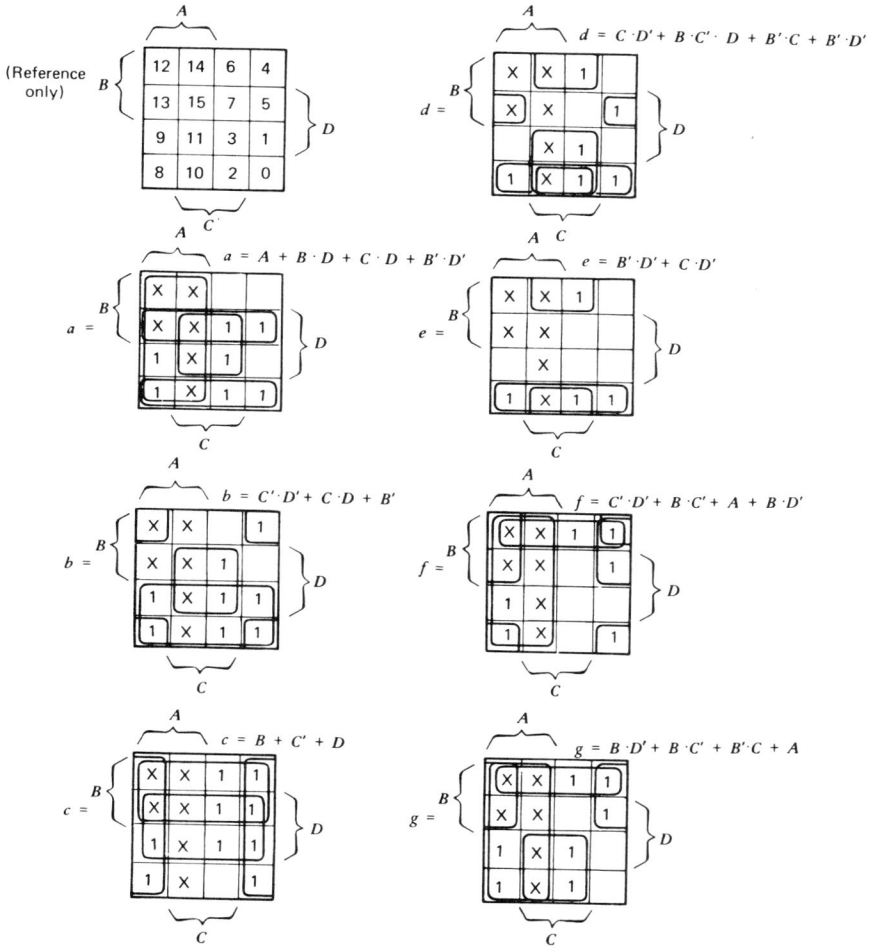

Fig. 3-8. Seven-segment display driver simplification.

3. Ones are then written into the correct minterm positions for each of the seven functions. This can be done from the truth table (Fig. 3-2*d*) or directly from the pictured displays (Fig. 3-2*b*)

4. Blocks of minterms are circled on each map until each one is included in at least one block.

5. Equations are written below each map with one term for each encircled block of minterms.

If we compare the resulting equations with unsimplified equations 3-6 through 3-12 stated earlier, we see that we have considerably reduced the number and size of the logic terms (and therefore the number and size of gates required).

3.8.3 Sharing Terms

If we look at the seven functions together, we see that there are several product terms in the final equations that are used in more than one function. For example, $C \cdot D$ appears in the equations for both a and b. The two-input gate required to generate $C \cdot D$ can therefore be shared by both functions. Notice how much easier it is to pick out this common term by looking at the diagrams than by looking at the equations. One simply has to look for similarly shaped circles.

On a multiple-function problem like this, it is well to remember the possibility of common terms when choosing terms to encircle since *there are often several equally good ways to choose the terms.* For example, segment a could just as well have been reduced as shown in Figure 3-9. Though the third term of the two resulting equations is different, both are correct and produce the same result. Though sharing a gate is possible with both choices in this case, it is often possible to save a gate by making the choice that allows sharing.

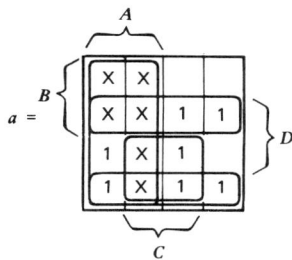

$a = A + B \cdot D + B' \cdot C + B' \cdot D$ Fig. 3-9

3.8.4 Mechanizing the Inverse of a Function

Since our product terms must cover all the minterms of the function, we sometimes find that the function is defined the hard way. For example, segment c is *on* when we display digits 0, 1, 3, 4, 5, 6, 7, 8, and 9. A simpler way of saying the same thing is that segment c is *off* only when we display a 2.

In Figure 3-10 we diagram the function c' instead of c; we can simply invert this function to produce c. This is the same equation for c as on Figure 3-8, written in a different form.

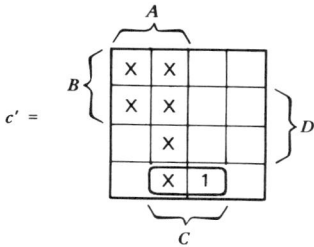

$c' = B' \cdot C \cdot D'$

$\therefore c = (B' \cdot C \cdot D')'$ Fig. 3-10

3.8.5 Other Diagram Formats

The Veitch diagram format used in this book is one of the many diagram layouts possible. It is chosen mainly because it is easy to read off the product terms when writing the simplified logic equation. Another method of labeling, called the Karnaugh map, is shown in Figure 3-11.

AB \ CD	00	01	11	10
00	0	1	3	2
01	4	5	7	6
11	12	13	15	14
10	8	9	11	10

Fig. 3-11.   Karnaugh map with minterm numbers.

Though all the examples so far have used the four-variable Veitch diagram, Figure 3-12 shows how diagrams can be made of functions of two to six variables. Actually, the diagramming technique's characteristic of positioning combinable minterms adjacent to each other does not work on a flat sheet of paper for more than four variables. If, however, we visualize stacking four-variable diagrams as shown in Figures 3-12e and g, we can have this characteristic, in space, for up to six variables. We can thus have solid groups of minterms one, two, or four deep if they are aligned between one plane and the next. Actually the diagrams must be drawn spread out in one plane as shown on Figures 3-12d and f to avoid the confusion of overlapping. Groups of minterms can simply be circled in the normal manner on the four-variable planes, then adjacent minterms or blocks of minterms can be joined with curved lines drawn between the "planes." If the drawing is done on a fairly large scale, the method is moderately workable (though unwieldy) for up to

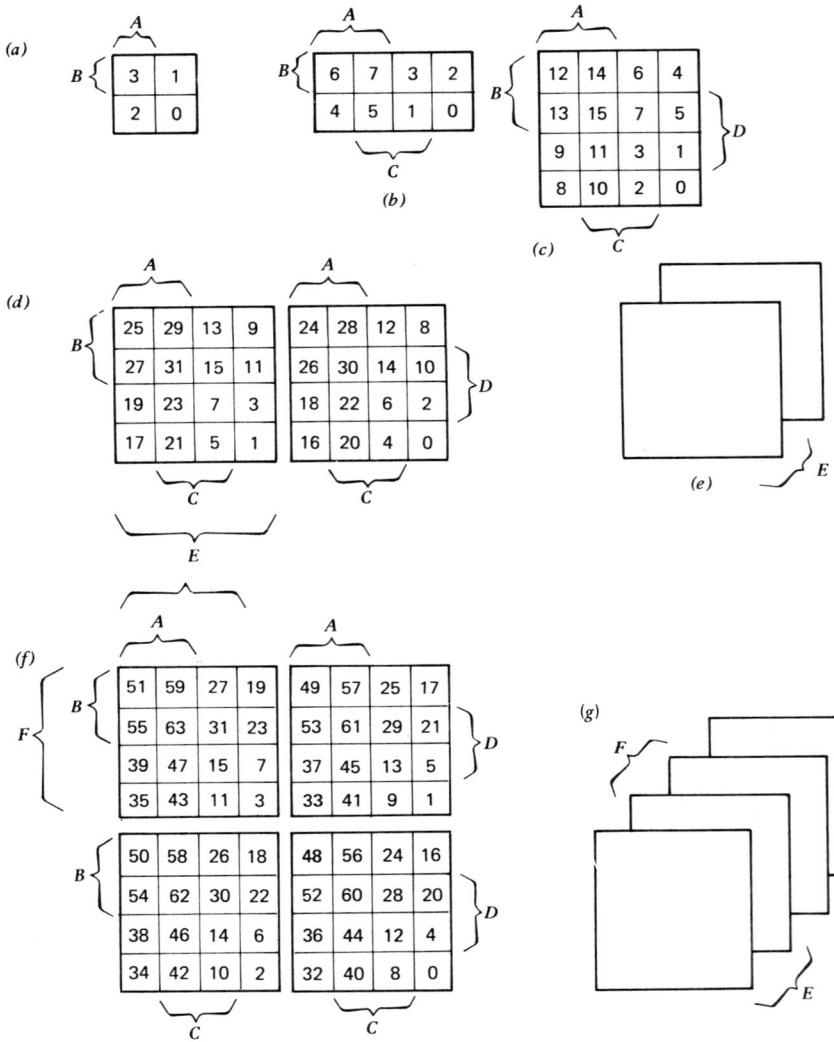

Fig. 3-12. Veitch diagrams for two through six variables (with minterm numbers): (a) two variables; (b) three variables; (c) four variables; (d) five variables; (e) visualization of (d); (f) six variables; (g) visualization of (f).

eight variables. Note that an eight-variable Veitch diagram is arranged like ā
Veitch diagram of 16 Veitch diagrams.

3.9 FACTORING LOGIC EQUATIONS

Logic equations can be factored just like normal algebraic expressions, and the
result can sometimes save gates. For example, the simplified segment e logic
equation

$$e = B'D' + CD' \qquad (3-22)$$

can be written

$$e = D' \cdot (B' + C) \qquad (3-23)$$

The gating required to mechanize this is shown in Figure 3-13. This is
somewhat simpler than the unfactored form shown in Figure 3-4b.

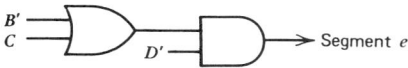

Fig. 3-13

The savings are particularly worthwhile where more can be factored. For
example,

$$f = A \cdot B \cdot C \cdot D + A \cdot B \cdot C \cdot E \qquad (3-24)$$

can be written

$$f = A \cdot B \cdot C \cdot (D + E) \qquad (3-25)$$

which requires *one* four-input gate and one two-input gate, instead of *two*
four-input gates and one two-input gate.

Another nice thing about factoring is that it leaves a function of fewer
variables, so it is often possible to factor out the "uninteresting" part of a
function, then use a simpler Veitch diagram to simplify the remaining
function.

3.10 DE MORGAN'S THEOREM

Most of the fundamental identities of Boolean algebra are fairly obvious, and
all can be easily verified by making truth tables. These identities are shown in
Table 3-1.

Because of the one-for-one relationship between logic equations and logic
gates, it is interesting to consider these identities as statements about logic

TABLE 3-1 BOOLEAN IDENTITIES

1. $A \cdot B = B \cdot A$ } (commutative law)
2. $A + B = B + A$
3. $(A + B) + C = A + (B + C)$ } (associative law)
4. $(A \cdot B) \cdot C = A \cdot (B \cdot C)$
5. $A(B + C) = AB + AC$ } (distributive law)
6. $A + B \cdot C = (A + B) \cdot (A + C)$
7. $A + A' = 1$
8. $A \cdot A' = 0$
9. $A + A = A$
10. $A \cdot A = A$
11. $(A')' = A$
12. $A + 1 = 1$
13. $A \cdot 0 = 0$
14. $(A + B)' = A' \cdot B'$ } (De Morgan's theorem)
15. $(A \cdot B)' = A' + B'$

gates. For example, the first and second identities in Table 3-1 simply say that the order in which the inputs are connected to a gate does not matter. Most of the other identities are equally uninteresting, but the last two are extremely important when related to gating. Number 15, for example, says that *a NAND gate* (as shown on Fig. 3-3d) *is the same as an OR gate with the inputs inverted.* This can be confirmed by simply making a truth table of the two functions, as shown in Table 3-2.

Actually De Morgan's theorem is quite reasonable. If we think of A and B as referring to the presence of two people in a room, it really does not matter whether we say "A *and* B are *not* both here" or "Either A is *not* here *or* B is *not* here." Therefore $(A \cdot B)'$ and $A' + B'$ are two different ways of saying the same thing.

TABLE 3-2 DE MORGAN'S THEOREM

A	B	$A \cdot B$	$(A \cdot B)'$	A'	B'	$A' + B'$
0	0	0	1	1	1	1
0	1	0	1	1	0	1
1	0	0	1	0	1	1
1	1	1	0	0	0	0

3.11 NAND/NAND LOGIC

The important result of De Morgan's theorem is that *the same physical gate circuit can perform either a NAND function or a NOR function, depending on how we define the logic levels.* When we talk about a NAND circuit we usually mean that it is a NAND if we define the logic as *"high true"* (high voltage = 1). If we redefine the logic as "low true" (low voltage = 1), the same circuit becomes a NOR gate. We can show this by making a truth table in terms of voltage levels where H = high (or +) voltage and L = low voltage. This defines the performance of a physical gate circuit. Now we will get two different truth tables, depending on whether we define our logic levels as "high true" or "low true," as shown in Figure 3-14. As can be seen, the same type of gate circuit can thus perform the NAND and the NOR functions.

Actually De Morgan's theorem simply points up the fact that the NAND function and the OR function are opposites: OR says that one or more of the inputs are *true*, while NAND says that one or more of the inputs are *false* (therefore they are *not* all true). Inverting the sense of the logic, then, changes a NAND to a NOR.

Note that the small circles ("bubbles") on the logic symbols indicate low true logic, and no "bubbles" indicates high true. The symbol on the left in Figure 3-14 is thus an AND gate with high true inputs and a low true output

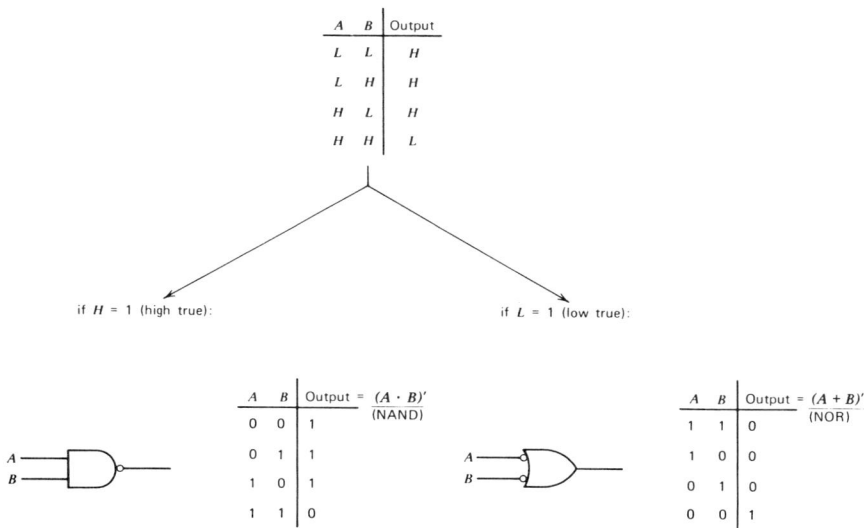

Fig. 3-14

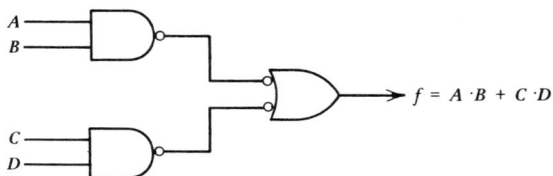

Fig. 3-15.   NAND/NAND gating—high true.

(therefore a high true NAND). The symbol on the right is an OR gate with low true inputs and a high true output (therefore a low true NOR). If we connect this type of gate in a two-level structure as shown in Figure 3-15 *we get the equivalent of an AND–OR structure even though all three gates are physically the same type of circuit!* The low true output of the first level AND causes the second level to act as an OR with high true output.

If we want to define our logic levels as low true, then the same two-level NAND/NAND gating gives us an OR–AND structure, as shown in Figure 3-16. Again, a truth table in H and L would show that the circuit is actually doing the same thing in both cases. The only difference is in our interpretation of the logic levels. We can make the same transformation on an equation by applying De Morgan's theorem.

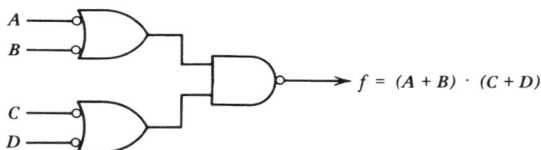

Fig. 3-16.   NAND/NAND gating—*low true.*

Actually, the statement of equality between the two gate symbols in Figure 3-14 is a statement, in terms of gates rather than equations, of De Morgan's theorem (see Fig. 3-17).

Fig. 3-17.   De Morgan's theorem for NAND.

It is easy to see, then, why NAND logic has become so popular. *Any* logic function can be mechanized by simply interconnecting *a single type of logic circuit.*

3.12 SOME NAND/NAND EXAMPLES

Let us look at several actual examples of NAND/NAND logic. The equation from Figure 3-8 for segment b would be mechanized with NAND gates as shown in Figure 3-18. Notice that the B' term is actually connected to B be-

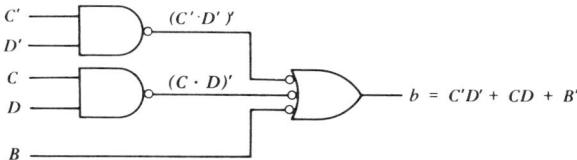

Fig. 3-18. Segment b.

cause our logic convention is high true, and the inputs to the second level gate must be low true to generate the OR function. Notice also that the three-input gate is drawn as a low true input OR strictly to make the logic structure easy to understand. Since it is also a NAND gate, it could just as correctly be drawn as such, as shown in Figure 3-19.

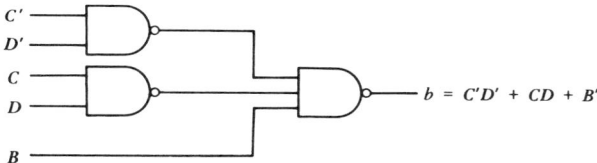

Fig. 3-19. Segment b.

It is important to see the two ways of drawing the NAND gate as interchangeable, since it is sometimes impossible to draw a circuit entirely in the easy-to-read form of Figure 3-18. We can look at a circuit and write an equation for any output by *working back from the output*. With high true NAND logic the output level is always an OR function with low true inputs. As we work back through the logic structure, the levels are alternately OR and AND functions for the odd and even levels. The logic convention can likewise be thought of as starting at high true at the output and alternating between high true and low true with each level.

Figure 3-20, for example, shows a four-level logic circuit. We can analyze this circuit and write the equation for f_1, by first writing down the low true OR inputs to the final gate, then working back, one level at a time, writing the equation directly. Drawing the NAND gate alternately as NAND and low true input OR is very helpful in this case.

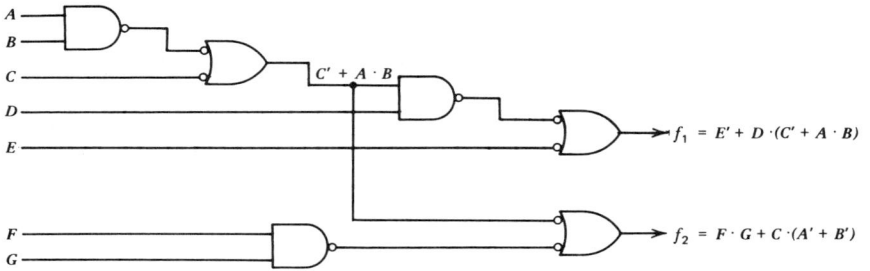

Fig. 3-20

The lower function f_2 is an example of how, when signals are used to generate several functions, it is sometimes impossible to draw the circuit symbols in the easily read form for all functions. The signal $C' + A \cdot B$ is an odd level in function f_1 and an even level in f_2. Therefore the two gates generating it must be imagined as being drawn in the opposite form to write function f_2 directly. If this appears confusing, we can simply transform the function $[C' + A \cdot B]'$ as an equation using De Morgan's theorem:

$$[C' + (A \cdot B)]' = C \cdot (A \cdot B)' = C \cdot (A' + B') \qquad (3\text{-}26)$$

Drawing the NAND gate in two different forms is strictly for convenience in designing and analyzing circuits. It is important to realize that there are always two ways of looking at a function.

3.13 NOR/NOR LOGIC

Since it is more natural to think in terms of AND/OR when describing logic functions, the NAND/NAND structure has generally proven to be the easiest to use. Some types of physical logic circuits produce the NOR function more naturally. It is important to remember, however, that they are called NOR only because it is customary to name the circuits using a high true logic convention. In using them, we can just as well make our logic convention low true, essentially changing them to NAND gates.

A NOR/NOR structure can, in some cases, be useful where the functions are best defined as a sum of maxterms, since two levels of NOR logic are equivalent to an OR/AND structure. In general, everything said about NAND gates has a dual that applies to NOR gates. For example, the high true NOR gate can be drawn in two equivalent ways effectively expressing De Morgan's theorem, as shown in Figure 3-21.

Fig. 3-21. De Morgan's theorem for NOR.

$(A + B)'$ = $A' \cdot B'$

Also, output functions can be analyzed by starting at the output and working back through the gating levels, considering them low input AND gates or OR gates for odd and even levels respectively. This is shown in Figure 3-22.

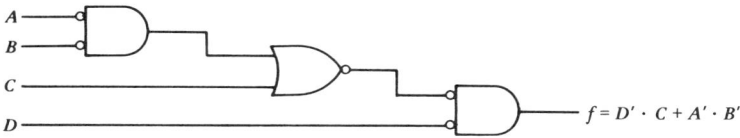

$$f = D' \cdot C + A' \cdot B'$$

Fig. 3-22. NOR/NOR logic.

3.14 MIXING LOGIC TYPES

In the interest of standardization, it is generally advisable to use the NAND/NAND structure. Mixing in other types usually produces degenerate results that are inefficient as a general mechanization scheme. In a few cases, however, gates can be saved by allowing use of a limited number of NOR gate types (e.g., two-input types only). Generally the only advantage of this is that inverters can sometimes be saved. For example, if the function we want is not AND/OR but simply an AND function, then we must invert the output of a NAND gate to generate the function. With a NOR gate, one level gives us an AND (with inverted inputs). This can be seen in Figure 3-23.

$f = A' \cdot B'$ Fig. 3-23. Using NOR to save inverters.

Often in decoding a function the primed version of the input variables is unavailable. In this case a NOR/NAND, which degenerates to NAND, is sometimes useful (see Figs. 3-24a and b). Another useful combination is shown in Figure 3-24c. Here we effectively make a 16-input AND gate from two 8-input NAND's.

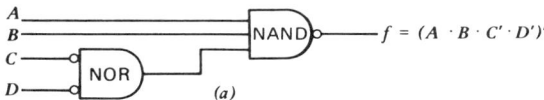

$f = (A \cdot B \cdot C' \cdot D')'$

(a)

Fig. 3-24. Useful degenerate NOR/NAND combinations: (a), (b) inverted inputs not available; (c) making large AND gates.

$$f = A' \cdot B \cdot C$$

(b)

$$f = 16 \text{ input AND}$$

(c)

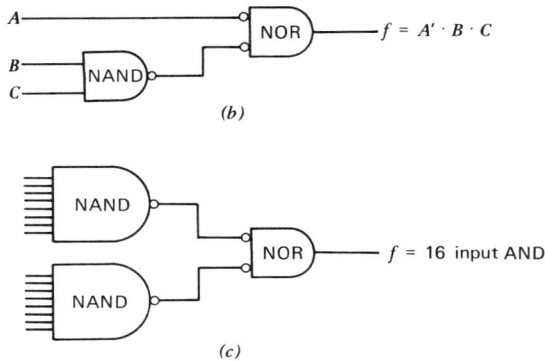

Fig. 3-24. (*Continued*)

Another useful small-scale gate circuit is AND/OR or AND/OR/IN-VERT. Because of the internal connection between logic levels, this circuit simply allows more logic per IC package. Usually the best procedure for using special circuits of this type (including NORs) is to examine the final logic diagram for opportunities to use them.

Groups of four 2-input and two 2-input NAND/NAND gates, for example, are candidates for replacement by packaged AND/OR circuits. Since there are, for example, two 2-input AND/OR circuits in a package, it makes no sense to use only one of the circuits as that would be using one package to replace three-fourths of a quad two-input-gate package. Likewise, with NOR circuits it is usually better not to use just one or two 2-input NORs since they come four in a package.

3.15 EXCLUSIVE OR

Another special function that is sometimes useful is the Exclusive OR, which is defined as shown in Figure 3-25. Exclusive OR means simply that *one and only one* of the inputs is true. Though the function can be generated with three 2-input NAND gates, the quad two-input Exclusive OR IC gives us four circuits in a package. It is thus worthwhile using the special Exclusive OR circuit to save packages in some cases. Since $A \oplus B$ can also be thought of as $A \neq B$, the circuit is often used to compare two signals.

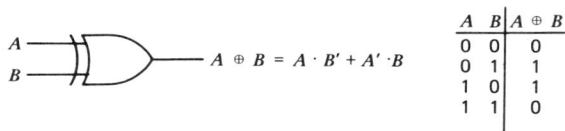

$$A \oplus B = A \cdot B' + A' \cdot B$$

A	B	$A \oplus B$
0	0	0
0	1	1
1	0	1
1	1	0

Fig. 3-25. Exclusive OR.

Yet another way to look at the function is as a *"controlled inverter"* circuit. If we think of one input as a control input, then if the control input is true, the other signal will be inverted. If the control input is false, then the other signal will not be inverted.

3.16 VIRTUAL OR AND AND

When a gate output is active in only one direction, we can tie several outputs together and actually perform a logic function. For example, open-collector TTL gates actively pull their output low, but require a "pullup" resistor to pull their output high. When several gate outputs are connected in parallel, the output goes low whenever any one or more of the individual gate outputs goes low. This is effectively a low true OR function made without a physical OR gate. It is therefore called a Virtual OR or sometimes Implied OR, Collector OR, Bus OR, or Dot OR.

Figure 3-26a shows how open-collector NAND gate outputs can be paralleled to generate an AND/OR/INVERT function. If either $A \cdot B$ or $C \cdot D$ goes true, one of the NAND gates pulls the output low. The dotted OR gate outline is shown simply to clarify the intended logic function, as there is no actual gate. If the gates were normal TTL or **CMOS (complementary metal oxide semiconductor)** gates, it would not be permissible to connect their outputs in parallel. If one output actively tries to go high and the other tries to pull low, they work against each other and produce an indeterminate result.

Some logic types, such as **emitter-coupled logic (ECL)** and **diode transistor logic (DTL)**, are normally active in only one direction, so Virtual ORs or ANDs can be done by tying together normal gate outputs. As with normal logic gates, an OR function becomes an AND if we define the logic as low true instead of high true. Thus, by simply defining the logic as low true, the circuit of Figure 3-26a can be redrawn as shown in Figure 3-26c with a Virtual AND function.

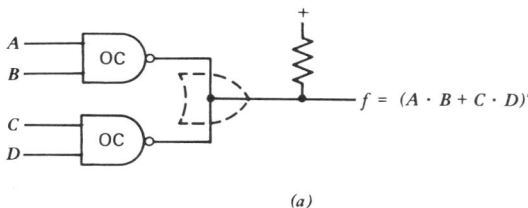

(a)

Fig. 3–26. Virtual OR: (a) NAND/virtual OR logic; (b) same function by NAND/NAND logic; (c) same function with low considered true.

$f = (A \cdot B + C \cdot D)'$

(b)

$f = (A' + B') \cdot (C' + D')$

Virtual AND

(c)

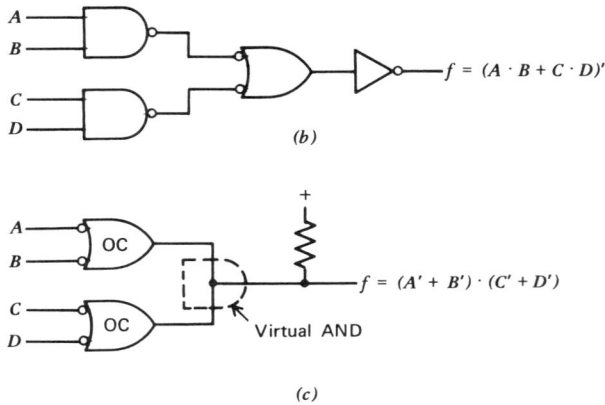

Fig. 3-26.   *(Continued)*

3.17   BUSING

Since virtual gates have no fixed number of inputs, they are often used in a *bus* arrangement to provide expandability. For example, if signals may come from any of a number of separate circuit modules, a single wire bus can go to each module. A Virtual OR function is formed along the bus because any of the modules can put its signal on the bus.

Usually, some sort of logical rule ensures that only one signal will be applied to the bus at a time. In this type of system, special circuits with **"tristate"** outputs are often used for improved speed. When a tristate circuit is enabled, it drives the bus actively high or low. When it is disabled, it looks like an open circuit. Thus there are three states: high, low and open.

EXERCISES

1. Make truth tables of the functions generated by the circuits shown in Figure 3-4*a* and *b* to see if they produce identical functions. If there are differences, will they affect the operation of the seven-segment display?

2. *(a)* Make a truth table like the one shown in Figure 3-2*d* for a seven-segment decoder to be drive by a **biquinary** code.
    (b) Simplify the seven logic equations with Veitch diagrams.
    (c) Mechanize the simplified equations with AND and OR gates and compute (14-pin) package count.
    (d) Mechanize the same equations with NAND/NAND logic.
    (e) Mechanize the same equations with NOR/NOR logic.
    (f) Mechanize with NAND/NAND logic using factoring wherever possible.

3. (a) Make a truth table for the common household circuit used to control a lamp with two switches (e.g., A = switch at the top of the stairs and B = switch at the bottom of the stairs).

(b) Derive a logic equation for the lamp signal.

(c) Mechanize the circuit using AND and OR gates.

(d) Mechanize the circuit using NAND/NAND logic.

(e) Mechanize the circuit using a special gate function described at the end of the chapter.

4. (a) Write a logic equation for a burglar alarm bell from the following inputs:

$$A \text{ = arming switch}$$
$$F \text{ = front door closed}$$
$$B \text{ = back door closed}$$
$$W \text{ = window closed}$$
$$M \text{ = motion detected in room}$$

(b) Mechanize the logic function using AND and OR gates.

(c) Mechanize the logic function using NAND/NAND.

(d) Mechanize the logic function using NOR/NOR.

5. (a) Write logic equations (from a truth table if desired) for a circuit that will produce four bits that indicate in BCD code which one of 10 inputs (I_0 thru I_9) goes true.

(b) Mechanize the equations using AND and OR gates and compute package count.

(c) Mechanize the equations using NAND/NAND logic.

(d) Mechanize the equations using NOR/NOR logic.

6. (a) Make a Veitch diagram of the function

$$f = A \cdot B \cdot C + B \cdot C' \cdot D + A' \cdot B \cdot C$$

(b) Write a simplified equation for f.

(c) Mechanize f using NAND/NAND logic.

7. (a) Write simplified equations for the functions shown in Figure 3-27.

(b) Mechanize the simplified equations using AND/OR gates.

(c) Mechanize the simplified equations using NAND/NAND gates.

(d) Mechanize the simplified equations using NOR/NOR gates.

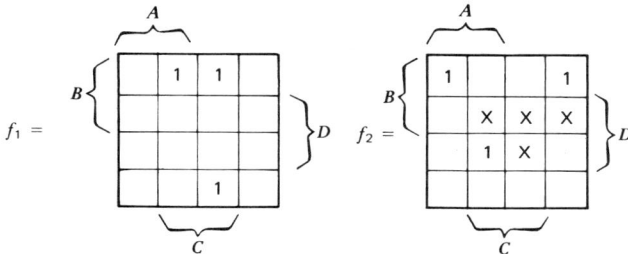

Fig. 3-27

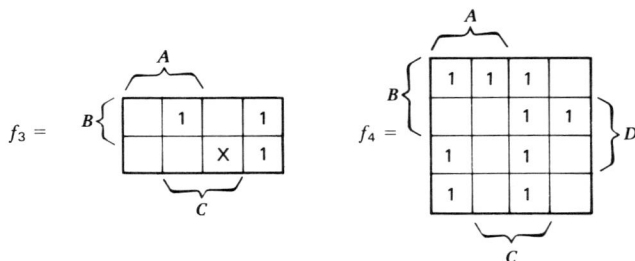

Fig. 3.27.   (Continued)

8.   (a)   Diagram the following (four-variable minterm form) equations:

$$f_1 = m_1 + m_2 + m_8 + m_{13}$$
$$f_2 = m_0 + m_1 + m_2 + m_6 + m_8$$
$$f_3 = m_5 + m_7 + m_9 + m_{10} + m_{14}$$
$$f_4 = m_0 + m_2 + m_4$$

   (b)   Make a truth table of the four functions.
   (c)   Write four simplified equations from the diagrams.
   (d)   Mechanize using NAND/NAND logic.

9.   (a)   Make Veitch diagrams of the four functions defined by the truth table in Figure 3-28.
   (b)   Write simplified logic equations.
   (c)   Mechanize the equations using AND/OR logic and compute package count.
   (d)   Mechanize the equations using NAND/NAND logic.

Input				Outputs			
A	B	C	D	F_1	F_2	F_3	F_4
0	0	0	0	1	0	0	1
1	0	0	0	0	1	0	1
1	1	0	0	0	1	1	1
1	1	1	0	1	1	0	1
1	1	1	1	1	1	1	0
0	1	1	1	0	0	0	0
0	0	1	1	0	0	1	1
0	0	0	1	0	0	1	0

Fig. 3-28

10. (a) Make a truth table of a majority function of four variables. This function should be true whenever three or more of the inputs are true.
    (b) Diagram the function.
    (c) Mechanize it with NAND/NAND logic.
11. (a) Write simplified logic equations for the functions shown in Figure 3-29.
    (b) Mechanize them with NAND/NAND logic.
    (c) Compute (14-pin) IC package count.

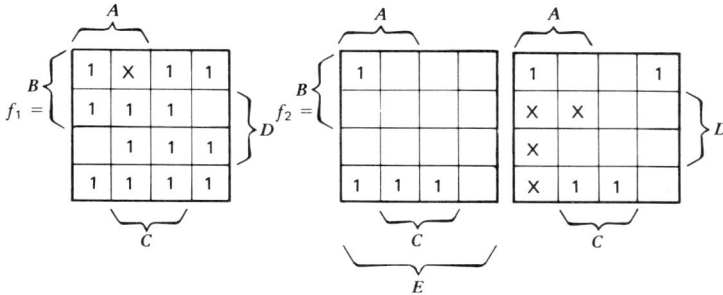

Fig. 3-29

12. (a) Make a truth table of a function that is true when two or more of four variables are true.
    (b) Diagram the function.
    (c) Mechanize it with NAND/NAND logic.
    (d) Compute package count.
13. (a) Simplify the following equation:

$$f = SH1 \cdot WH2 \cdot AM' + AM \cdot WO' + WH1' \cdot WO \\ + WH2' \cdot WO + WH1 \cdot WH2' \cdot AM'$$

    (Note that the four variables are $WH1$, $WH2$, AM, and WO.)
    (b) Mechanize it with AND/OR gates.
14. (a) Make a Veitch diagram of the following function:

$$f = A \cdot (B + C + D) + A \cdot B' \cdot C + A \cdot C'$$

    (b) Write the equation in simplified form.
    (c) Mechanize it with NAND/NAND logic.
15. (a) Write a simplified logic equation for the function shown in Figure 3-30.
    (b) Mechanize it with AND/OR logic and compute package count.
    (c) Mechanize it with NAND/NAND logic.
    (d) Mechanize it with NOR/NOR logic.

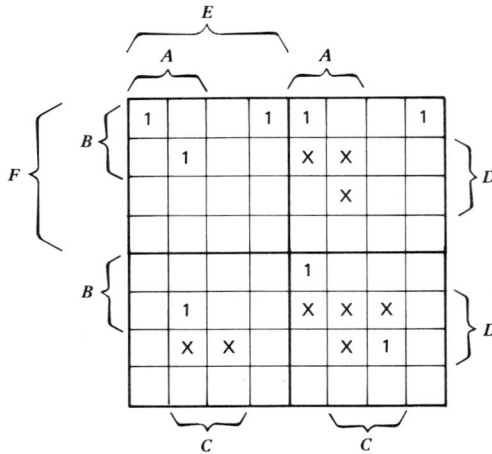

Fig. 3-30

16. (a) Make a truth table of a function of four variables (*A, B, C, D*) that is true whenever any two adjacent variables (*A* and *B*, *B* and *C*, *C* and *D*, *D* and *A*) are true. Only the 10 *BCD* input combinations will occur, so the rest are "don't cares."
   (b) Diagram the function.
   (c) Write a simplified equation for the function.
   (d) Mechanize the function with NAND/NAND logic.
17. (a) Mechanize the seven-segment display functions of Figure 3-8 with NAND/NAND logic.
   (b) Rewrite the equations on Figure 3-8, factoring any common terms.
   (c) Mechanize the logic from the factored equations using NAND/NAND logic.
18. (a) Mechanize the seat belt interlock equation (3-13) with NOR/NOR logic.
   (b) Do the same with NAND/NAND logic.
19. State the fundamental Boolean identities in Table 3-1 by drawing pictures of gates (in the same way that Fig. 3-17 expresses the fifteenth identity).
20. (a) Mechanize the Exclusive OR function with NAND/NAND logic.
   (b) Assuming the input variables are not available in complemented form, mechanize an Exclusive OR with just four 2-input NAND gates.
21. (a) Make a truth table for an Exclusive OR function of four variables.
   (b) Diagram the function.
   (c) Mechanize the function using NAND/NAND logic.
   (d) Mechanize the function using open collector NAND gates in parallel and inverting the Virtual OR output.
   (e) Mechanize the function using only three 2-input Exclusive OR gates and no primed input variables.

22. (*a*) Make a truth table for a function that goes true when two 2-bit binary num-
         bers are equal (i.e., $AB = CD$).
    (*b*) Write a logic equation for the function.
    (*c*) Mechanize it with NAND/NAND logic.
    (*d*) Compute package count.

BIBLIOGRAPHY

Calebotta, Steven, "CMOS—The Nearly Ideal Logic Family," *Digital Design,* July 1973, pp. 34–44.

Fronck, Donald, "Ring Map Minimizes Logic Circuit," *Electronic Design,* August 17, 1972, pp. 80–81.

Phister, M., *Logical Design of Digital Computers,* Wiley, New York, 1958, Ch. 31, "Boolean Algebra," and Chapter 4, "The Simplification of Boolean Functions."

Richards, R. K., *Digital Design,* Wiley, New York, 1971.

Su and Nam, "Computer Aided Synthesis of Multiple Output, Multi-Level NAND Networks With Fan-In and Fan-Out Constraints," *IEEE Transactions on Computers,* December 1971, pp. 1445–1455.

Texas Instruments Staff, *Designing With TTL Integrated Circuits,* McGraw-Hill, New York, 1971.

Texas Instruments Staff, *The Integrated Circuits Catalog,* Texas Instruments Inc., Dallas, Tex., 1972, available through distributors for $4.95.

Veitch, E. W., "A Chart Method for Simplifying Truth Functions," *American Mathematical Monthly* **59,** 1952, pp. 521–531.

CHAPTER FOUR

COMBINATIONAL LOGIC II
MSI and LSI Logic Design

4.1 DESIGNING FOR MSI AND LSI

In Chapter 3 we developed those techniques of traditional logic design that still seem to be useful. Before the advent of ICs all logic design was done in this one-gate-at-a-time manner. In modern logic design the problem is more complicated. Certain logical combinations are available as standard products that are much better bargains than discrete gates. For this reason, though we must often resort to them, *the techniques of the preceding chapter must be considered a last resort*—to be used only when higher levels of integration cannot be used. Because the standardized MSI and LSI circuits seldom fit our needs exactly, it is usually necessary to use discrete logic to adapt the circuits to the application.

Though we cover some techniques for using MSI and LSI circuits to mechanize logic equations, the real chance to use MSI and LSI is at the system design level—before we get to such details as logic equations. If we define a logic function without considering *available* MSI circuits, we may end up writing equations that must be mechanized by discrete gates, since they do not fit the available MSI circuits.

For this reason, *it is essential to memorize the contents of the IC manufacturers' catalogs*. Only by knowing what MSI and LSI circuits are available can we design a system to use available components. Only after we have a detailed system block diagram, showing MSI and LSI circuit types as boxes, should we start to write logic equations.

Useful as they are, logic equations tend to obscure the function being performed. For example, the seven-segment display driver used repeatedly as an example in the previous chapter is available as *a single MSI IC*. Once we start writing equations, the chances of discovering this are practically nil. The equations on Figure 3-8 do not suggest a seven-segment display driver at all. However, the original system problem statement of "driving a seven-segment display" certainly *does* suggest it. For this reason, we should *postpone writing logic equations as long as possible*. The entire system should be fully defined in terms of MSI and LSI blocks first.

4.2 THE DIGITAL MULTIPLEXER/SELECTOR

One of the most generally useful MSI circuits is the digital multiplexer. This circuit is also sometimes called a "selector" because it functions very much like a selector switch. An eight-input multiplexer, for example, "selects" one of eight inputs to appear on the output. A three-bit binary code determines which input will appear on the output. Figure 4-1 shows a logic diagram and truth table for such a circuit. Multiplexers also come packaged as quad two-input (four 2-input multiplexers in one package), dual four-input, and 16-input (in a larger, 24-pin package).

Here we can see the advantage of knowing the standard ICs and designing them in. If our system requires that eight bits of data input come from one of four sources, our block diagram design will simply show four dual four-input multiplexer boxes. If we had written logic equations, we would have to write eight separate equations of the form

$$Y = I_0 \cdot A' \cdot B' + I_1 \cdot A' \cdot B + I_2 \cdot A \cdot B' + I_3 \cdot A \cdot B \qquad (4-1)$$

If we mechanized these equations with gates, we would use at least 10 gate packages. Though we *might* recognize from the equations that the resulting circuit fits the multiplexer function, it is certainly less clear that the data will come "from one of four sources."

If we know our MSI circuits well, we can work much more intelligently at the system level. For example, since it is no more expensive to select one of four inputs than one of three, we can often add features to the system at no extra cost. It is for this reason that a "design specification" should include a block diagram with details to the MSI circuit level. The portion of the logic that must be done with discrete gates can simply be covered by a catch-all block marked "control logic."

Logic Diagram

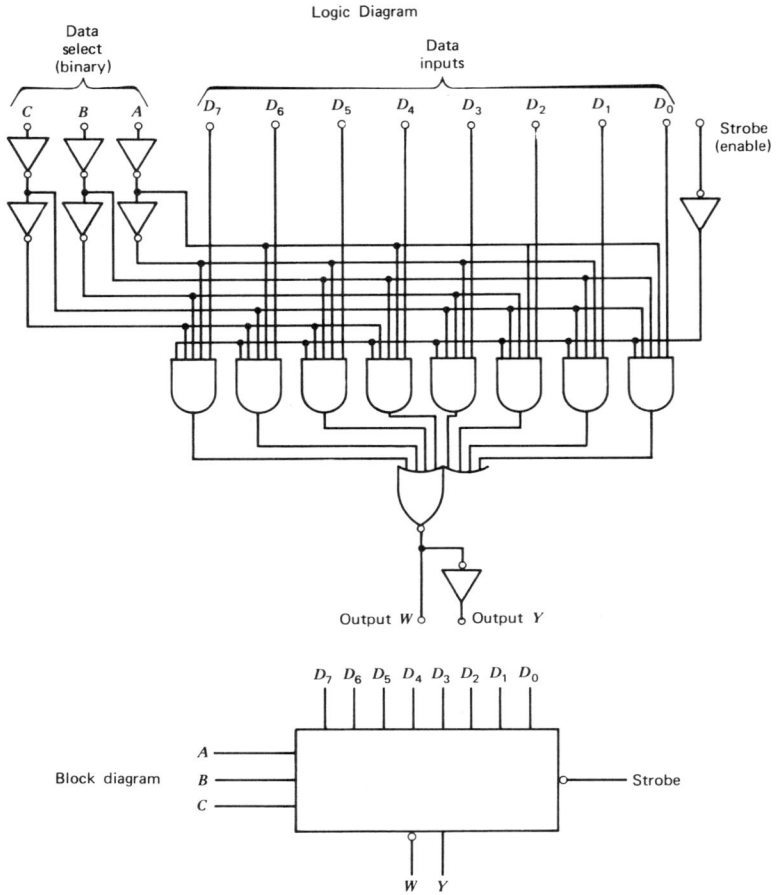

Block diagram

Truth Table

				Inputs									Outputs	
C	B	A	Strobe	D_0	D_1	D_2	D_3	D_4	D_5	D_6	D_7	Y	W
X	X	X	1	X	X	X	X	X	X	X	X	0	1
0	0	0	0	0	X	X	X	X	X	X	X	0	1
0	0	0	0	1	X	X	X	X	X	X	X	1	0
0	0	1	0	X	0	X	X	X	X	X	X	0	1
0	0	1	0	X	1	X	X	X	X	X	X	1	0
0	1	0	0	X	X	0	X	X	X	X	X	0	1
0	1	0	0	X	X	1	X	X	X	X	X	1	0
0	1	1	0	X	X	X	0	X	X	X	X	0	1
0	1	1	0	X	X	X	1	X	X	X	X	1	0
1	0	0	0	X	X	X	X	0	X	X	X	0	1
1	0	0	0	X	X	X	X	1	X	X	X	1	0
1	0	1	0	X	X	X	X	X	0	X	X	0	1
1	0	1	0	X	X	X	X	X	1	X	X	1	0
1	1	0	0	X	X	X	X	X	X	0	X	0	1
1	1	0	0	X	X	X	X	X	X	1	X	1	0
1	1	1	0	X	X	X	X	X	X	X	0	0	1
1	1	1	0	X	X	X	X	X	X	X	1	1	0

When used to indicate an input, X = irrelevant.

Fig. 4-1. Eight-input multiplexer/selector (74151).

4.2.1 Use of Multiplexers to Mechanize Logic Functions

In addition to serving as a system component to select signal sources, the digital multiplexer can be very useful for generating and simplifying logic functions. If we look at the equation for the output of a four-input multiplexer, for example,

$$Y = I_0 \cdot A' \cdot B' + I_1 \cdot A' \cdot B + I_2 \cdot A \cdot B' + I_3 \cdot A \cdot B \qquad (4\text{-}2)$$

we see that we have a separate input for each of the four combinations of A and B. We can thus "factor out" A and B from a function of n variables and end up, instead, with four separate functions of $n - 2$ variables to be connected to the inputs of the multiplexer. Similarly an eight-input muliplexer can eliminate three variables, and a 16-input multiplexer eliminates the four variables connected to its address inputs.

If we have a function of four variables and we eliminate three of them with an eight-input multiplexer, we have to connect eight functions of only *one* variable to the inputs. The only possible functions of one variable, C, are: C, C', 1, or 0. Since these are all available with no additional gating, *we can generate any function of four variables with only an eight-input multiplexer.*

As an example, let us try

$$f_1 = A' \cdot B' \cdot C' \cdot D' + A' \cdot B' \cdot C \cdot D + A' \cdot B \cdot C' \cdot D + A' \cdot B \cdot C' \cdot D'$$

$$+ A \cdot B' \cdot C' \cdot D + A \cdot B' \cdot C \cdot D' + A \cdot B \cdot C' \cdot D' + A \cdot B \cdot C' \cdot D \qquad (4\text{-}3)$$

If we use an eight-input multiplexer with A, B, and C as address inputs, then we can find the input functions required by simply making a table of the eight multiplexer input address combinations. This is shown in Table 4-1.

TABLE 4-1   MULTIPLEXER INPUTS TO
GENERATE F_1

Input	Address	Other Variables (Residues)
I_0	$A' \cdot B' \cdot C'$	D'
I_1	$A' \cdot B' \cdot C$	D
I_2	$A' \cdot B \cdot C'$	$D + D'$
I_3	$A' \cdot B \cdot C$	
I_4	$A \cdot B' \cdot C'$	D
I_5	$A \cdot B' \cdot C$	D'
I_6	$A \cdot B \cdot C'$	$D' + D$
I_7	$A \cdot B \cdot C$	

The "Other-variables" column of this table is taken directly from the logic equation (4-3). For each product term of the equation, an entry is made on the table. For example, for the first term, $A' \cdot B' \cdot C' \cdot D'$, we enter D' in the $A' \cdot B' \cdot C'$ row. For the next term we enter D in the $A' \cdot B' \cdot C$ row. For the third term we enter D in the $A' \cdot B \cdot C'$ row, and for the fourth term we enter D' in that same row.

The multiplexer inputs (Fig. 4-2) can be read directly from this table, since each row corresponds to a multiplexer input. Since $D + D' = 1$, we simply connect a true logic level (+ supply) to inputs I_2 and I_6. Since there is nothing (0) in rows 3 and 7, we connect a false logic level (ground) to these inputs. The resulting circuit is shown in Figure 4-2. Note that *the single multiplexer package shown in Figure 4-2 generates a logic function that would require four discrete gate packages to mechanize* even after simplification. As a bonus, the enable input can be used as a final AND with the whole function and the inverted output is also available. In addition, only one variable (D) has to be available in complemented form.

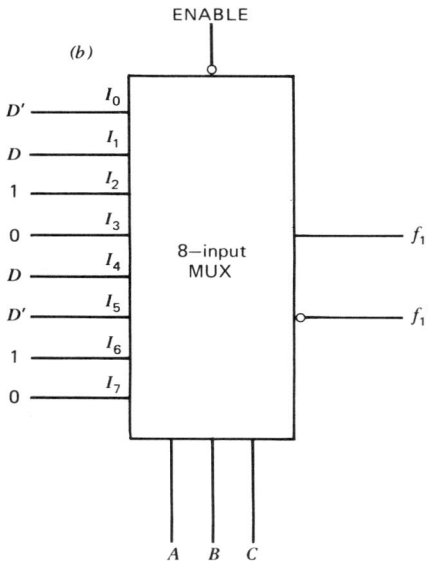

Fig. 4-2.   A multiplexer as a logic function generator.

Figure 4-3 shows how to work directly from the Veitch diagram rather than from the equation. The numbers in Figure 4-3a correspond to multiplexer input numbers. The pattern of 1's in each of the eight rectangles indicates whether to connect D, D', 1, or 0 to that input of the multiplexer.

Though the savings in this example is three IC packages, this is a particularly difficult function of four variables. Remember that many easy functions of

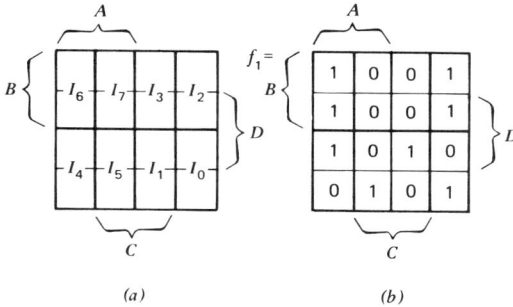

Fig. 4-3. Veitch diagram method: (a) multiplexer input numbers; (b) equation 4-3.

four variables can be done with one or even one-half package using discrete gates. So the multiplexer method is indicated only for the more difficult functions. Also, common product terms cannot be shared between two functions using this method. Since a considerable amount of sharing of common terms is often possible where many functions of the same variables are generated, one should not be overly anxious to discard the gate approach. In general, the choice of method can be made by examining the gate approach and remembering that *any* of the kinds of functions shown in Table 4-2 can be done with *one* IC. This gives us the maximum number of discrete gate packages we should use to generate a function.

TABLE 4-2 FUNCTIONS REQUIRING ONLY ONE
MULTIPLEXER

Function	Multiplexer Type
Any four functions of same two variables	Quad two-input
Any two functions of same three variables	Dual four-input
Any function of four variables	Eight-input
Any function of five variables	16-input
	(or two 8-input)

4.2.2 Simplifying Logic Functions with Multiplexers

Since a multiplexer essentially "removes" the variables connected to its address input from a function, it is sometimes useful for simplifying functions. Gates are used to generate the "residue" functions that are connected to the multiplexer inputs.

We can minimize the residue function generation by choosing the variables that will go to the multiplexer address inputs so the residue functions will be the trivial ones or functions that appear repeatedly. Since the residue functions can be used in generating many different functions, we try to use the variables that seem to appear in many different combinations as the multiplexer address.

As an example, let us look again at the seven-segment digital display problem. The seven functions of four variables are defined by equations 3-6 through 3-12 in Chapter 3. We could, of course, generate these seven functions with seven 8-input multiplexers as just described. However, the simplified logic functions of Figure 3-8 can be mechanized with $5\frac{1}{4}$ conventional gate packages (nine 2-input three 3-input, and four 4-input gates), so the use of eight-input multiplexers is not justified.

However, if we consider the possibility of *simplifying* the functions by using dual four-input multiplexers, our savings will be significant. Because of the possibility of sharing residue functions, we want to reduce all seven functions by the same two variables. We choose the two variables from Veitch diagrams of the functions of Figure 3-8. Depending on which two variables we choose to connect to the multiplexer address, our residue functions (to be connected to the four multiplexer inputs) will be found from various subdivisions of the Veitch diagram as shown in Figure 4-4. Each subdivision can be thought of as a little two-variable Veitch diagram for the required residue function. If we

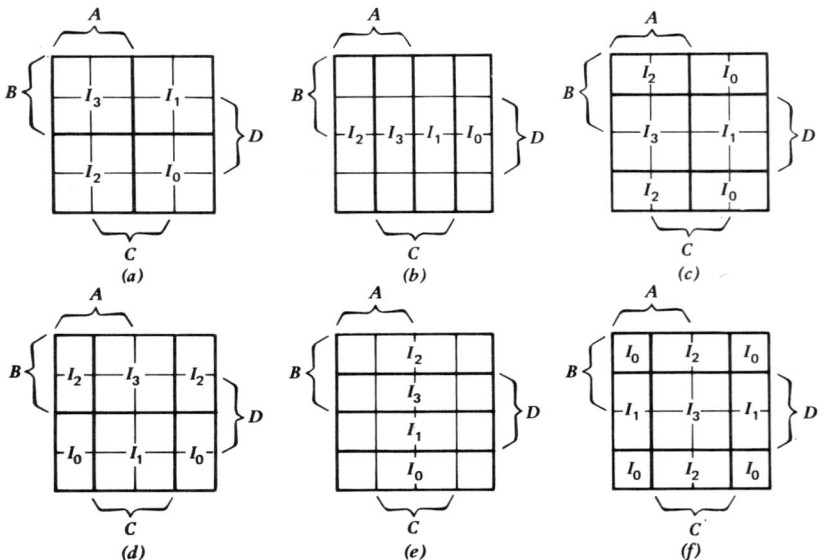

Fig. 4-4.   Six possible ways to reduce function by two variables (I_n = multiplexer input numbers): (a) A and B; (b) A and C; (c) A and D; (d) B and C; (e) B and D; (f) C and D.

look at Figure 3-8 with these possible subdivisions in mind, we can see that the best choice is to eliminate variables B and D. With this choice, each residue function is represented on the Veitch diagram by a horizontal bar. Since this gives us one or two X's ("don't cares") in each function, the chances of ending up with trivial functions or functions that can be shared are very good.

We can now make a table of the the required input functions directly from the Veitch diagrams, as shown in Figure 4-5a. The seven-segment display Veitch diagrams from Figure 3-8 are redrawn with circled terms for this method in Figure 4-5b. The residue functions are simply read off the Veitch diagram rows in their simplest form—just as though each row were a Veitch diagram in two variables. As with any multiple function problem, we try to choose common terms where possible.

Input	MUX ADR	a	b	c	d	e	f	g
I_0	$B' \cdot D'$	1	1	C'	1	1	C'	$A + C$
I_1	$B' \cdot D$	$A + C$	1	1	C	0	A	$A + C$
I_2	$B \cdot D'$	0	C'	1	C	C	1	1
I_3	$B \cdot D$	1	$A + C$	1	C'	0	C'	C'

(a)

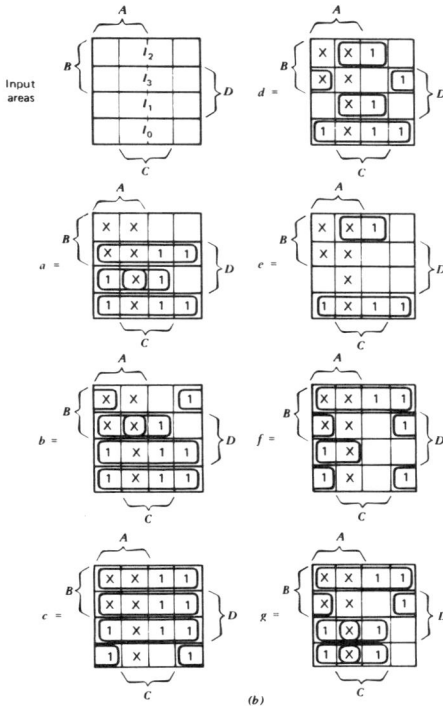

Fig. 4-5. (a) Residue functions. (b) Veitch diagrams used to generate the table of functions.

As the table of Figure 4-5a shows, there is only one nontrivial residue function $(A + C)$, and that can be generated by a single two-input NAND gate. The required circuits for the seven functions are shown in Figure 4-6. Notice that, since the multiplexer circuits come two to a package and there are seven functions, we have done the easiest segment (c) with a conventional three-input gate. The circuit shown requires only $3 + \frac{1}{3} + \frac{1}{4} = 3\frac{7}{12}$ IC's packages.

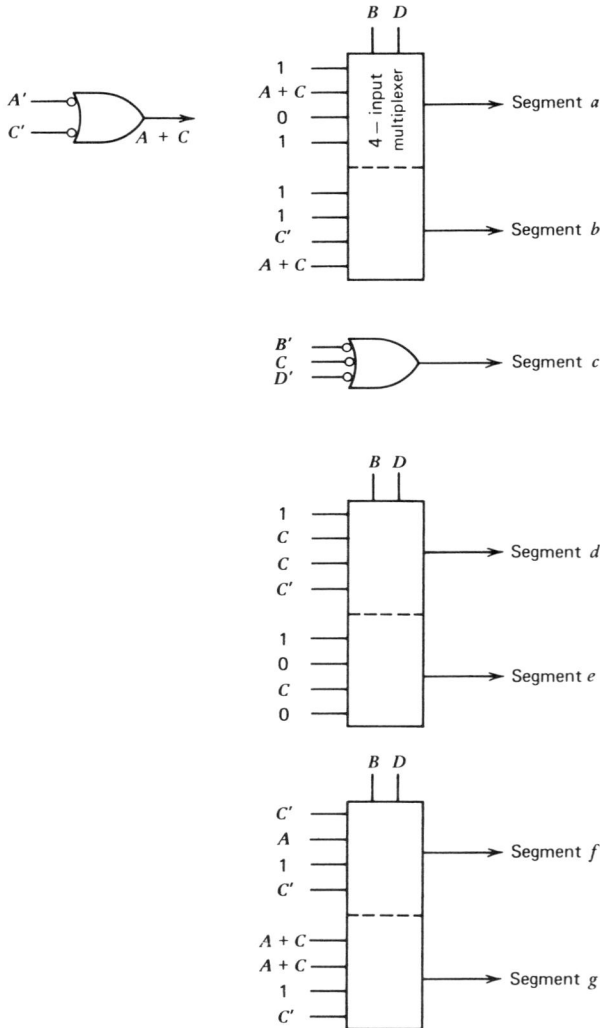

Fig. 4-6.   Seven-segment display driver—multiplexer logic example.

This is considerably less than the 5¼ packages required for a conventional gating mechanization.

4.2.3 Simplifying Functions of Many Variables

The method used in the preceding example is also useful for simplifying the mechanization of functions of a large number of variables. As we saw in Chapter 3 (see Fig. 3-13), the Veitch diagram method becomes messy with functions of more than four variables and downright horrendous for more than six. By using multiplexers, the size of the problem can be greatly reduced. For example, by using a 16-input multiplexer to "remove" four of the variables, a function of eight variables becomes 16 separate and much easier to handle functions of four variables. In addition to making the problem solution *much* easier, the number of IC packages required to mechanize the function can be greatly reduced. Of course, multiplexers can also be used to generate the residue functions, so particularly complicated functions can be done by using a second level of multiplexers to generate the more complicated residue functions.

4.2.4 An Absolute Maximum

Though our example showed that we can often generate residue functions of two variables with very few gates, we can easily calculate the maximum number of packages required in the very *worst case* as follows. There are 16 functions of two variables. Six of these are trivial: A, A', B, B', 1, and 0. This leaves $A' \cdot B'$, $A' \cdot B$, $A \cdot B'$, $A \cdot B$, and $(A' \cdot B) + (A \cdot B')$ (Exclusive OR), plus their complements. All 10 of these functions can be generated with five 2-input NAND gates and five inverters, or a total of $1¼ + \frac{5}{6} = 2\frac{1}{12}$ packages.

Of course, most real functions will not require *all* of these functions, but the important thing is that we can set an *absolute upper limit*. We can thus do *any* function of six variables with a 16-input multiplexer plus a maximum of $2\frac{1}{12}$ gate packages. This approach should thus be considered for any function of six variables that requires more than three gates to mechanize by traditional methods.

4.2.5 Multiplexer Trees

Though the maximum size multiplexer IC available is eight inputs (or 16 inputs in a 24-pin package), we can make any size multiplexer by simply interconnecting several multiplexers in a "tree" structure. Figure 4-7 shows, as

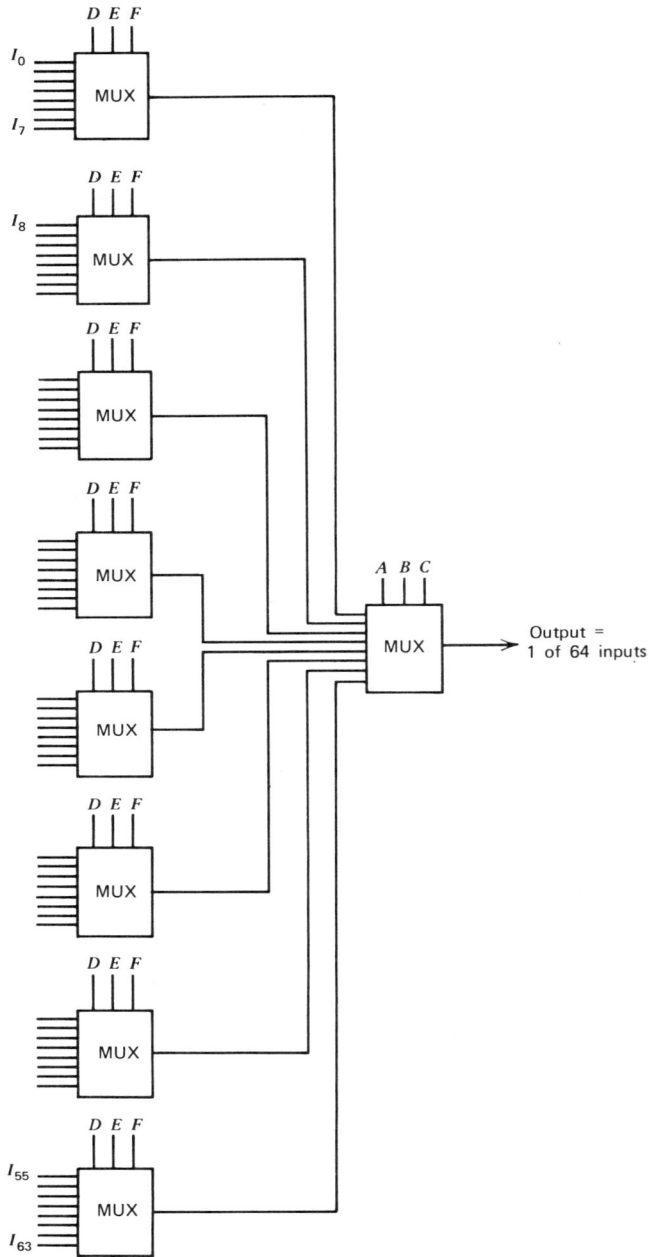

Fig. 4-7. Sixty-four-input multiplexer tree.

an example, a 64-input multiplexer made up of nine 8-input multiplexers. Each first-level multiplexer selects one of its eight inputs depending on address bits *D, E,* and *F.* The second-level multiplexer then selects a first-level multiplexer output, as defined by bits *A, B,* and *C. The final result is that one of the 64 inputs,* as defined by six-bit code *A, B, C, D, E, F, appears on the output.*

Smaller trees can be made by simply omitting some of the first-level multiplexers. *Trees can be made up to any size desired* simply by adding multiplexer levels. For example, eight 64-input multiplexers like the one shown in Figure 4-7 could each drive an input of a third-level eight-input multiplexer to select from a total of $8 \times 64 = 512$ inputs.

This ability to build larger multiplexers extends our capacity to generate logic functions with multiplexers. For example, the 64-input multiplexer of Figure 4-7 can generate any function of seven variables with its nine IC packages. Using the method of Section 4.2.4, however, we can generate any function of seven variables even more economically. Since we can generate any required residue function of two variables with a maximum of $2\frac{1}{12}$ packages, we can use a 32-input multiplexer ($4 + 1 = 5$ packages) to remove the other five variables for a total of $5 + 2\frac{1}{12} = 7\frac{1}{12}$ *packages to do any function of seven variables.* Using the 64-input multiplexer (nine packages) plus a maximum of $2\frac{1}{12}$ packages to generate residue functions, we can generate *any function of eight variables with a maximum of $11\frac{1}{12}$ packages.*

These techniques should be used whenever a discrete gate mechanization of a function requires more ICs. As the seven-segment decoder example showed, we can often get much better results than the maximums shown on the table by factoring out certain variables that carry more information.

TABLE 4-3 MAXIMUM PACKAGE COUNT FOR ANY FUNCTION

Number of Variables	Maximum Number of ICs	Multiplexer Type
2 (four functions)	1	Quad two-input
3 (two functions)	1	Dual four-input
4	1	Eight-input
5	2^a	16-input
6	$4\frac{1}{12}^a$	16-input + residue gates
7	$6\frac{1}{3}^a$	32-input + residue gates
8	$11\frac{1}{12}$	64-input + residue gates

[a] Since 16-input multiplexers require twice the board space, they are counted as two.

The savings with this technique can be very significant (a worst-case function of five variables, for example, requires *18 gate packages* to mechanize). However, real-world functions of many variables, as they appear in simple control logic, usually are so simple that discrete gates are the most economical solution. With experience we can learn to spot functions that should be done with multiplexers by looking at their Veitch diagrams or equations.

4.3 DECODERS/DEMULTIPLEXERS

The opposite of a multiplexer is a demultiplexer, or decoder. *In a decoder, a binary input address determines which one of many outputs will go low.* Figure 4-8 describes a decimal decoder. Other MSI circuits are available with two 4-output decoders (2 address bits common), or a 16-output decoder in a larger, 24-pin package. Other decoders are available, for specialized applications, with high-current or high-voltage outputs, seven segment display driver outputs, excess-three-coded inputs, and so on.

The truth table on Figure 4-8 shows that the decoder generates 10, low true minterms of four, high true, variables. If we have many functions of the same four variables, we can often save gating by "predecoding" the variables with an MSI decoder. For example, the seven-segment display driver problem we have been using as an example can be mechanized (with low true outputs) with only $2\frac{7}{12}$ gate packages plus a decoder, or a total of $3\frac{7}{12}$ packages.

Figure 4-9 shows the circuitry required. The equations are mechanized directly from the truth table in Figure 3-2d by connecting a minterm input for each 0 on the truth table. Note that, since there are many more 1's than 0's on the table, much larger gates are needed to generate the functions using one input per 1 on the table. We are effectively mechanizing the primed functions directly as the canonical minterm form, with the minterms generated by the MSI decoder. The only exception is that we have simplified the e' gating by utilizing the fact that all four minterms constituting the d' function appear in the e' function. If we did not do this, we would have to use an entire IC package to generate e'. This is because an eight-input gate would have to be used to get the required six minterms (since the standard gate sizes are two-, three-, four-, and eight-input only).

This solution, coincidentally, requires the same number of packages as the previous solution of the same problem with multiplexers. It has some slight advantages, however. The ICs are somewhat cheaper since all but one are discrete gates. Also, the complement of the input variables is not required, and the input signals are connected only to the inputs of the decoder, so they are not heavily loaded, as in the multiplexer solution. If we had needed high true

Logic Diagram

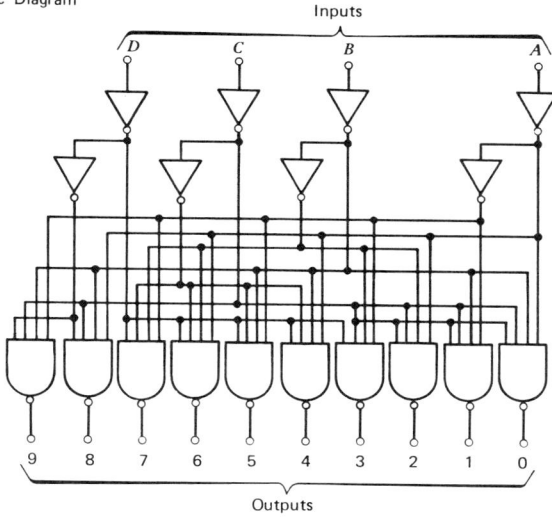

Inputs

Outputs

Block Diagram

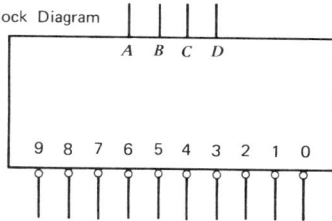

S5442/N7442
BCD
Input

Truth Table

Decimal
output

D	C	B	A	0	1	2	3	4	5	6	7	8	9
0	0	0	0	0	1	1	1	1	1	1	1	1	1
0	0	0	1	1	0	1	1	1	1	1	1	1	1
0	0	1	0	1	1	0	1	1	1	1	1	1	1
0	0	1	1	1	1	1	0	1	1	1	1	1	1
0	1	0	0	1	1	1	1	0	1	1	1	1	1
0	1	0	1	1	1	1	1	1	0	1	1	1	1
0	1	1	0	1	1	1	1	1	1	0	1	1	1
0	1	1	1	1	1	1	1	1	1	1	0	1	1
1	0	0	0	1	1	1	1	1	1	1	1	0	1
1	0	0	1	1	1	1	1	1	1	1	1	1	0
1	0	1	0	1	1	1	1	1	1	1	1	1	1
1	0	1	1	1	1	1	1	1	1	1	1	1	1
1	1	0	0	1	1	1	1	1	1	1	1	1	1
1	1	0	1	1	1	1	1	1	1	1	1	1	1
1	1	1	0	1	1	1	1	1	1	1	1	1	1
1	1	1	1	1	1	1	1	1	1	1	1	1	1

Fig. 4-8.   Decimal decoder.

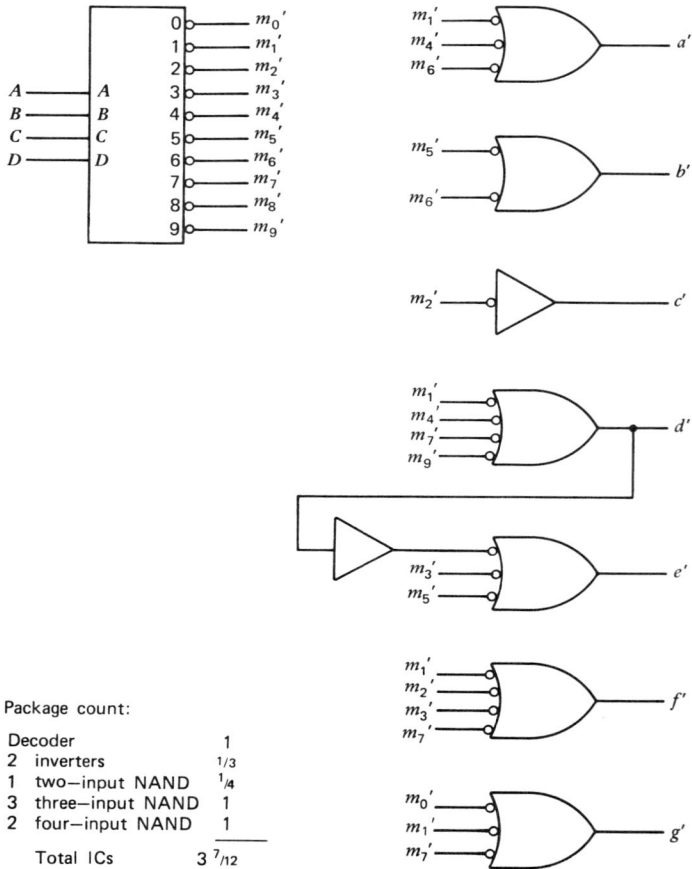

Fig. 4-9.   Seven-segment decoder predecoding example.

outputs, however, we would have needed seven-additional inverters using this approach.

In general, predecoding is a useful tool for simplification *where there are many functions of the same variables.* Loading of the input signals with too many gate inputs is often a problem in this type of situation and predecoding can alleviate it. Instead of a strict sum of minterms approach, we can often use predecoding in conjunction with Veitch diagram simplification. In writing the simplified equations from the diagram, we simply remember that the decoded product terms are available and use them where convenient. This approach is useful only on certain types of functions that can be recognized when writing equations from the Veitch diagrams.

If many large product terms using the same variables seem to be required, then it is worth examining this approach—with the decoder decoding those variables. Often, when three or four bits indicate one of many different modes of operation, functions of this sort result. Whether to use multiplexers or a decoder to handle the mode bits depends largely on how many functions there are. The multiplexer requires less other gating, yet one multiplexer is required for each function, so it is preferred if there are not too many functions. The outputs of a *single* decoder, however, can be used in generating a large number of functions, so it is often more efficient where there are a large number of functions. There are even some cases where the two circuit types can be used together: multiplexers are used to remove variables, and decoders are used to generate the more complex residue functions.

4.3.1   The Decimal Decoder as a Demultiplexer

The decimal decoder shown in Figure 4-8 is often used as an eight-output demultiplexer. If the D input is thought of as a low true enable, then the circuit can be thought of as an eight-output decoder with a common enable (outputs eight and nine are not used in this mode). The three-bit address on inputs A, B, and C thus selects an output, which will be true or false depending on input D. The circuit can thus be thought of as the opposite of a multiplexer—a demultiplexer in that it works something like a selector switch in reverse.

Figure 4-10 shows why these circuits are called multiplexers and demultiplexers. If we want to send many signals over a single path, we can use a switch at each end so that one signal is sent at a time. This is called *multiplexing* and is illustrated with switches in Figure 4-10a and with MSI circuits in Figure 4-10b. Note that the block diagram for the decimal decoder is drawn with the A, B, and C address inputs at the side in this application to better show its function. Also, the D input is drawn with a "bubble" since a low input causes a low output from one of the eight outputs.

4.3.2   Building Demultiplexer/Decoder Trees

Now that we have seen the duality between multiplexers and demultiplexers, let us reconsider the multiplexer tree of Figure 4-7. We can use the same principle with demultiplexers to make a tree as shown in Figure 4-11. If the D input on the left of Figure 4-11 is held low, then one of the eight outputs of the decoder will go high, enabling one of the eight output decoders. Address bits D, E, and F determine which output of that decoder goes low. We thus have one of the 64 outputs going low, as selected by the six-bit address. If we want

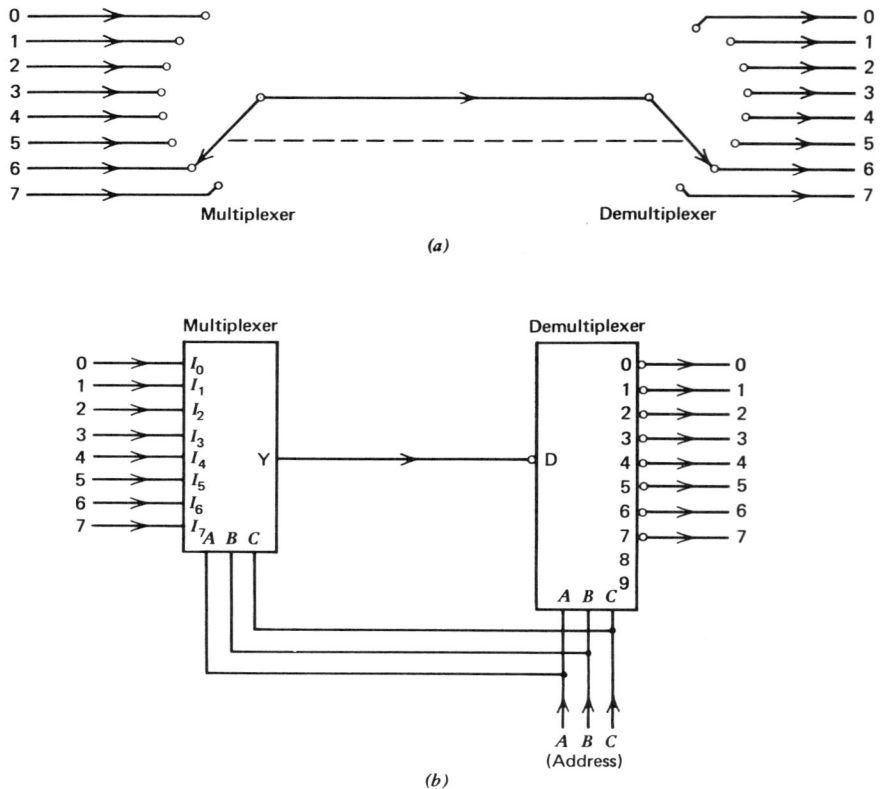

Fig. 4-10. Multiplexing eight signals over one path: (a) using mechanical switches; (b) using MSI logic circuits.

to make an 80-output decoder, we simply add two more output decoders and use the D input of the first decoder as an address bit. We will thus have 10 enables available. Note that this is impossible if we are demultiplexing, since we need input D for the data input.

As with the multiplexer trees, we can build up any size tree we want simply by adding more levels. For example, if each of the 64 outputs is connected to the D inputs of 64 more decoders, we can make a $64 \times 8 = 512$-output decoder/demultiplexer.

4.3.3 Submultiplexing/Demultiplexing

A practical application of large decoder and multiplexer trees is to activate, or sense the state of, one of a large number of separate devices in a large system.

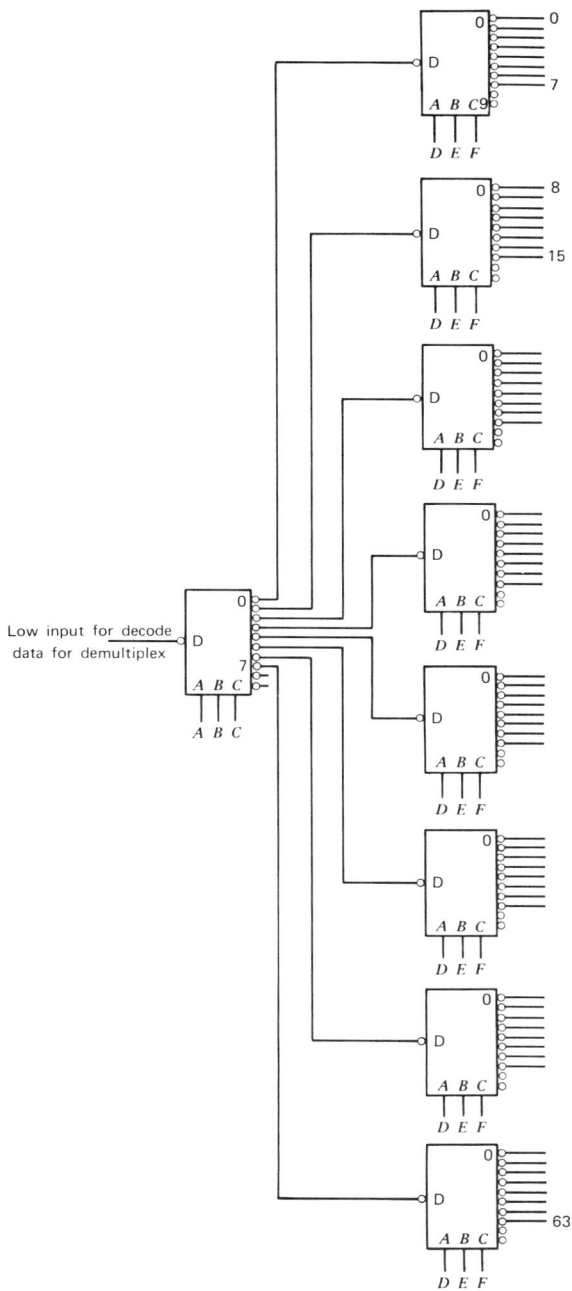

Fig. 4-11. Sixty-four-output demultiplexing/decoding tree.

One example might be a 1000-input multiplexer to sense the state of 1000 telephone lines. Rather than have one multiplexer with 1000 input wires coming to one place from 1000 places, it is more efficient to disperse the final multiplexing circuits. For example, if the telephone line circuitry is arranged on printed circuit cards, with 16 circuits on each card, we could put a 16-input multiplexer on each of the 16 line modules. The same four-bit address goes to each of the line modules, and a single, "submultiplexed" output comes from each line module to a centrally located, 64-input multiplexer tree like that shown in Figure 4-7. We can thus collect data from $64 \times 16 = 1024$ points through only 64 wires. Each of the 64 wires is said to be "submultiplexed." We could similarly distribute control pulses to 1024 destinations from a 64-output decoder.

The idea is that the tree need not be all in one place, but can be distributed throughout the system. Besides saving interconnections, this makes it possible to have only a small tree in the basic system and expand the size in modules of 16 as line circuits are added.

4.4 MULTIDIMENSIONAL ADDRESSING

Sometimes we want to address many points, but the basic module does not include in one place the eight or more addresses handled by MSI decoders. If, for example, there is one address per module, it is impractical to put the final level of a tree on the module. It is also impractical to have a centralized tree with 1000 outputs with a wire to each module. Another approach would be to decode separately a common 10-bit address with a 10-input gate on each module. This is inefficient, however, since we end up using 1000 ten-input gates to do the decoding. Also, the 10 address lines would each have to drive 1000 gates.

The best solution is to use a smaller gate on each module and do some predecoding with MSI decoders. For example, with a three-input gate on each module, we can use three decimal decoders for predecoding 1000 addresses. The predecoder will have only $3 \times 10 = 30$ output wires. In this case, we can think of the three decimal decoders as units, tens, and hundreds digits of a decimal number. Each of the 1000 modules has a decimal address from 000 to 999. For example, module number 132 has as inputs to its three-input select gate: output number 1 from the "hundreds" decoder, number 3 from the "tens" decoder, and number 2 from the "units" decoder. These three signals will all go true only when the BCD address 132 is input to the predecoders. Each predecoder output in this case drives 100 gates, so decoder-drivers should be used. This decoding scheme can be called three-dimensional. Since there are 10 signals in each of the three dimensions, we can address $10 \times 10 \times 10 = 10^3$ points.

If we decrease the number of dimensions to two, then we will need $\sqrt{1000}$ = 32 signals in each dimension. We will thus have 32 select wires in the X dimension and 32 in the Y dimension. We have thus simplified the select gates at the expense of slightly more complicated preselect logic. (Fig. 4-12a shows an 8×8 = 64-output, two-dimensional, matrix.) Each preselector output now drives 32 select gates, so we have reduced the loading by a factor of 3. Though 10 packages are required for preselect logic versus three for the three-dimen-

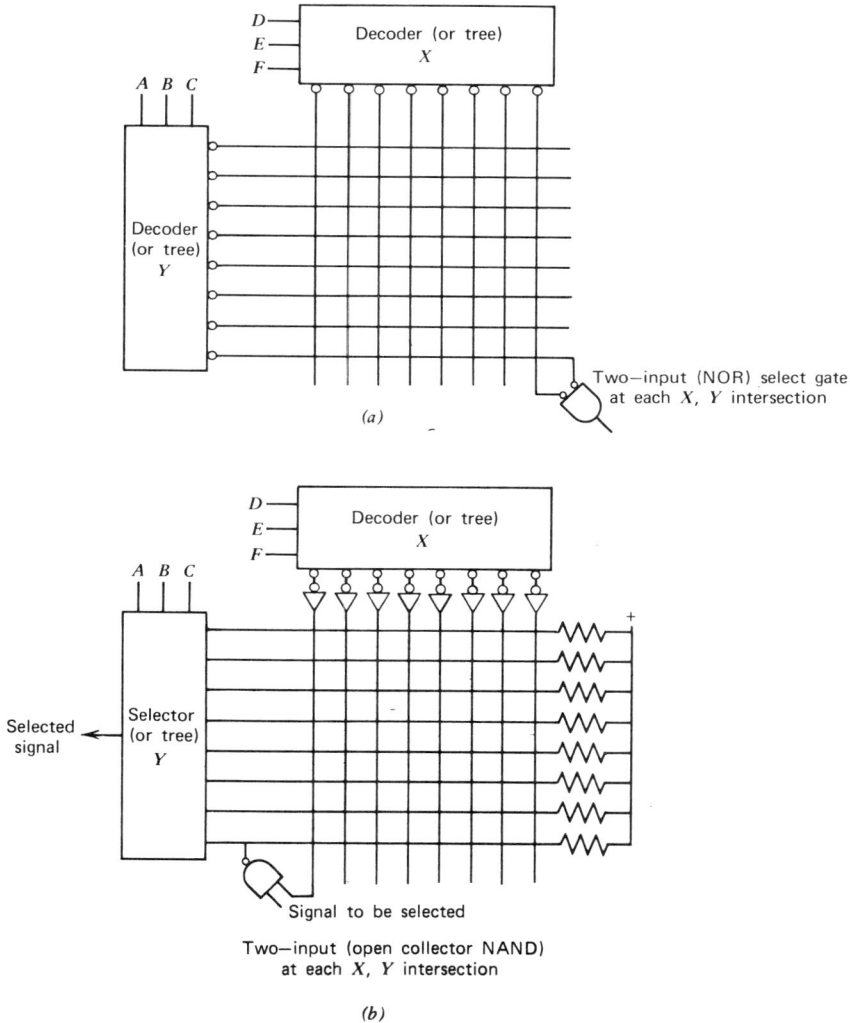

(a)

(b)

Fig. 4-12. *X-Y* matrix selection: (a) output selector; (b) input selector.

sional solution, the decrease in the select gates from 1000 three-input gates (333 packages)* to 1000 two-input gates (250 packages)* more than offsets this. In fact, the most economical approach from an *IC package* count point of view would be the full 1024-output tree. This approach, however, is less desirable because of the interconnection problem.

We can conclude, then, that the two-dimensional approach is the best overall since the $2 \times 32 = 64$ select wires and 10 preselect ICs can easily fit on one central module.

If we had to select a larger number, three dimensions would be justified to reduce the number of wires. For example, with $3 \times 24 = 72$ wires we could select $24^3 = 13,824$ points.

Figure 4-12b shows how the same matrix principle can be be used to select input points. Here, an open-collector gate is used to put the signal selected by decoder X on each Y line. The Y selector picks the correct input. In Figure 4-12b any of $8 \times 8 = 64$ inputs can be selected; but by using trees, of course, as in the above output selection example, we can select one of 1024 points with a 32×32 matrix.

Of course we could use unequal X and Y dimensions such as 128 X's and eight Y's ($128 \times 8 = 1024$); but this is usually less desirable (note that there would be $128 + 8 = 136$ select wires). In fact we can state as a general rule: *Always divide the dimensions in such a way that the cost of each dimension is equal.*

This rule applies in many situations where capability varies as the product of the dimensions and cost, varies as the sum. In the two-dimensional case, the costs of the X and Y dimensions are equal when their sizes are equal, since the circuits in each are identical. If, however, one dimension were an expensive sense amplifier and the other a simple logic decoder output (as is sometimes the case in RAMs), then we would probably use unequal dimensions—divided in inverse proportion to the cost per circuit. Of course, as in the example, modularity and pin limitations can force a choice that is not strictly an equal cost division.

Although multiplexer and demultiplexer circuits are sometimes useful at the logic equation mechanization level, their greatest utility is at the system design level. Whenever signals must be selected from, or directed to, many sources, or when mode of operation must be switched, the system design should be "*adapted to*" the capabilities of MSI multiplexers and demultiplexers.

4.5 OTHER MSI COMBINATIONAL LOGIC CIRCUITS

The catalogs of the IC manufacturers are filled with detailed descriptions of many other MSI circuits. *These catalogs are the digital designer's most im-*

---

* This assumes we can use the other gates in the package elsewhere.

*portant tool.** The available circuits should be learned well and scanned and rescanned during the system design phase. If maximum use is made of the complex circuits available, we only need to "fill in the cracks" with traditional logic design techniques.

An important job that used to be performed by logic designers was design of logic circuits to perform binary arithmetic operations. This job has now been essentially taken over by the IC manufacturers. A one-chip programmable processor (the subject of Chapter 7) can handle all arithmetic operations and can be programmed to combine them in any way. If a programmed processor is too slow, we can buy quite a selection of MSI arithmetic circuits to do the job. The following are some selected examples:

1.   A *four-bit adder* circuit that generates the binary sum of two, 4-bit numbers. A carry input and output makes it possible to expand to any word length. For example, four 4-bit adder circuits add 16 bits (e.g., 74283).

2.   A *look-ahead carry generator,* which can be used with four of the above adder circuits to give faster operation on a 16-bit addition. For longer words, an additional circuit is used for each 16 bits (e.g., 74182).

3.   A four-bit binary *magnitude comparison* circuit, which compares two binary numbers A and B and produces three outputs: $A = B$, $A > B$, and $A < B$. Expansion inputs allow expansion to any word length in a tree structure. For example, up to 24 bits can be compared using six circuits in a two-level tree. A three-level tree can compare up to 120 bits or can be partially implemented to compare 32 bits with 10 packages (e.g., 7485).

4.   A multiple-function *arithmetic logic unit* (ALU), which can perform four-bit addition, subtraction, Exclusive OR, OR, NOR AND, NAND, or comparison—as specified by a control code. It also stores the result and shifts it to the right or left, with or without carry. The word length can be expanded to any size with just one circuit required for each four bits. For very high-speed operation, the same carry look-ahead circuit described above can be used (e.g., 745281).†

5.   A 4 × 4 binary *multiplier,* which multiplies two 4-bit binary numbers and produces a four-bit output. Two circuits together can produce an eight-bit product. Expansion is not so easy with these circuits. For example, an 8 × 8 bit multiplier with 16-bit output requires 16 packages—only eight of which are multipliers (e.g., 74284).

Besides arithmetic, standard MSI packages are available to handle several other standard, combinational logic operations. Here are a few examples:

1.   An eight-input, three-output *priority encoder,* which produces a bi-

---

* Texas Instruments' excellent catalog can be purchased at a very low price. Signetics and Advanced Micro Devices also have excellent catalogs.

† It is interesting to note that this circuit costs *about the same* as a single IC gate circuit cost 12 years ago, and it performs 16-bit addition in about the same time it took a signal to propagate through *one gate* at that time.

nary output code indicating which of eight inputs is low. If more than one input is low, the highest numbered input has priority. Enable inputs and outputs make it possible to cascade circuits to handle more inputs, but external gating is required to combine the output bits. For example, for 16 inputs, two encoder circuits and one quad two-input NAND gate are required. Another version of the circuit encodes 10 inputs into BCD but is not expandable. This circuit is useful for encoding small keyboard arrays, handling computer priority interrupts, and so on (e.g., 74148).

2. A nine-bit **parity checker**/*generator,* which can generate or check either odd- or even-parity error-checking codes. These circuits can be connected in a two-level tree to check up to $9 \times 9 = 81$ bits with $9 + 1 = 10$ packages. With a partially completed tree, unused second-level inputs can be used, so, for example, $9 + 9 + 7 = 25$ bits can be checked with three packages. In addition, several parity checkers can be connected to different groups to generate a **"hamming" code**, which indicates not only that there is an error, but which bit is in error (e.g., 745280).

3. A *six-bit binary-to-BCD converter,* which converts a six-bit binary number to a seven-bit BCD number. Three circuits can be connected together to convert eight binary bits, and 16 packages will convert 16 bits. As with the BCD-to-binary converter below, the expansion structure is a complex, multilevel structure (e.g., 74185).

4. A six-bit *BCD-to-binary converter,* which actually has half of the circuitry required to convert two 4-bit BCD decodes into a seven-bit binary number. Six circuits can be interconnected to convert three BCD decodes to a nine-bit binary number. Each additional decode beyond this point requires about twice as many circuits (e.g., 74184).

The preceding circuits are listed mainly to indicate the kinds of special MSI combinational logic circuits available. The logic equations for these circuits are very complex, so *their applicability must be recognized at the system design level.* Once we start making truth tables or writing logic equations, it is too late to use this type of circuit. Quite often the system specification is affected by the initial knowledge of available standard circuits.

4.6 READ-ONLY MEMORIES

A **read-only memory (ROM)** is a good example of what follows standardization; we might call it "customized standardization." When Henry Ford created the Model T, he brought the price way down by making thousands of identical automobiles. He said that you could have the Ford "in any color you want—as long as it's black." Now we are at a greater level of sophistication where computers and automatic machinery make it possible to order a "customized" car, with any options you want. We still get the same benefits of standardization, but *the standard process is just more sophisticated.*

Read-only memories represent the ultimate in customized standardization; for example, we can actually specify a 1 or a 0 in any of 16,384 bit locations. The number of possible combinations is thus $2^{16,384}$, or about 10^{5000}. For comparison, it has been calculated that only 10^{63} electrons could be packed in a sphere the size of the earth.

This fantastic customization is made possible by using computers to generate automatically a customized **mask** for one masking step of the normal IC process. As hundreds of chips are made at a time, the mask has hundreds of repetitions of the custom pattern of 16,384 light or dark spots. Since the mask generation is automatic, the IC manufacturer simply has to load a deck of punched cards containing the information and push a button! The light and dark spots on the mask cause oxide to be selectively etched away or left there, to define 1's and 0's.

The 16,384-bit ROM used as an example has a 128×128 $X\!-\!Y$ array of points that either are or are not conducting (see Fig. 4-13). The circuit has eight output pins and 11 address input pins, so it is organized as 2048 words of eight bits each. Thus, for each binary combination of the input pins, a different, custom specified, eight-bit output is produced. Operation is similar to the $X\!-\!Y$ selection matrix of Figure 4-12*b* except the outputs are selected here in eight groups of 16.

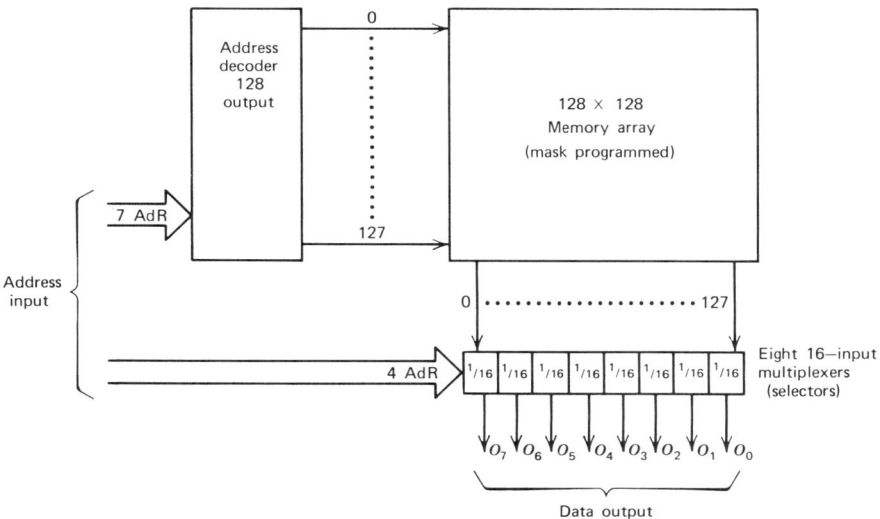

Fig. 4-13. A 16,384 bit ROM.

TABLE 4-4 ROM'S AS FUNCTION GENERATORS

Logic Functions/ Variables	Mounted Gate IC Cost Multiple (1978)	ROM Organization
Any 4 of 8 variables	~2	256 × 4 bits = 1,000
Any 8 of 8 variables	~2	256 × 8 bits = 2,000
Any 4 of 9 variables	~2	512 × 4 bits = 2,000
Any 8 of 9 variables	~2	512 × 8 bits = 4,000
Any 4 of 10 variables	~2	1024 × 4 bits = 4,000
Any 8 of 10 variables	~2	1024 × 8 bits = 8,000
Any 4 of 11 variables	~2	2048 × 4 bits = 8,000
Any 8 of 11 variables	~3	2048 × 8 bits = 16,000
Any 4 of 12 variables	~3	4096 × 4 bits = 16,000

We thus have another maximum to add to our maximum package count list (Table 4-3). It is unfair, however, to count a ROM as only one package. The customization increases the cost somewhat, and large ROMs, like the 16,384 bit unit we have been using as an example, require a 24-pin package, which is equivalent to two small ICs in space used on the circuit board. Although the cost of ROMs is many times greater than that of gate ICs, if we include the hidden costs we discussed in Chapter 2, one ROM mounted in a system costs about the same as 2 to 4 mounted gate ICs today. Future reductions should decrease this ratio to two or three in the near future. Read-only memories are available in a variety of sizes and organizations. We can tabulate the capabilities of various ROMs for mechanizing logic functions as shown in Table 4-4.

All of the ROMs in the table occupy a single 24-pin package (except the 1000 and four-bit 2000, which have 16 pins), so it is clear that *really complex logic functions should not be done with discrete gates.*

4.7 PROGRAMMABLE READ-ONLY MEMORIES

Even though custom mask preparation is totally automated, mask programming of ROMs is practical only when we can use a minimum of 100 or so identical devices. The reason, of course, is that even though only one masking step is customized, a special production run is effectively done for each mask pattern. Another disadvantage of mask-programmed ROMs is that it is virtually impossible to make changes if an error is found in the design.* It is thus

---

* Though it is possible to add AND gates to remove, or OR gates to add, additional minterms.

essential that we use another type of ROM for small production runs and for initial checkout and pilot production.

The **programmable read-only memory (PROM)** is a completely standard device that can be manufactured by the millions. To program the device, the customer (or sometimes the distributor) simply plugs it into a "PROM Programmer," which writes the desired bit pattern into the device. Figure 4-14 shows a simple programmer that uses a punched paper tape to read in the desired program. Some PROMs are *reprogrammable* in that they can be erased electrically, or by exposure to ultraviolet light, and reprogrammed repeatedly. Since PROMs are usually pin-compatible with an equivalent mask programmed ROM, *high-volume systems can be made with PROMs until enough field experience has been accumulated to justify commitment to volume production of mask-programmed ROMs.* The ROM masks are then generated from the *same* punched tape or cards used for the PROMs and plugged into the same circuit board locations previously filled by PROMs.

System checkout can be further simplified by using a ROM simulator (see Fig. 4-15). This device has IC-shaped plugs that plug into the ROM socket in the system being checked out. A RAM is used to simulate the ROMs so single words can be changed, without reprogramming the entire memory, by simply pressing buttons on the control panel to indicate the desired new word and its address. Programs are read in with punched tape with the same format used on the PROM programmer.

If we compare a PROM to discrete gating as a method of mechanizing complex logic, we see that a programmed PROM performs the function of a complex, interconnected group of gates. The difference is that an inexpensive

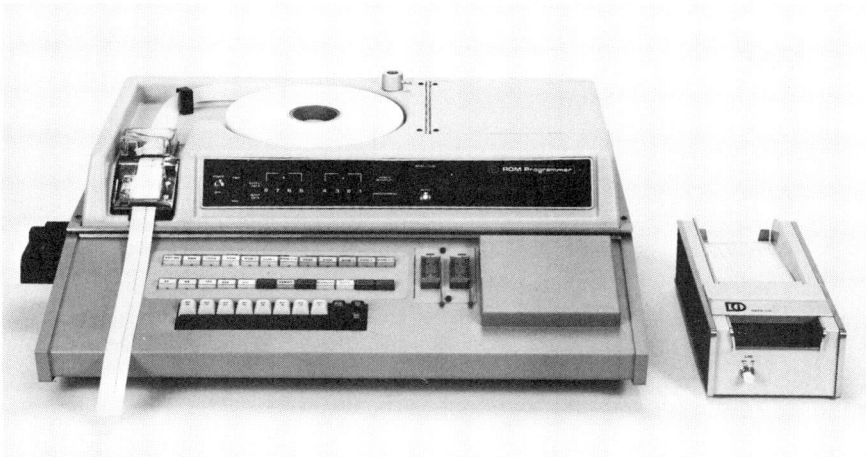

Fig. 4-14. PROM programmer.

Fig. 4-15.   ROM simulator.

PROM programmer can automatically handle the "interconnection." This is a little like having your own, briefcase-sized, programmable, automatic, wire wrap machine to interconnect gates (these machines cost hundreds of thousands of dollars and weigh tons). The standardized punched tape or card format makes it possible to use computer-aided design programs to produce, modify, and document the tape or cards. The same tape or cards can then be used to load a ROM simulator and program an automatic tester that can completely test a ROM or PROM in seconds by checking the outputs for each possible input code.

4.8   ROM/PROM LOGIC SIMPLIFICATION

Simplification of logic equations, as discussed in Chapter 3, is meaningless when ROMs are used. In fact, if the function is simplified, it must essentially be expanded to canonical minterm form to be mechanized on a ROM. This expansion can be done directly on the truth table, as illustrated on Figure 4-16. Since each output represents one function, we work with one function at a time. For each term of the equation for the function we write one or more 1's on the truth table. If all the input variables are represented in the term, then we write only a single 1. If n variables are missing, then we must write 2^n 1's to mechanize the term (though some of the 1's may already be there

CUSTOMER: _____   THIS PORTION TO BE COMPLETED BY SIGNETICS

P.O. NO.: _____   PART NO.: _____

YOUR PART NO.: _____   S.D. NO.: _____

DATE: _____   DATE RECEIVED: _____

INPUTS

ABCDEF	Word	GH = 00 OUTPUT				Word	GH = 01 OUTPUT				Word	GH = 10 OUTPUT				Word	GH = 11 OUTPUT			
		O_4	O_3	O_2	O_1		O_4	O_3	O_2	O_1		O_4	O_3	O_2	O_1		O_4	O_3	O_2	O_1
000000	0	1	0	0	0	64	1				128					192				
000001	1	1	0	0	0	65	1				129					193				
000010	2	1	0	0	0	66	1				130					194	1			
000011	3	1	0	0	0	67	1				131					195				
000100	4	0	0	0	0	68					132					196	1			
000101	5	0	0	0	0	69					133					197	1			
000110	6	0	0	0	0	70					134					198	1			
000111	7	0	0	0	0	71					135					199				
001000	8	0	0	0	0	72					136					200				
001001	9	0	0	0	0	73					137					201				
001010	10	0	0	0	0	74					138					202	1			
001011	11	0	0	0	0	75					139					203	1			
001100	12	0	0	0	0	76	1				140					204	1			
001101	13	0	0	0	0	77	1				141					205	1			
001110	14	0	0	0	0	78	1				142					206	1			
001111	15	0	0	0	0	79	1				143					207	1			
010000	16	0	0	0	0	80					144					208				
010001	17	0	0	1	0	81			1		145					209				
010010	18	0	0	0	0	82					146					210	1			
010011	19	0	0	1	0	83			1		147					211				
010100	20	0	0	0	0	84					148					212	1			
010101	21	0	0	1	0	85			1		149					213	1			
010110	22	0	0	0	0	86					150					214	1			
010111	23	0	0	1	0	87			1		151					215				
011000	24	0	0	0	0	88					152					216				
011001	25	0	0	0	0	89					153					217				
011010	26	0	0	0	0	90					154					218	1			
011011	27	0	0	0	0	91					155					219				
011100	28	1	0	0	0	92	1				156	1				220	1			
011101	29	0				93	1				157					221	1			
011110	30	1				94	1				158	1				222	1			
011111	31	0				95	1				159					223	1			
100000	32	1				96	1				160	1	1			224	1	1		
100001	33	1				97	1				161	1	1			225	1	1		
100010	34	1				98	1				162	1	1			226	1	1		
100011	35	1				99	1				163	1	1			227	1	1		
100100	36	0				100					164		1			228	1	1		
100101	37	1				101	1				165	1	1			229	1	1		
100110	38	0				102					166		1			230	1	1		1
100111	39	1				103	1				167	1	1			231	1	1		
101000	40	0				104					168					232				
101001	41	0				105					169					233				
101010	42	0				106					170					234	1			
101011	43	0				107					171					235				
101100	44	1				108	1				172	1				236	1			
101101	45	0				109					173					237	1			
101110	46	1				110	1				174	1				238	1			
101111	47	0				111					175					239				
110000	48	0				112					176					240				
110001	49	0		1		113			1		177					241				
110010	50	0				114					178					242	1			
110011	51	0		1		115			1		179					243				
110100	52	0				116					180					244	1			
110101	53	0		1		117			1		181					245	1			
110110	54	0				118					182					246	1			
110111	55	0		1		119			1		183					247				
111000	56	1				120	1				184	1				248	1			
111001	57	1				121	1				185	1				249	1			
111010	58	1				122	1				186	1				250	1			
111011	59	1				123	1				187	1				251	1			
111100	60	1				124	1				188	1				252	1			
111101	61	1				125	1				189	1				253	1			
111110	62	1				126	1				190	1				254	1			
111111	63	1				127	1				191	1				255	1			

Fig. 4-16.   1024 bit ROM truth table/order blank.

from other terms). Figure 4-16 illustrates a ROM mechanization of the following equations:

$$f_1 = 0_1 = A \cdot B' \cdot C' \cdot D \cdot E \cdot F' \cdot G \cdot H \tag{4-4}$$

$$f_2 = 0_2 = B \cdot C' \cdot F \cdot G' \tag{4-5}$$

$$f_3 = 0_3 = A \cdot C' \cdot G \tag{4-6}$$

$$f_4 = 0_4 = A' \cdot B' \cdot C \cdot D' \cdot E \cdot F \cdot G \cdot H + A \cdot B' \cdot C' \cdot F + E \cdot F' \cdot G \cdot H$$

$$+ B \cdot C \cdot D \cdot F' + A \cdot B \cdot C + D \cdot E' \cdot G \cdot H + A \cdot C \cdot D \cdot F'$$

$$+ B' \cdot C' \cdot D' \cdot G' + A' \cdot C \cdot D \cdot H \tag{4-7}$$

The first three functions are, of course, too simple to mechanize economically on a ROM, but they illustrate how we can locate the squares on the truth table for a term. By writing the states of the input variables along the left edge and top of the table, we make it possible to locate the words defined by the term by a process of elimination. For example, f_1 is simply a single minterm: word number 230. On the other hand, f_2, which has only four variables, indicates $2^4 = 16$ minterms. We simply find the eight places on the left edge where $B = 1$, $C = 0$, and $F = 0$ and write 1's in the two columns where $G = 0$.

Because of the orderly nature of the truth table representation of ROM-mechanized logic functions, computer-programmed design automation is quite simple. A simple computer program can thus convert logic equations directly into punched cards or paper tape for ROM or PROM programming.

Since ROMs can mechanize *any* function, they represent a tremendous amount of excess capability if used to mechanize most real-world logic functions as they naturally occur. To make full use of a ROM's capabilities, the input and output variables normally have to be encoded efficiently so that each variable has a very high information content. The more compactly coded the inputs and outputs of the ROM are, the more complicated the logic equations will be. However, since the ROM size is determined *only* by the number of inputs and outputs, the result is a savings in hardware.

Our example of the seven-segment display driver is a perfect example. Since the seven-segment signals can be easily derived from a four-bit BCD code, the seven signals have less information content than the four bits of the BCD code. One way, then, of reducing the number of ROM output bits on a system with a seven-segment display output would be to use an external BCD-to-seven-segment decoder connected to four ROM output bits instead of using seven ROM outputs and doing the decoding in the ROM.

A more extreme example would be where the logic must route eight bits of data to one of four locations. Instead of using eight inputs and $8 \times 4 = 32$ outputs to do this in the ROM, we should simply have two bits out of the

ROM control the address inputs to four separate dual four-output demulti-plexer ICs. We thus *concentrate the truly random logic functions in the ROM* by using MSI or gates to do the simple special functions.

With practice we can learn to look at all the inputs for groups that can be externally encoded and the outputs for groups that can be externally decoded. *Any group of variables that is mutually exclusive* (that is, only one can go true at a time) *can be encoded.* For example, we press only one key of a 16-key keyboard at a time; we can thus reduce the number of ROM input pins required from 16 to only four by just encoding them externally.

Another example of an encodable input is the 12 rows of the punched cards used for computer input and output. Only one of the first seven rows on the card is ever punched. We can thus encode these seven rows into three bits with an MSI encoder circuit. Figure 4-17 shows two approaches to the problem of converting the punched card into an eight-bit **ASCII** code. Without en-coding, our 12 inputs give us $2^{12} = 4096$ words, so a $4096 \times 8 = 32{,}768$-bit ROM is needed. Figure 4-17b shows how the use of one MSI encoder can reduce the ROM size to $2^8 = 256$ words or 256×8 bits $= 2048$ bits. Actually, since there are only 128 valid codes, we could further compress the code to eight bits (the five remaining bits appear only in certain combinations),

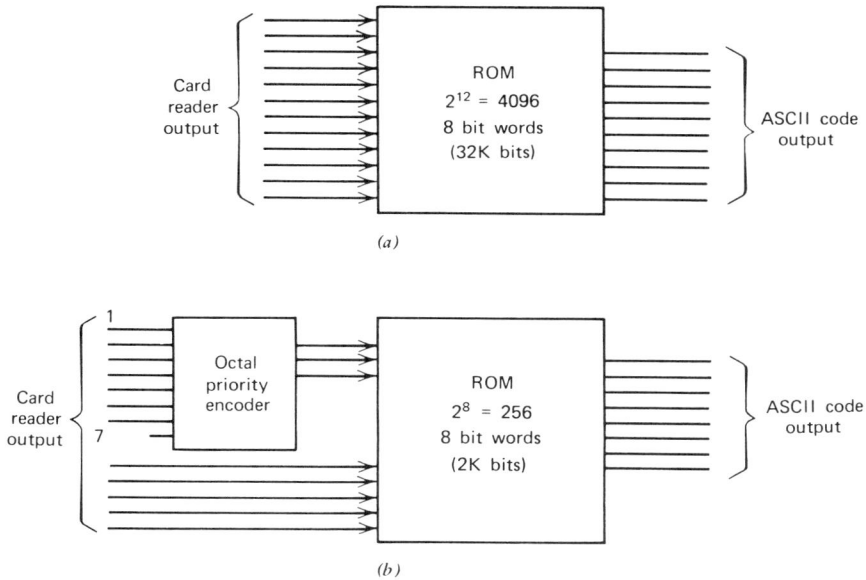

(a)

(b)

Fig. 4-17. Encoding ROM inputs—punched card code converter (a) without en-coding; (b) with encoding.

making it possible to use a 1024-bit ROM. In this case, however, the larger ROM costs less than the encoding logic would. If we want to build a code converter in the reverse direction (ASCII to Hollerith), we can similarly use a decoder to produce the first seven outputs. The savings from decoding outputs is not so great, however, since the *ROM size is directly proportional to the number of outputs while it doubles for each additional input bit.*

Multiplexers are often useful for reducing the number of ROM inputs. If, for example, there are several modes of operation, and different inputs are of interest for each mode, we can use a multiplexer to select only the bits of interest. If the mode of operation is not nicely encoded, we can sometimes even use a separate ROM to produce the multiplexer address as shown in Figure 4-18. We can even use outputs of a ROM to select *its own* inputs (as long as those inputs do not enter into the selection equation).

In general, the classical "gates and flip-flops" logic design approach will not result in logic functions that are sufficiently densely encoded to make the use of ROMs or PROMs feasible. Instead, we must define the system from the start with the idea of concentrating the logic in a ROM block on the block diagram. Since this block can perform any logic function, the only thing we need worry about in the system design phase is minimizing the number of inputs and (to a lesser degree) outputs to and from the ROM block. This philosophy often leads us to a "microprogrammed" approach, where the ROM is actually thought of as containing something like a computer program. This approach is discussed further in Chapter 6.

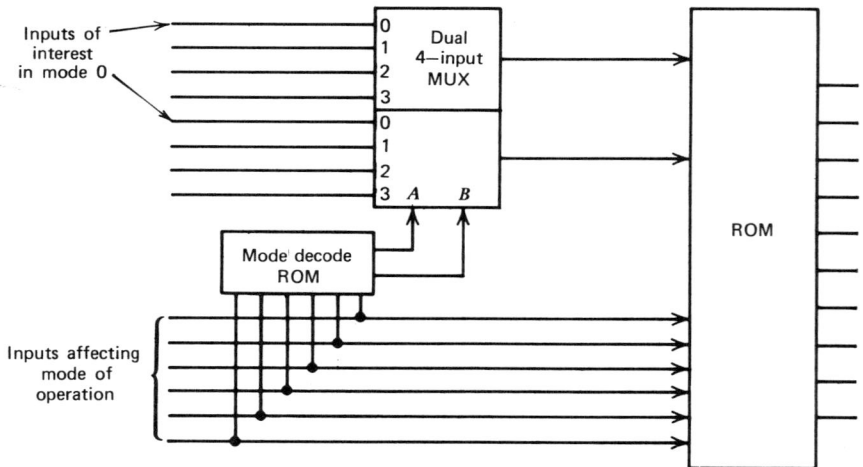

Fig. 4-18.   Using a ROM to select pertinent inputs.

4.9 PROGRAMMABLE LOGIC ARRAYS

The **programmable logic array (PLA)** is a solution to the input encoding problem we just discussed with ROMs. *With a PLA we specify not only an output bit pattern for each word but also the word address.* This makes it possible to have more inputs for the same-size ROM. For example, a 96-word × 8-bit = 768-bit PLA can have 14 inputs.* If all input combinations defined a word (as they do on a ROM), 14 inputs would define $2^{14} = 16,384$ words, or a $16,384 \times 8 = 131,072$-bit ROM. With a PLA we can define the 96 words in terms of the 14 inputs. Since we can specify 1, 0, or X ("don't care") on each of the 14 inputs, we essentially perform the input code compression within the word decoding portion of the chip.

Just as with the ROM, *we can look at a PLA as a memory,* as a *code converter,* or as a *logic function generator.* As a *code converter* we can do the punched-card-to-ASCII code conversion shown in Figure 4-17b *directly* with a 96-word PLA with 12 inputs and eight outputs. Each of the 96 words we define will correspond to a *valid* punched card code.

As a memory a PLA can, internally, perform the same function as the extra ROM and multiplexer in Figure 4-18. Because we are not limited to simple, easily mechanized, input reducing functions, we can usually do a much better job of making use of input redundancies.

If we think of the PLA as a logic function generator,† we see that it is actually mechanized as an AND/OR logic structure where the 96-word addresses are decoded by 96 14-input AND gates, and each output is a logical OR of any of the 96 AND gates. Each output can thus be defined as *a sum of logical product terms.*

As a simple example, Figure 4-19a shows a small PLA layout with six inputs, 13 product terms (words), and four outputs. The dots on the regular matrix at the top can be thought of as AND inputs to the vertical "product term" line. These are either connected to the inverted, or uninverted, input or not connected at all (for a "don't care") as specified by the customer. The smaller matrix at the bottom is the output OR matrix. Each dot can be thought of as an OR input to generate one of the outputs. Both of these matrices are specified by the customer on a truth table as shown in Figure 4-19b. As with the ROM, the truth table data are punched onto computer cards and used to automatically generate a special mask, which is used in one step of the manufacture of the PLA.

It is easy to see from Figure 4-19a how the matrix arrangement makes very efficient use of the chip. In fact, custom LSI random logic ends up wasting so

---

* The National Semiconductor DM 7575, for example.

† The PLA is so flexible that the 14-input, eight-output one used as an example could be programmed to do the job of *any* of the special MSI logic circuits described earlier in this chapter.

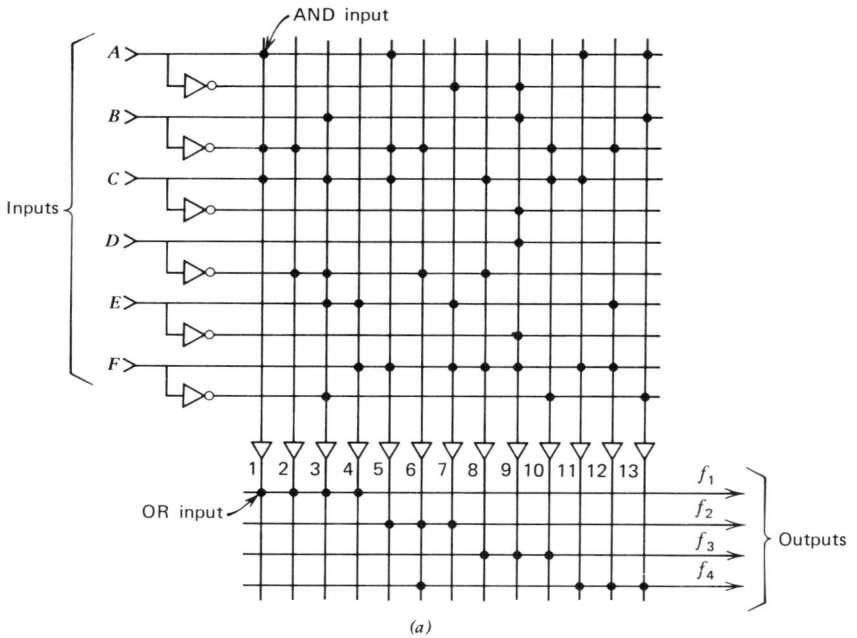

Fig. 4-19. (a) Programmable logic array (simplified). (b) Truth table for ordering such a PLA.

Product	Inputs							Outputs			
Term No.	A	B	C	D	E	F		f_1	f_2	f_3	f_4
1	1	0	1	X	X	X		1	0	0	0
2	X	0	X	0	X	X		1	0	0	0
3	X	1	1	0	1	0		1	0	0	0
4	X	X	X	X	1	1		1	0	0	0
5	1	0	1	X	X	1		0	1	0	1
6	X	0	X	0	X	X		0	1	0	1
7	X	0	X	X	1	1		0	1	0	0
8	X	X	1	0	X	1		0	0	1	0
9	0	1	0	1	0	1		0	0	1	0
10	X	0	1	X	X	0		0	0	1	0
11	1	X	1	X	X	1		0	0	0	1
12	X	0	X	X	1	1		0	0	0	1
13	1	X	1	X	X	0		0	0	0	1

(b)

much space on the random interconnections (e.g., see Fig. 7-3) that the chip area per gate is up to 100 times the area per intersection of a PLA matrix. Of course, a ROM has this same advantage; in fact, the word decoding on a ROM is done with an exhaustive matrix like the one shown in Figure 4-19a, but with fewer inputs.

To demonstrate how to order a PLA to mechanize logic functions, let us use our tiny PLA example of Figure 4-19 to mechanize the following equations:

$$f_1 = A \cdot B' \cdot C + B' \cdot D' + B \cdot C \cdot D' \cdot E \cdot F' + E \cdot F \qquad (4\text{-}8)$$

$$f_2 = A \cdot B' \cdot C \cdot F + B' \cdot D' + B' \cdot E \cdot F \qquad (4\text{-}9)$$

$$f_3 = C \cdot D' \cdot F + A' \cdot B \cdot C' \cdot D \cdot E' \cdot F + B' \cdot C \cdot F' \qquad (4\text{-}10)$$

$$f_4 = A \cdot C \cdot F + B' \cdot E \cdot F + B' \cdot D' + A \cdot C \cdot F' \qquad (4\text{-}11)$$

Because the sum-of-products form of the equations fits the logic of the PLA exactly, we can simply enter the functions directly on the truth table of Figure 4-19b. We start by counting the product terms to be sure they fit the circuit. There are four terms in f_1, three in f_2, three in f_3, and four in f_4, for a total of 14. Since the circuit can handle only 13 product terms, we must first combine two terms. Noting that the second term of f_2 and the third term of f_4 are both $B' \cdot D'$, we use the same product term for both functions. We can now begin to mechanize the functions by entering each term, in order, on the truth table (Fig. 4-19b). The first product term, $A \cdot B' \cdot C$, is mechanized by writing 101XXX on the first row of the table indicating that the first AND gate should be connected to A, B', and C and should have no connection to D, E, and $F(X = $ don't care). A 1 in column f_1 indicates that the OR gate for output f_1 should be connected to this term. We simply continue this way until we have filled out a row for each term. The dots in Figure 4-19a show the circuit that would result from the table of Figure 4-19b.

Actually equations 4-8 through 4-11 could be mechanized with only 10 product terms, as shown in Figure 4-20. Since there is no advantage in reducing the product terms below the number provided by the PLA being used, we stopped at 13 in the example. However, let us now reduce the terms futher. *First we look for identical product terms;* for example, the second terms of f_1 and f_2 and the third term of f_4 are all $B' \cdot D'$, so there is no need to generate the same term three times. Instead we can use the same term for all three outputs as shown on Figure 4-20. If we still have too many product terms after combining common terms, we can *examine each function for combinable pairs of terms that differ only by the inversion of one variable.* For example, the first and last terms of f_4 can be combined:

$$A \cdot C \cdot F + A \cdot C \cdot F' = A \cdot C(F + F') = A \cdot C \qquad (4\text{-}12)$$

Product term No.	Inputs							Outputs			
	A	B	C	D	E	F		f_1	f_2	f_3	f_4
1	1	0	1	X	X	X		1	0	0	0
2	X	0	X	0	X	X		1	1	0	1
3	X	1	1	0	1	0		1	0	0	0
4	X	X	X	X	1	1		1	0	0	0
5	1	0	1	X	X	1		0	1	0	1
6	X	0	X	X	1	1		0	1	0	0
7	X	X	1	0	X	1		0	0	1	0
8	0	1	0	1	0	1		0	0	1	0
9	X	0	1	X	X	1		0	0	1	0
10	1	X	1	X	X	X		0	0	0	1

Fig. 4-20.   Simplified truth table.

The only thing we care about is reducing the number of product terms since the number of variables in each term is irrelevant. We reduce the number of product terms only until they fit the limitations of the PLA, since there is no advantage in reducing them further.

When more than one PLA is used, even more complex logic functions can be mechanized. For example Figure 4-21 shows two 14-input PLAs used to generate 14 functions of 18 variables. Examination of the product terms shows that none contains all variables. In fact, inputs A, B, C, D, O, P, Q, and R occur in only a few places, and A, B, C, or D never occur in the same product term with O, P, Q, or R. We can therefore generate only product terms that do not contain O, P, Q, or R in PLA 1, and terms that do not contain A, B, C, or D in PLA 2. Since functions f_7 and f_8 contain terms from both groups, *we connect outputs from both PLAs together for these functions, forming a Virtual OR.* This allows us to use terms from both PLAs for these functions. The other functions contain product terms strictly from one group or the other so they do not have to be paralleled. Notice that this trick can also be used with ROMs when some of the variables appear in only a few terms.

When many PLAs are used, it is useful to put functions that share the same product terms in the same package, even when the inputs to all the PLAs are the same. Again, "Virtual ORing" can be used for some functions to make it possible to use product terms from more than one PLA.

As with ROMs, it is often most economical to decode mutually exclusive outputs outside of the PLA with more economical circuits. Even though PLAs essentially encode their own inputs, there is a limit to the amount of input re-

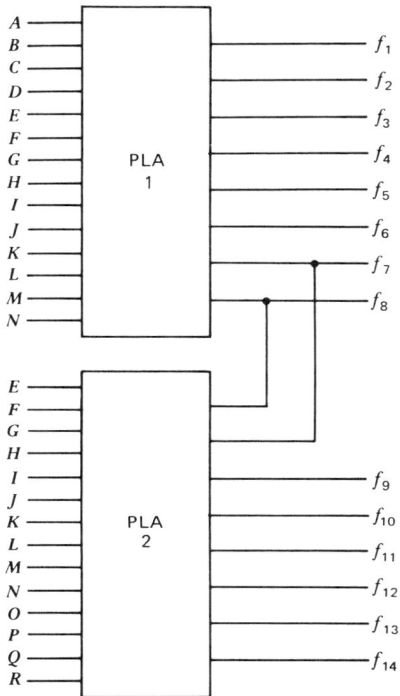

Fig. 4-21. Multiple PLAs.

dundancy they can handle. It is still often desirable to do some encoding of inputs before connecting them to a PLA. Ideally, almost all of the product term lines in a PLA should be used, and each should define a different output. If highly redundant inputs, such as unencoded keyboards, are used, this is usually impossible. Sometimes it is even desirable to use a multiplexer scheme like that shown in Figure 4-18 for PLA inputs. Even though PLA product terms can ignore certain variables, these "don't cares" are fixed. With an extra level of logic ahead of the PLA, the "don't cares" can be made a function of inputs that define the input format. This makes it possible to use the same PLA product term (word) for several input formats, instead of a separate word for each.

4.10 SUMMARY

In Chapters 3 and 4 we have developed a wide range of techniques for mechanizing logic functions. Each technique can be thought of as a tool and each has its place. The job of an engineer is to mentally reach into his extensive "bag of tricks" and come out with the proper tool for the job. Though any tool will work, the key is to select the right one.

If, for example, we want to select one of eight inputs, we could do it with discrete gates, or a ROM, or a PLA; but the really nice fit is an MSI multiplexer/selector. Likewise, if we want to mechanize many complex, nonstandard, logic functions, it is pretty hard to beat a ROM or a PLA. Simple logic functions and small nonstandard functions are, likewise, best done by discrete gating. Each logic type has its special niche.

TABLE 4-5 IC PACKAGES REQUIRED TO MECHANIZE LOGIC FUNCTIONS (SUMMARY)

Number of Variables	Equivalent Packages	Structure
2 (any four functions)	1	Quad 2-input multiplexer
3 (any two functions)	1	Dual 4-input multiplexer
4	1	8-input multiplexer
5	2^a	16-input multiplexer
6	2 to $4\frac{1}{2}^a$	16-input multiplexer + residue gates
7	$4\frac{1}{3}$ to $6\frac{1}{3}^a$	32-input multiplexer + residue gates
8	$9\frac{1}{2}$ to $11\frac{1}{2}$	64-input multiplexer + residue gates
8 (any four functions)	1 (cost ≈ 2)^b	1,024 bit (256×4) PROM or ROM
8 (any eight functions)	2^a (cost ≈ 2)^b	2,048 bit (256×8) PROM or ROM
9 (any four functions)	1 (cost ≈ 2)^b	2,048 bit (512×4) PROM or ROM
9 (any eight functions)	2^a (cost ≈ 2)^b	4,096 bit (512×8) PROM or ROM
10 (any four functions)	2^a (cost ≈ 2)^b	4,096 bit (1024×4) PROM or ROM
10 (any eight functions)	2^a (cost ≈ 2)^b	8,192 bit (1024×8) PROM or ROM
11 (any four functions)	2^a (cost ≈ 2)^b	8,192 bit (2048×4) PROM or ROM
11 (any eight functions)	2^a (cost ≈ 3)^b	16,384 bit (2048×8) PROM or ROM
12 (any four functions)	2^a (cost ≈ 3)^b	16,384 bit (2048×4) PROM or ROM
12 (most 12 functions)	2^a (cost ≈ 3)^b	12 inputs 12 out **FPLA** or PLA
13 (most 10 functions)	2^a (cost ≈ 3)^b	13 inputs 10 out FPLA or PLA
14 (most eight functions)	2^a (cost ≈ 3)^b	14 inputs 8 out FPLA or PLA
17 (most 10 functions)	4^a (cost ≈ 3)^b	17 inputs 10 out FPLA or PLA

^{a} One 24-pin package counts as two since it is more costly and occupies the space of two 16-pin ICs on a circuit board.
^{b} Base on cost in 100 quantities in 1978.

Though we orient the system design around making use of higher levels of integration, it always seems to turn out that there are still a lot of little jobs, joining the big pieces together, that are best done with discrete gates. As an aid in determining at what point we should *not* use discrete gating, *Table 4-5 summarizes what can be done with the various other approaches we have discussed.* For any given function we can compare this table to the number of packages required to mechanize the same function with discrete gates.

EXERCISES

1. (a) Mechanize the following function of four variables using only an eight-input multiplexer:

$$f = m_0 + m_3 + m_5 + m_6 + m_9 + m_{10} + m_{12} + m_{15}$$

   (b) Mechanize the same function, in simplest form, using NAND/NAND logic.
   (c) Compare the IC package count for the two approaches.

2. (a) Mechanize the following function of four variables using only an eight-input multiplexer:

$$f = m_0 + m_1 + m_2 + m_3 + m_{11} + m_{12} + m_{14} + m_{15}$$

   (b) Mechanize the same function, in simplest form, using NAND/NAND logic.
   (c) Compare the IC package count for the two approaches.

3. (a) Mechanize a **Gray code**-to-BCD converter, as defined in Figure 4-22 by

Inputs				Outputs			
A	B	C	D	f_1	f_2	f_3	f_4
0	0	0	0	0	0	0	0
0	0	0	1	0	0	0	1
0	0	1	1	0	0	1	0
0	0	1	0	0	0	1	1
0	1	1	0	0	1	0	0
0	1	1	1	0	1	0	1
0	1	0	1	0	1	1	0
0	1	0	0	0	1	1	1
1	1	0	0	1	0	0	0
1	1	0	1	1	0	0	1

Fig. 4-22

the truth table, using two dual four-input multiplexers and residue gates. (Outputs for codes not on table are "don't cares"—use Veitch diagram.)

(b) Mechanize the same function with four 8-input multiplexers.

(c) Mechanize the same function with a quad two-input multiplexer plus residue gates.

(d) Mechanize the same function with NAND/NAND logic gates.

4. (a) Mechanize the biquinary-to-seven-segment-display decoder of exercise 2, Chapter 3, using four 8-input multiplexers.

(b) With two dual four-input multiplexers and residue gates.

(c) With a quad two-input multiplexer and residue gates.

(d) With a decimal decoder and NAND gates.

(e) Compare package count for the various approaches.

5. (a) Mechanize the decimal priority encoder of exercise 5, Chapter 3, with four 8-input multiplexers.

(b) With two dual four-input multiplexers plus residue gates.

(c) With a quad two-input multiplexer plus residue gates.

(d) With a decimal decoder and NAND gates.

(e) Compare package count for the various approaches.

6. (a) Mechanize functions f_2 and f_4 from exercise 7, Chapter 3, with two 8-input multiplexers.

(b) With a dual four-input multiplexer.

7. (a) Mechanize the functions described in exercise 9, Chapter 3, with four 8-input multiplexers.

(b) With two dual four-input multiplexers.

(c) With a quad two-input multiplexer.

(d) Compare package count for the various approaches.

8. Mechanize the function of exercise 12, Chapter 3, with an eight-input multiplexer.

9. Mechanize the function of exercise 13, Chapter 3, with an eight-input multiplexer.

10. (a) Mechanize the function of exercise 15, Chapter 3, with a 16-input multiplexer.

(b) With an eight-input multiplexer and residue gates.

11. Mechanize the function of exercise 22, Chapter 3, with an eight-input multiplexer.

12. Make a chart, similar to Figure 4-4, of the four possible ways to reduce functions of four variables with an eight-input multiplexer.

13. (a) Mechanize a BCD-to-Gray-code converter (exchange inputs for outputs on the truth table for exercise 3) using four 8-input multiplexers.

(b) With two dual four-input multiplexers and residue gates.

(c) With a quad two-input multiplexer and residue gates.

(d) With a decimal decoder and NAND gates.

(e) With NAND/NAND logic.

(f) (Compare package count for the various approaches.

14. (a) Mechanize a seven-segment-to-BCD converter (switch inputs for outputs on the truth table in Fig. 3-2d) by using four 8-input multiplexers and residue

gates (make eight 4-variable Veitch diagrams). Use multiplexers to generate residue functions if it results in a savings.

(b) What size ROM is required to mechanize this function (actual minimum size, ignoring industry standard sizes)?

(c) What size PLA is required?

(d) Make a truth table for ordering a PLA (though size is unrealistically small).

15. Draw a 16-input multiplexer tree made with eight-input multiplexers and a two-input NAND gate.

16. Draw a 40-input multiplexer tree made with six 8-input multiplexers.

17. Draw a 40-output demultiplexer tree made with six decimal decoders.

18. (a) Draw a 40-output decoder tree made with five decimal decoders.

(b) With four decimal decoders and a quad two-input NAND gate.

19. Design a 16,384-input multiplexer using an X–Y matrix with a 16-input, tristate, submultiplexer on each of 1024 circuit modules. The central multiplexer module has two decimal decoder/drivers and nine 8-input multiplexers and has only 100 connector pins available.

20. For a 2048-output decoding/demultiplexing matrix, make a table of the following factors for one, two, three and four equal dimensions:

Number of Matrix Dimensions	1	2	3	4
Size of Each Dimension				
Number of gate packages (total)				
Number of predecoder packages				
Wires from predecoder				
Loads on predecoder outputs				

21. (a) Make a similar table for a matrix of two dimensions where the X dimension is twice as large as the Y dimension.

(b) For a matrix of three dimensions of size ratio $2:2:1$.

22. Make a circuit with eight inputs and eight outputs that will have one of the eight output wires low corresponding to the highest priority of the 10 input lines that is low. Use only two of the special MSI circuits discussed in this chapter.

23. (a) Make a truth table for a (unrealistically small) ROM to convert BCD to seven-segment display driver signals.

(b)   What is the ROM size in bits?

(c)   Draw a diagram of the ROM, similar to Figure 4-19a, including dots to indicate the 2^n word decode and dots for the specified word outputs.

24.   What size ROM would it take to make:

(a)   A quad two-input multiplexer?

(b)   A dual four-input multiplexer?

(c)   An eight-input, three-output priority encoder?

25.   What size PLA would be required to make:

(a)   A quad two-input multiplexer?

(b)   A dual four-input multiplexer?

(c)   An eight-input, three-output priority encoder?

25.   What size PLA would be required to make:

(a)   A quad two-input multiplexer?

(b)   A dual four-input multiplexer?

(c)   An eight-input three-output priority encoder? (Include number of inputs, outputs, and product terms.)

26.   (a)   How many bits is the large ROM shown on Figure 4-18?

(b)   How many bits would it have to be if the input bits had not been preselected with a multiplexer?

BIBLIOGRAPHY

Carr, W. N. and J. P. Mize, *MOS/LSI Design and Application,* McGraw-Hill, New York, 1972, Chapter 8, "Programmable Logic Arrays," and Ch. 7, "Memory Applications."

Fairchild Semiconductor Staff, *The TTL Applications Handbook,* Fairchild Semiconductor, Mountain View, Calif., 1973.

Intel Staff, *The Intel Memory Design Handbook,* Intel Corp., Santa Clara, Calif., 1973, Chapter 2, "Read Only Memories."

National Semiconductor, *How to Design With Programmable Logic Arrays,* Application Note # AN-89, National Semiconductor Corp, Santa Clara, Calif., 1973.

Texas Instruments Staff, *The TTL Data Book,* Texas Instruments, Dallas, Tex., 1973.

Uimari, David, "PROM's—a Practical Alternative to Random Logic," *Electronic Products,* January 21, 1974, pp. 75–91.

Priel, Vry, and Phil Holland, "Application of a High Speed Programmable Logic Array," *Computer Design,* December 1973, pp. 34–96.

Yau and Tang, "Universal Logic Modules and Their Application," *IEEE Transactions on Computers,* Vol. c-19, No. 2, February 1970.

CHAPTER FIVE

SEQUENTIAL LOGIC
Design

5.1 MEMORY ELEMENTS

In Chapter 4 we saw that combinational logic and ROM are just different forms of the same thing. An eight-input AND gate, for example, can be thought of as a 2^8 = 128-word $\times$ 1-bit ROM programmed so the output is false for all "words" except the one defined by all "1's" inputs. Thus logic gates and ROMs are really the same thing functionally.

In this chapter we introduce the changeable memory. If we change the circuitry inside a ROM *so we can change its bit patterns at will*, we can then use it to remember things. The PROM is changeable but not so easily as a true read/write RAM.* To change the contents of a RAM, we simply present the desired bit pattern to the data input and pulse the *write clock* input; the addressed word is immediately changed. This ability to remember is, of course, very important in solving most real world problems. Most problems consist of performing combinational operations on various data, but seldom do all of the data occur at the same instant. Memory allows us to bring all the data together at the instant the computation is made. Also, since a RAM can do the logic itself, just like a ROM, we can do different problems with the same

---

* Actually ROMs are random-access memories since they can be randomly accessed, but RAM has come to mean a *read/write* memory.

hardware (as a computer does), and even make the hardware adapt itself to the problem.

Just as we have a logic gate as the basic combinational element and a ROM as a large combinational array, we have a flip-flop as a basic memory element (basically a one-bit RAM) *and a RAM as a large memory array*. Essentially, gates and ROMs are analogous to flip-flops and RAMs, but with flip-flops and RAMs we have the added dimension of *time*. Both devices have a **clock** input, and *their states change only after a transition of the clock*.

5.2   FLIP-FLOP TYPES

Because of the similarity of memory elements to combinational logic elements, it is not surprising that we still use the truth table, the Veitch diagram, and the logic equation in our design procedures. In fact, we can define our basic memory elements in a format almost exactly like the one we used to represent our basic logic elements. Figure 5-1 shows the basic flip-flop types in a format almost identical with that of the gate definitions on Figure 3-3.

In both cases a logic equation and truth table define the output as a function of the inputs. With the memory elements the output does not change as a function of the inputs *until the clock transition*. Therefore, a superscript notation is used to indicate that the output during clock period $n + 1$ is a function of the inputs during the previous clock period n.

For a D (delay) flip-flop, the equation is quite simple, for it tells us that the *input (D) is "stored" in the flip-flop when the clock occurs and will appear on the output (Q) during the next ($n + 1$) clock time*. The D flip-flop is thus very much like a single-bit RAM. It is very useful for data storage and other special applications.

The other three types of flip-flops defined on Figure 5-1 are still just one-bit storage elements, but instead of simply storing the input, they change state in response to the inputs by various logical rules. Since they hold their previous state in spite of the clock, unless an input goes true, they often simplify the combinational logic functions required to control them in control applications.

The T (toggle) flip-flop, for example, simply stays in its previous state if the T input is false before the clock. If the T input is true, the output changes to the opposite state (toggles) on the clock. The T flip-flop is thus useful, for example, in binary counters where we want each bit to invert every time there is a carry from the lower order bits. If we compare the equation for the T flip-flop output with the one for the D flip-flop, we see that the T flip-flop is identical in operation to a D flip-flop with an Exclusive OR gate on its input, as shown on Figure 5-2. As a matter of fact, *any of the flip-flop types can be*

(a)

Data → D   Q → $Q^{n+1} = D^n$

Clock → C

D^n	Q^{n+1}
0	0
1	1

(b)

→ T   Q →

→ C $Q^{n+1} = (T' \cdot Q + T \cdot Q')^n$

T^n	Q^{n+1}
0	Q^n
1	$(Q')^n$

(c)

→ S   Q →

→ C $Q^{n+1} = (S + R' \cdot S' \cdot Q)$

→ R $R \cdot S \neq 1$

R^n	S^n	Q^{n+1}
0	0	Q^n
0	1	1
1	0	0
1	1	X

(d)

— J   Q

— C $Q^{n+1} = (J \cdot Q' + K' \cdot Q)^n$

— K

J^n	K^n	Q^{n+1}
0	0	Q^n
0	1	0
1	0	1
1	1	$(Q')^n$

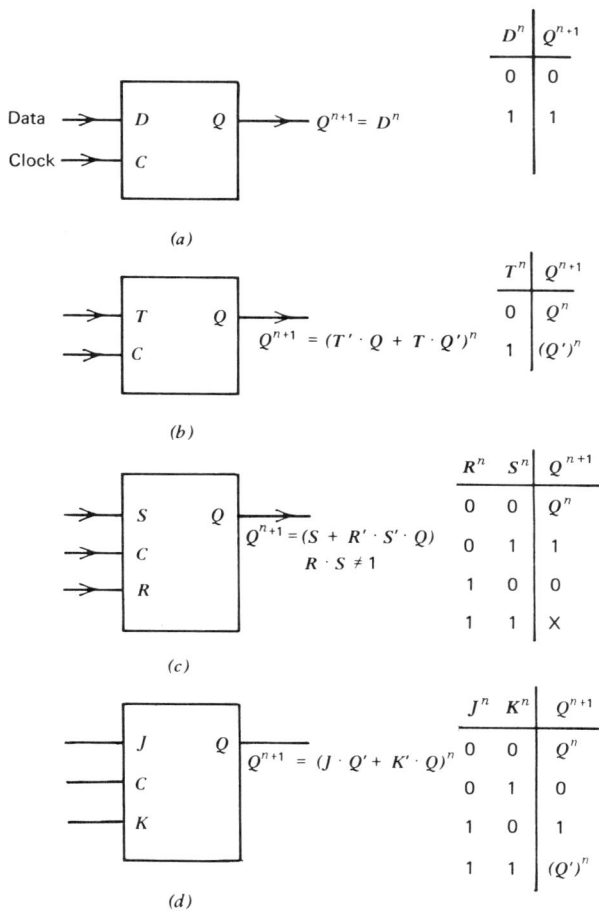

Fig. 5-1. (a) D (delay) flip-flop. (b) T (toggle) flip-flop. (c) R–S (reset-set) flip-flop. (d) J–K flip-flop.

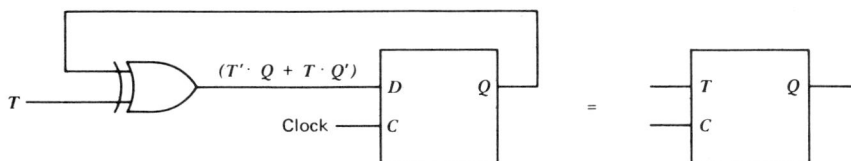

T —⊕— $(T' \cdot Q + T \cdot Q')$ → D   Q

Clock — C

=   — T   Q —

       — C

Fig. 5-2. Making a T flip-flop.

made from a D flip-flop by simply adding a few combinational logic gates, as
indicated by the equations on Figure 5-1.

The *R–S* flip-flop sets after the *S* input is true and resets after the *R* input is
true. Its output is undefined if both *R* and *S* are true. It is possible to define a
Set Overrides Reset (SOR) or a Reset Overrides Set (ROS) flip-flop—meaning
simply that it will set or reset respectively if both the *R* and *S* inputs are
true.

The *J–K* flip-flop (Figure 5-1*d*) sets after *J* is true and resets after *K* is true.
It is thus similar to an *R–S* flip-flop except that *if J and K are both true, the
output changes to the opposite state (toggle).* It can thus be used as a *T* flip-
flop by simply tying the *J* and *K* inputs together. Since the *J–K* flip-flop can
essentially do the job of both the *R–S* and the *T* flip-flop, the *R–S* and the *T*
flip-flop are seldom seen. The choice, therefore, is between *J–K* flip-flops for
small counters and control or *D* flip-flops for data storage applications.
Actually, *J–K* flip-flops can even do the job of a *D* flip-flop with the addition
of a single inverter, as shown on Figure 5-3.

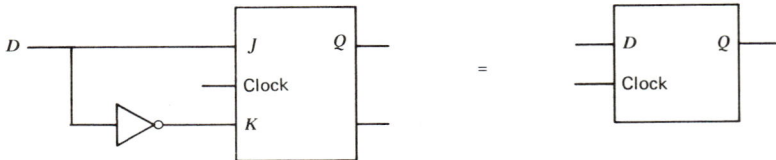

Fig. 5-3.   Making a D flip-flop.

Actual integrated circuit flip-flop symbols are slightly more complicated
than those shown in Figure 5-1. They are packaged together as "dual *D*" or
"dual *J–K*" flip-flop packages. Also, they usually have **asynchronous**
PRESET and **CLEAR** inputs that allow immediate set or reset without
waiting for the clock. These inputs are used only for certain special jobs, such
as setting initial conditions, since there is danger of "race conditions," as
explained in Chapter 6. As with logic gate inputs, a "bubble" on an input
means low true. *With flip-flops, however, it is not just an arbitrary
convention.*

For example, in the flip-flop shown in Figure 5-4, a low input on the
CLEAR or PRESET input sets the flip-flop to 0 or 1 no matter how we define
our logic levels. Also, the "bubble" on the clock input indicates that the flip-
flop output will change state on the *negative-going* edge of the clock signal,
whether that is defined as true or false. (The flip-flops in Figure 5-1 would all
change on the positive-going edge of the clock since they have no "bubble.") If
flip-flops are always drawn with the inputs and outputs arranged as shown in
Figure 5-4, logic diagrams are much more readable. Note that most flip-flops

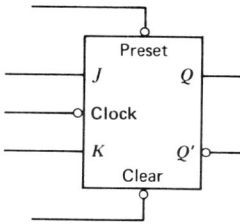

Fig. 5-4. An actual J–K flip-flop.

have an inverted output Q', which is drawn coming out at the lower right. Since the CLOCK input of all flip-flops in a system are usually connected to one signal, they are often left off the logic diagram for simplicity. Power and ground connections also are often shown separately from the actual logic diagram.

Another memory element type, called a *latch,* is often described on data sheets with a truth table like the one for the D flip-flop in Figure 5-1a. It is definitely not like a D flip-flop, however, because the output changes as soon as the clock goes high and does not "latch" until the clock falls (if the input changes while the clock is high, the output follows it). Because of this characteristic, *a latch is not usable in the* **synchronous** logic we shall be discussing.

5.3 REGISTERS AND COUNTERS

Flip-flops are often grouped together conceptually as *registers* or *counters.* A *register* is thought of as storing *data,* and often the content of the register has some physical interpretation, such as a character code or a number used in computation. The register itself contains essentially no logic and consists simply of a group of flip-flops. Combinational logic is often used at the input to a register to select the source of data to be held in the register or to perform code conversion or arithmetic logic.

Flip-flops are also often grouped together as *counters.* Counters are used for controlling sequences of operations. The outputs of the flip-flops that make up the counter are spoken of together as the *state* of the counter. The sequences of states can be straightforward, such as counting a certain number of clocks or repetitions of some operation, or they can be complex. Complex sequences are often controlled at several points by variable input conditions. In either case, a counter can be defined by simply developing combinational logic for each flip-flop input. The next $(n + 1)$ state of the flip-flops is determined by their inputs before the clock (n) as shown on Figure 5-1. We thus have a two-step problem:

1. Define the counters required in the system and a desired sequence of states.

2. Design combinational logic for the flip-flop inputs to produce that sequence of states.

As we found before, the truly creative part of the design process is in the earliest stages. Once we have defined the counters in the system and their states, we follow a fairly cut-and-dried procedure for finding the required combinational logic.

An important tool we can use at the system design level is the idea of subdivision of the complex system into a collection of smaller subsystems. The counters and registers in a system can be represented simply as blocks in a block diagram. These blocks can be made to correspond to MSI package types where possible, but, even where discrete flip-flops will be used, the block can normally be given a name describing its function. Instead of thinking of all the sequence control as one counter, it is usually possible to *split the problem into several "dimensions"*—each controlled by a separate counter.

Often only one of the counters must have a significant amount of variability to its sequence. This counter is often called the *control-state counter* and can also be shown as a simple block on the block diagram. Other counters can often be simple, resettable, binary counters. The only way to discover whether a sequence has more than one "dimension" is to look at the basic problem for similarities.

For example, if a system handles several "words" of data one "bit" at a time, *word count* can be thought of as one dimension and *bit count* another. If it does several different operations on the data in some sequence, then the *control-state counter* can control those sequences. Our sequential control is thus divided into *three dimensions* (see Fig. 5-5) handled by three counters: the bit counter, the word counter, and the control-state counter. Aside from subdividing the problem into three simpler problems, this division also saves hardware (just as it did in the decoders in Section 4.4). Normally, for a particular control state some logical operation is performed. This operation normally stays the same throughout the bits of the word. *The logic that controls the type of operation can thus ignore the bit counter state.* Other logic in

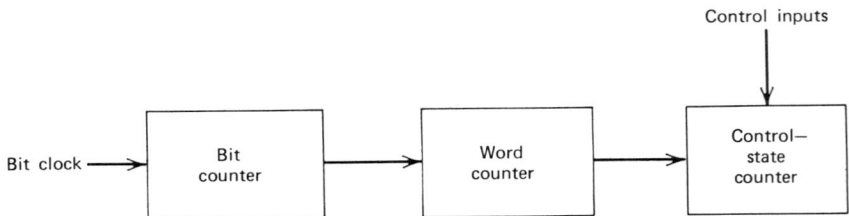

Fig. 5-5. Cascaded counters.

the system is concerned with storing or reading bits from correct memory or register bit locations. This logic can often ignore the type of logical operation being performed. *By dividing the problem into dimensions, we reduce the combinational logic in various parts of the system to functions of fewer variables.* The logic is thus simpler and easier to mechanize.

Another way to look at this division of the sequencing into dimensions is as a way to save flip-flops. We want to encode mutually exclusive groups of possibilities, just as we encoded mutually exclusive ROM inputs in Section 4.16. While the result in Section 4.16 was fewer ROM inputs, here the result is fewer flip-flops. Just as with the ROM inputs, ideally we would like to have a unique, logically different mode of operation for each unique combination of flip-flop states in the system. Where we have defined several counters, this would mean that the state of each counter is meaningful during each mode of operation. Although this is not always practical, we should always be aware of the possibility of using a counter for more than one function.

As an example of multiple use of a counter, suppose that our system of Figure 5-5 must input several words, compute their mathematical function, then output several words. If the computation is done in three steps, we might assign states to the control-state counter as follows:

> Input
> Computation Step 1
> Computation Step 2
> Computation Step 3
> Output

The control-state counter would thus require three flip-flops to represent the five states. Since the word counter has no meaning during the three computation steps, it is being wasted. If, however, *we use the word counter to keep track of the computation steps,* we can eliminate two of the "computation" control states. Our control-state counter therefore needs only three states: input, computation, and output.

Sometimes counters have so many uses that it is difficult to give them a simple name. For example, computer instruction execution is sometimes controlled by a "major" and "minor" control-state counter. The exact use of these two counters depends on the instruction being executed, but they represent two "dimensions" of instruction execution, making it possible to take advantage of the similarities between various instructions.

We can carry the analogy between saving control flip-flops and encoding ROM inputs one step further by observing that making multiple use of a single counter is similar to the trick we used in Figure 4-19 to make multiple use of ROM inputs by selecting the signals to be applied to them. Of course there is a certain logic cost in making multiple use of flip-flops. Discrete combinational logic is often increased when control states are more compactly encoded, so we must be sure when we cleverly save a flip-flop that we have not added more in combinational logic than we saved in flip-flops. This is particularly true since MSI counters come packaged in groups of four flip-flops. It really costs about the same for a 16-state counter as for a five-state counter.

5.4 DESIGNING COUNTERS WITH FLIP-FLOPS

Though MSI circuits are available for binary and a few other count sequences, it is sometimes desirable to design special counters with flip-flops and gates. For example, the count sequence shown in Figure 5-6a is sometimes useful because only one bit changes for each change of state. To design a counter for this sequence we must derive logic equations for the inputs of flip-flops A, B, and C.

If we want to design the counter using D flip-flops, then the equation on Figure 5-1a shows us that the D flip-flop input during clock time n determines the output during the next clock $n + 1$. We thus need logic that makes *the D input true whenever the next state of the flip-flop should be 1.* For example, flip-flop A must be 1 after states 2, 6, 7, and 5. We simply enter 1's in these positions of the Veitch diagram for D_A in Figure 5-6c. Simplifying in the normal way, we obtain an equation for the input of D flip-flop A. We similarly derive equations for D_B and D_C and mechanize the counter as shown in Figure 5-6f. The equation for D_c happens to be the Exclusive OR of A' and B (see Section 3.15), so it is mechanized here with an Exclusive OR gate.

Designing the same counter with J-K flip-flops is done by a similar procedure except that we must design logic for the J and the K inputs of each flip-flop.

Since we "don't care" if the J input is true when a flip-flop is already set, *we can immediately put X's in half of the squares for each Veitch diagram.* For example, if flip-flop A is set, we "don't care" about J_A, so we can put X's in the $A = 1$ half of the J_A diagram. We likewise "don't care" about the K input of a flip-flop that is already reset,* so we can, for example, put X's in the $A = 0$ half of the K_A diagram. These X's can always be written in immediately before starting to solve the problem. If there are unassigned states, of course, X's can also be put in those positions of all of the diagrams.

We next examine the states, one at a time, and note the change that is re-

---

* We can write X in all states where the flip-flop output is 0 because if K is true, it has no effect since the flip-flop is already reset. If the flip-flop is about to set, J will be true, but again $K = 1$ changes nothing since, if both J and K are true, the flip-flop will still set since the J-K flip-flop toggles when $J = K = 1$. (Note that this is not true with an R-S flip-flop; therefore we can write one fewer X.)

quired of each flip-flop to go to the next state. *For each 0 → 1 change (set) we write a 1 on a J Veitch diagram; for each 1 → 0 change (reset), we write a 1 on a K Veitch diagram square corresponding to the previous state.* For example, *from* state 0 we must set flip-flop C to advance to the next state so we write a 1 on the J_C diagram in the lower right corner (state 0). Next we

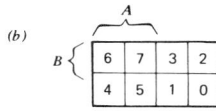

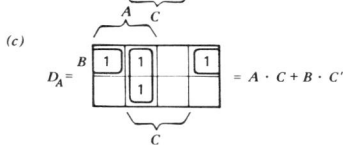

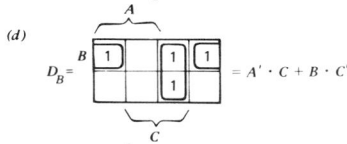

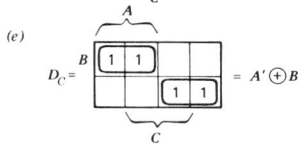

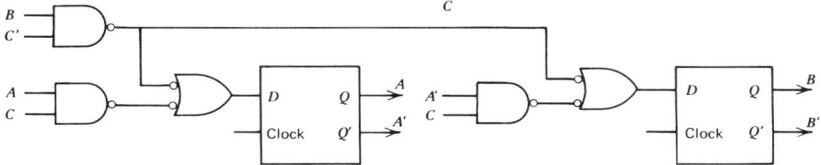

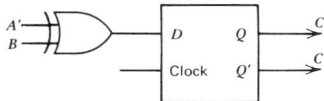

Fig. 5-6. (a) Reflected Gray code counter. (b) State numbers for reference. (c) D_A diagram. (d) D_B diagram. (e) D_C diagram. (f) D flip-flop mechanization.

(g) $J_A =$ ... $= B \cdot C'$

(b) $K_A =$ ... $= B' \cdot C'$

(i) $J_B =$ ... $= A' \cdot C$

(j) $K_B =$ ... $= A \cdot C$

(k) $J_C =$ ... $= A \cdot B + A' \cdot B' = A' \oplus B$

(l) $K_C =$ ... $= A' \cdot B + A \cdot B' = A \oplus B$

(m)

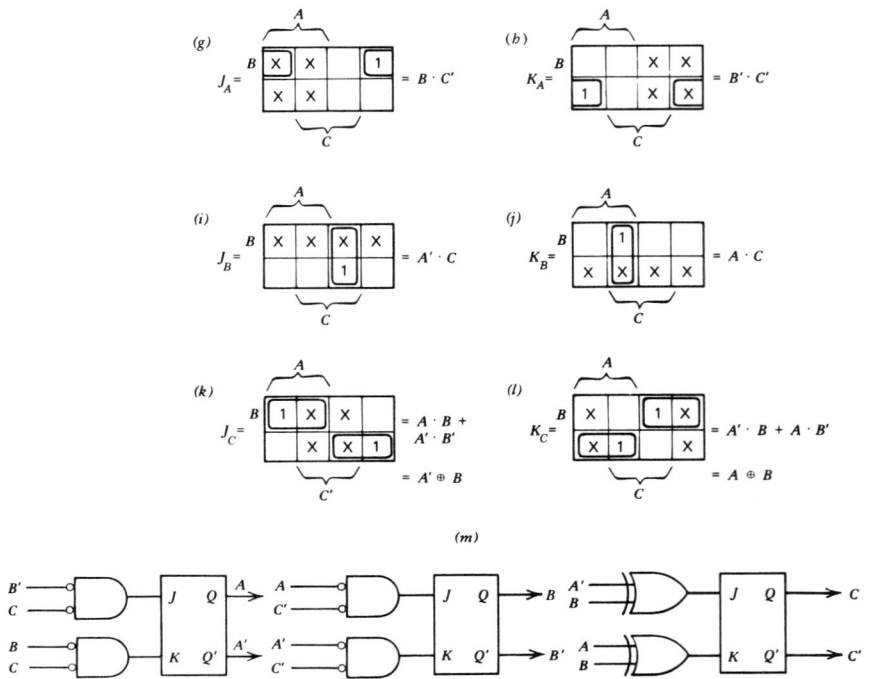

Fig. 5-6 (continued).   (g–l) J–K flip-flop design. (m) J–K flip-flop mechanization.

must set B to move out of state 1, so we write a 1 on the J_B diagram in the state 1 position. The next state is state 3, and we must reset flip-flop C to move out of this state, so we write a 1 on the K_C Veitch diagram in the state 3 position. We continue through the count sequence till all the state changes are mechanized.

When all Veitch diagrams are complete, we simply write the J and K equations in simplest form and mechanize them in the normal manner, as shown in Figure 5-6m. In this case we can save some gating by using NOR gates with inverted inputs to generate the AND functions required for J_A, K_A, J_B, and K_B (as was discussed in Section 3.13.) Examination of the equations for J_C and K_C reveals that they are actually Exclusive OR functions (see Section 3.15), so they are mechanized here using Exclusive OR gates.

If we design the counter with a T flip-flop, the procedure would be similar except that we would write a 1 in the T Veitch diagram for the state before any $0 \rightarrow 1$ or $1 \rightarrow 0$ change.

5.5. MOEBIUS COUNTERS

The reflected Gray code counter just designed requires a fairly significant amount of logic to mechanize. Some count sequences, however, result in very simple logic. For example, we can make an eight-state counter with the same characteristic of changing only a single bit for each state change, by using the sequence shown in Figure 5-7a. This type of counter is called a *"moebius counter"* or "twisted tail ring counter." We can confirm that the hardware mechanizations shown in Figures 5-7b and c are correct by just inspecting the count sequence. The state of each flip-flop is equal to the previous state of the flip-flop to the left. The state of the leftmost flip-flop is equal to the *complement* of the state of the rightmost flip-flop. MSI circuits called shift registers are available containing four flip-flops which can be so connected.

We can thus make a counter with the desired characteristics with a single MSI package instead of the minimum of three packages required with the sequence of Figure 5-6. Even with discrete flip-flops, we use fewer packages with this sequence despite the fact that one more flip-flop is used. Using the same principle, we can make a moebius counter of any size. The three-bit and two-bit moebius sequences are shown in Figures 5-7d and e. Note that these sizes do not waste flip-flops, since three flip-flops are the theoretical minimum for a six-state counter and two are the minimum for a four-state counter. Since any size (n) moebius counter has $2n$ states, very large counters are impractical.

The moebius sequence is often useful for control-state counters. When a moebius counter is mechanized, with J–K flip-flops, as in Figure 5-7c, *each state change is controlled by a different input*. We can thus control each state change independently by adding an AND (or NOR) gate at the input controlling that state change. For example, in Figure 5-7c the first state change $(0 \rightarrow 8)$ can be blocked by gating J_A. Subsequent state changes can be controlled by gating J_B, J_C, J_D, K_A, K_B, K_C, and K_D, respectively.

Another useful application of the moebius counter is multiphase clock generation. The eight flip-flop outputs in Figure 5-7b or c generate eight, interleaved, 50% duty cycle clocks. We can see by examining the count sequence in Figure 5-7a that (+) transitions $(0 \rightarrow 1)$ come on successive clock times in the following order: A, B, C, D, A', B', C', D'. We can thus generate an eight-phase clock by clocking a four-bit moebius counter at eight times the desired clock frequency. In general, an n-bit moebius counter generates $2n$ clock phases.

Another advantage of moebius counters is that *any state or consecutive group of states can be decoded with* a single two-input gate. To decode a single state we simply look at the adjacent 1 and 0—for example, state $14 = C \cdot D'$. We can confirm on a Veitch diagram that the other three minterms included in

Binary state	A	B	C	D
0	0	0	0	0
8	1	0	0	0
12	1	1	0	0
14	1	1	1	0
15	1	1	1	1
7	0	1	1	1
3	0	0	1	1
1	0	0	0	1
	0	0	0	0

(a)

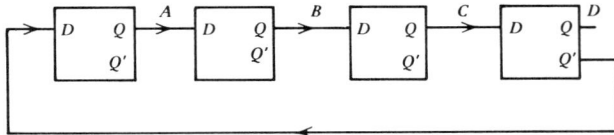

(b)

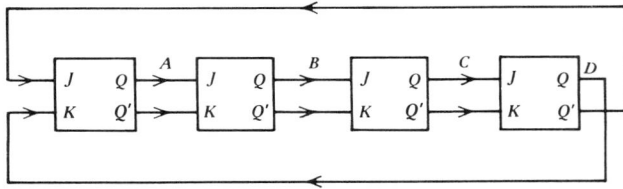

(c)

A	B	C
0	0	0
1	0	0
1	1	0
1	1	1
0	1	1
0	0	1
0	0	0

(d)

A	B
0	0
1	1
1	1
0	1
0	0

(e)

Fig. 5-7. (a) Four-bit Moebius counter ($\div 8$). (b) D flip-flop mechanization. (c) J-K flip-flop mechanization. (d) Three-bit Moebius count sequence ($\div 6$). (e) Two-bit Moebius count sequence ($\div 4$).

116

this function are "don't cares." To decode a consecutive group of states we simply look at the last 1 of the first state of the group and the first 0 of the last state of the group—for example, states 8, 12, and 14 $= A \cdot D'$. Any consecutive group of states can thus be decoded with a two-input gate.

By varying only slightly from the moebius count sequence we can often design other count sequences with minimum logic. For example, Figure 5-8 shows a three-state counter produced by simply changing the K_A input of a four-state counter to make it reset sooner.

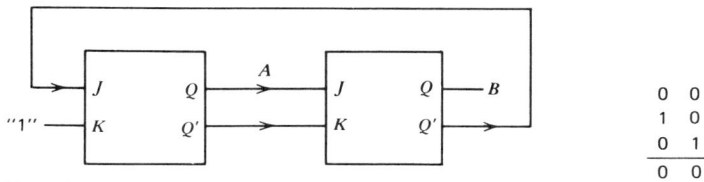

Fig. 5-8. A $\div 3$ counter.

5.6 DEFINING CONTROL-STATE COUNTERS

Except for very small ($\div 2$, $\div 3$, $\div 4$) counters and very special applications, most counters are made by simply connecting MSI counters for the desired binary count cycle length. In most systems, however, at least one counter must do more than simply count in a binary sequence. Often the sequence is so complicated that it looks more like a register than a counter in that it responds more to external inputs than to its own state.

The first step in designing such a sequential circuit is to make a *state diagram*. A state diagram (see Fig. 5-9) has a circle representing each state. Inside each circle is written the binary identity of the state and the "meaning" of the state of system operation. Arrows from the state point to all possible "next" states. Logic functions written next to each line show the conditions required for that state change. A state diagram thus completely defines the operation of a control-state counter.

The state diagram is generated one state at a time from the requirements of the problem. The binary state assignments are made only after the complete diagram is drawn with only the "meanings" of the states and entry conditions written in. The first state is always the "initial" state the system is usually set into when power is first turned on. From the initial state we draw an arrow for each possible "next" state with the logical condition for the state change to occur written next to each line. Likewise, from each "next" state we draw lines and write conditions until we have defined all states required for the system.

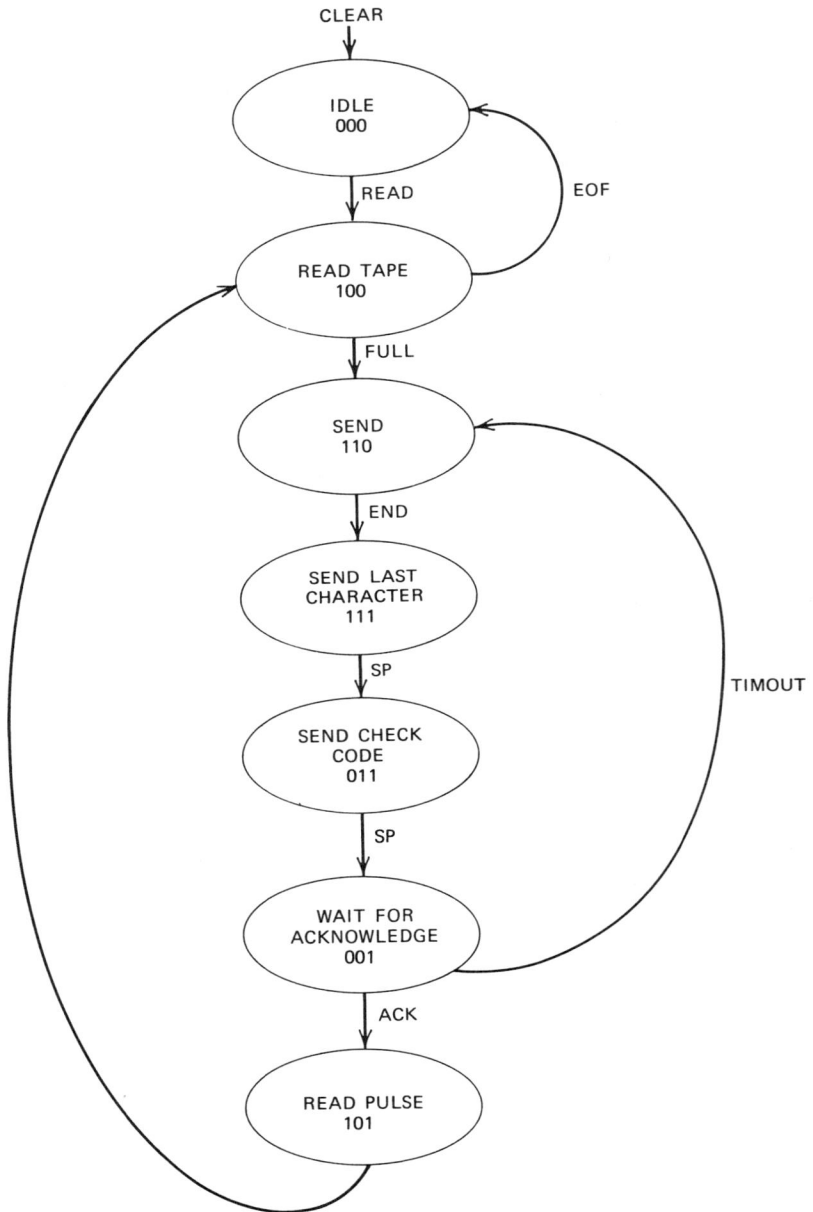

Fig. 5-9. State diagram for data transmitter.

Each time a new state is created we must determine whether anything must really be *"remembered"* by the state change. For example, if a toggle switch must cause the mode of operation to change, we may not need a new state since the logic can simply use the switch output to determine mode. If the switch is a momentary push button, however, a state change is required to "remember" the button was pushed. *State changes thus usually "remember" transient conditions that will disappear during the actual operation and will not be "remembered" elsewhere in the system.*

We must be careful when we define a new state that it is not just another name for a previously defined state. *It must either define a new operation or at least have different exit conditions than any other state.* Another thing to watch for is the possibility of the system's *"hanging up."* If we enter a state and the conditions for leaving the state are never met, we will stay in that state forever! Most state diagrams link back to the initial state at the completion of an operation.

5.7   AN EXAMPLE

Figure 5-9 defines the control states for a system that reads data records from magnetic tape into a RAM and sends them over a telephone line with a check character. It waits between records for a signal acknowledging correct reception. If no ACKnowledge is received within a ⅓-sec TIMOUT, the message is sent again. This cycle is repeated indefinitely until the ACKnowledge is received. The system also has a bit counter, a word counter, and another control-state counter to control tape operations.

Following the diagram down from the initial (IDLE) state, we see that pressing the READ button puts us in the READ TAPE state. If **End of File (EOF)** is detected, meaning there is nothing more on the tape, we return to the IDLE state. When the RAM is FULL of data from the tape, we go to the SEND state and shift out information until END indicates the RAM is empty. This puts us in the SEND LAST CHARACTER state, which is necessary since the data are sent out, one bit at a time, from a register that is loaded by the RAM. When the bit counter indicates that the last bit has been sent, the stop bit state (SP) of the bit counter puts us in the SEND CHECK CODE state. When the last bit of the check code has been sent, SP advances us to the WAIT FOR ACKNOWLEDGE state. If an ACKnowledge is received, we enter the READ PULSE state for one clock time to simulate the pressing of the READ button and cause another record to be read from the tape in the READ TAPE state. The sequence thus repeats until all records on the tape are sent and EOF returns the system to the IDLE state. If ACKnowledge is not received within ⅓ sec of entering the WAIT FOR ACKNOWLEDGE state, the TIMOUT signal causes the record to be sent over again.

We have thus described all the states and conditions for changing state, in Figure 5-7, in much the same way as we would generate that state diagram from our knowledge of the system requirements. Note that it was important to recognize the need to return to previously defined states instead of defining new ones when it came time to READ TAPE again and SEND again after TIMOUT. Also, if we did not provide the TIMOUT, the system could hang up in the WAIT FOR ACKNOWLEDGE state.

5.8   STATE ASSIGNMENT

Once we have defined the required states, we assign binary codes to the states. Since these codes correspond to the three flip-flops that make up the counter, the assignments will have a strong effect on the logic required to mechanize the state counter and the other logic in the system.

Logic in the rest of the system can be minimized by *making the counter flip-flops correspond to actual control signals* required. For example, on the state assignment on Figure 5-9, the middle flip-flop is set during the three states in which we want to enable the sending of data (110, 111, and 011). This means that we can use this middle flip-flop output *directly* as a "send enable" instead of having to use gating to generate a function that goes true during these three states.

The logic of the counter itself can be minimized by remembering that *certain count sequences are easier to generate than others.* All the state changes (except the TIMOUT) on Figure 5-6 require a *change of state of only one flip-flop.* This means that the gating for each conditional state change needs to appear on only one flip-flop input. The count sequence used on Figure 5-9 primarily follows a straight line from top to bottom with the exception of three *"branches."* To simplify counter and decode logic, the first six states are given the "moebius" sequence described in Section 5.5. Even though we have added a state outside of the sequences (READ PULSE) and have two branches that do not follow the sequence, we can expect our logic to turn out simpler because we used this sequence as a framework. Also, by making the initial state 000, we make it possible to initialize the counter with the asynchronous CLEAR flip-flop inputs.

5.9   FLIP-FLOP CONTROL-STATE COUNTER MECHANIZATION

We can mechanize the control-state counter of Figure 5-9 with the same technique used earlier for a counter with no control. *The essential difference is that, instead of writing in 1's on the Veitch diagram as required for state changes, we write in the logic function required for the change.* We use J–K flip-flops because they always give simpler results in applications where the counter must be stopped (since no signal is required to keep the flip-flops in their present state).

We start by making an oversized Veitch diagram (Fig. 5-10a) and identifying the states on it by name. This diagram is used as a reference to locate the states on the six other diagrams used to find the logic functions required for the six flip-flop inputs in the counter. Since the 010 state is not used, we can put an X ("don't care") in the upper right corner of all seven diagrams. Next we fill in four X's on each of the six J and K diagrams since we "don't care" about the J input when a flip-flop is set, and we "don't care" about the K input in states where it is already reset.

We now make entries on the diagrams for each state change. In the IDLE state we want to set flip-flop A if READ is true, so we enter READ in the lower right corner of the J_A diagram. Since there are no more changes from the IDLE state, we move on to the READ TAPE state. Since we must set B if FULL is true, we enter FULL in the lower right corner of the J_B diagram. We also want to reset A from this state if EOF goes true, so we enter EOF in the lower left corner of the K_A diagram. We continue in this way mechanizing each state change arrow on Figure 5-9 by writing in the change conditions on the proper diagrams. If more than one arrow affects one flip-flop in the same state (as on J_A), we simply use a "$+$" symbol and write in both functions.

The next step is to write simplified logic equations from the six diagrams, using X's wherever possible. We are actually mechanizing six functions of *10 variables,* but, since seven of the variables interact so little, we can use three-variable diagrams for the simplification. Since we assigned the states so that most state changes required only one flip-flop to change, there are no common functions in the example; but it is sometimes possible to change the terms in such a way that the same gate can be used by more than one function.

Figure 5-11 shows a mechanization of the counter using NAND/NAND logic (with two NORs where convenient). A total of $5\frac{2}{3}$ IC packages are needed.

Actually, the design just completed is somewhat wasteful since it assumes that any of the control signals can occur at *any* time. This is a good, conservative assumption if we have no intimate knowledge of how the control signals are generated. Some of the gating can be eliminated, however, if we examine some of the control signals. TIMOUT, for example, is generated by a counter that counts *only* while we are in the WAIT FOR ACKNOWLEDGE state. We can therefore eliminate the associated gating in J_A, J_B, and K_C (e.g., $J_B = A \cdot C' \cdot \text{FULL} + \text{TIMOUT}$). EOF is generated only when we are reading tape, so the equation for K_A can be simplified to $K_A =$

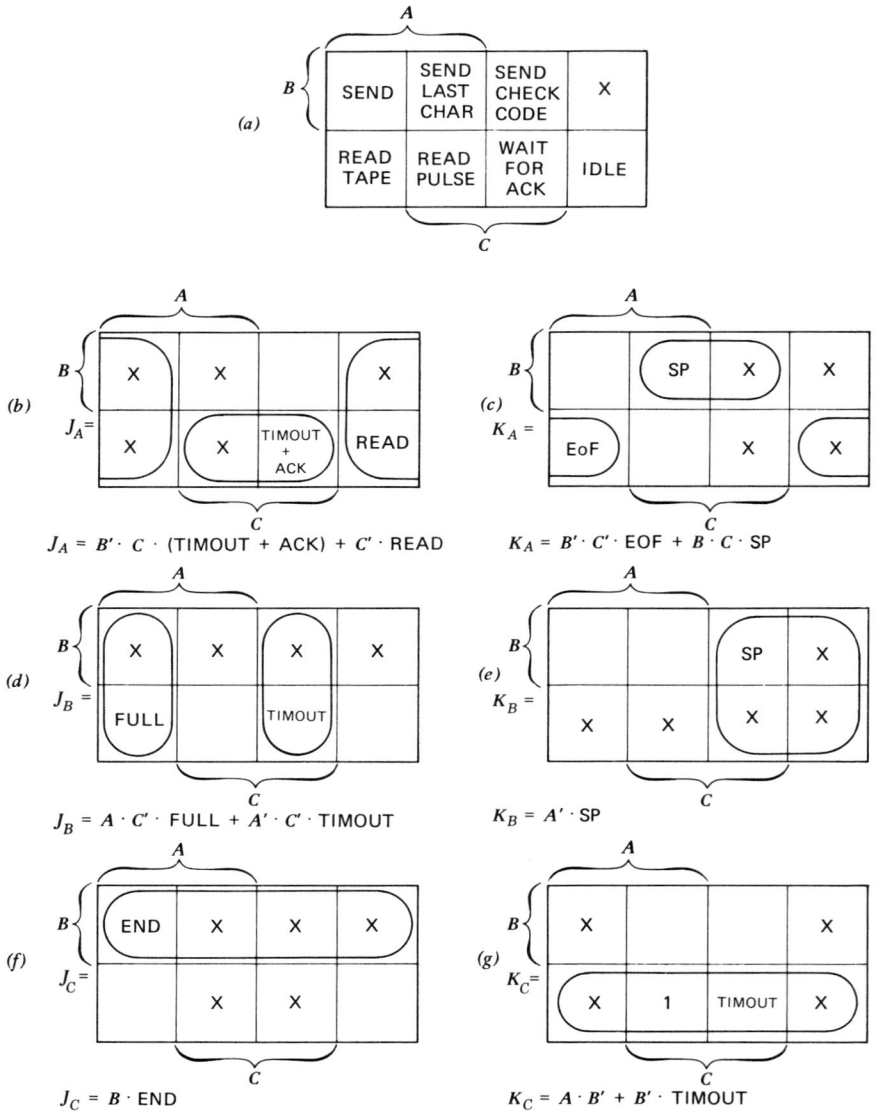

Fig. 5-10. Data transmitter control-state counter. (a) States for reference. (b) J_A. (c) K_A. (d) J_B. (e) K_B. (f) J_C. (g) K_C.

$J_A = B' \cdot C \cdot (\text{TIMOUT} + \text{ACK}) + C' \cdot \text{READ}$

$K_A = B' \cdot C' \cdot \text{EOF} + B \cdot C \cdot \text{SP}$

$J_B = A \cdot C' \cdot \text{FULL} + A' \cdot C' \cdot \text{TIMOUT}$

$K_B = A' \cdot \text{SP}$

$J_C = B \cdot \text{END}$

$K_C = A \cdot B' + B' \cdot \text{TIMOUT}$

122

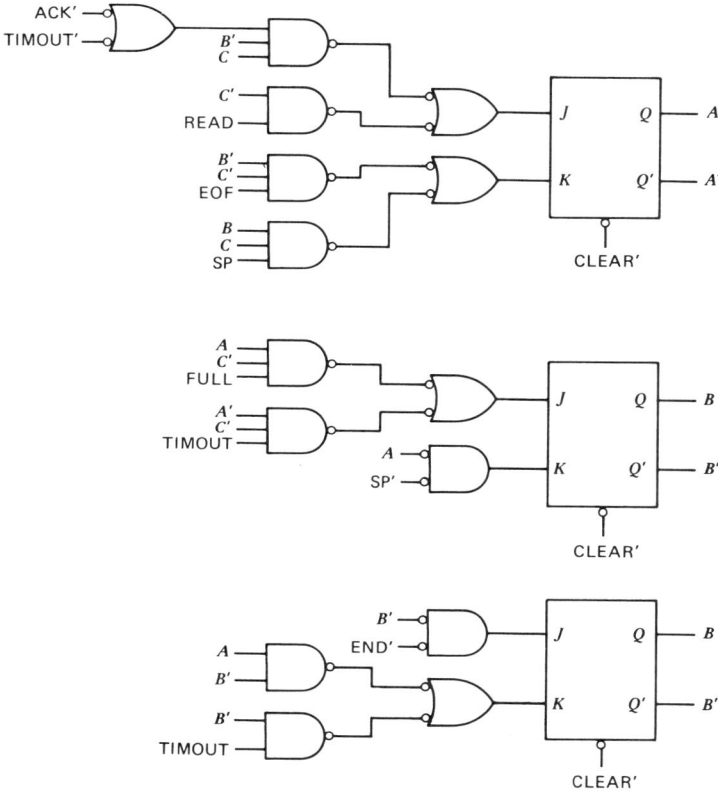

Fig. 5-11.   Mechanization of data transmitter control-state counter.

$EOF + B \cdot C \cdot SP$. Since control signals often come from subsystems designed by others or at a different time, it is often better to waste a few gates and eliminate the possibility of undesired state changes than to make assumptions about the control signals.

5.10   USE OF MSI COUNTERS

The availability of MSI counters, with four flip-flops per package, greatly simplifies most counter designs. Figure 5-12a shows a very general type of MSI

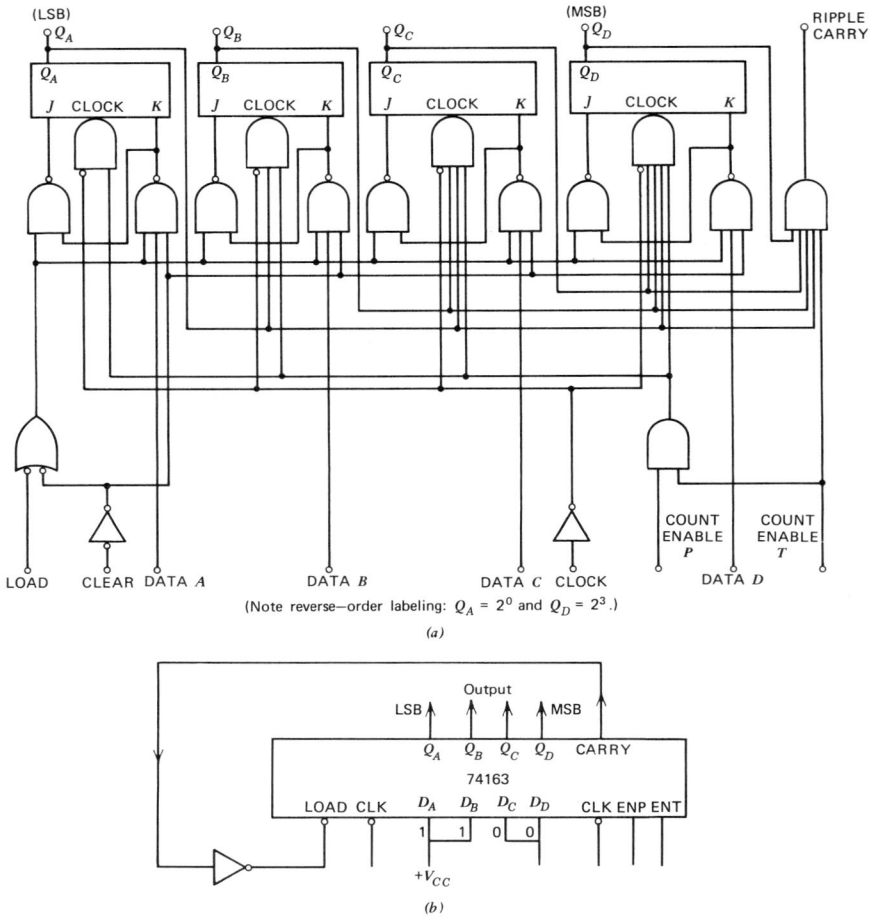

Fig. 5-12. Synchronous, presettable binary counter (74163): (a) logic diagram; (b) as an n state counter (n = 13).

binary counter. When the LOAD' input goes low, the counter is synchronously preset, on the next clock, to any state as defined by four DATA inputs. A CARRY output and COUNT ENABLE inputs make *it possible to cascade as many circuits as desired to make large counters.* By connecting the CARRY output to the LOAD' input as shown in Figure 5-12*b, we can make ANY length counter.* For example, the figure shows a counter connected to

* A CARRY output from the first package on the sixteenth count enables one count on the second package. When the second package reaches the sixteenth count, the CARRY output enables one count of the third package, and so on.

count 13; since binary 0011 = 3 is connected to the DATA inputs, the count cycle will jump from the sixteenth count (CARRY) to the third count.* The total cycle length is thus 16 − 3 = 13. Any length count cycle can be set up by simply setting up the desired jump on the DATA inputs. This technique also works for large counters made up of several ICs. The final CARRY output is simply inverted and connected to all of the LOAD' inputs.

Since we can make any length synchronous count sequence with four bits per package, *the usefulness of discrete flip-flop counters is limited to special count sequences and very small counters* (÷2, ÷3, and ÷4 only). Even control-state counters can often be done with an MSI counter.

Any control-state counter can be thought of as a linear counting sequence with "jumps" or "branches." For example, Figure 5-9 counts from top to bottom and has three "branches" back. *If we assign the states in a binary sequence, and use LOAD' and CLEAR' to do the branches, we can make MSI* control-state counters using the counter IC shown in Figure 5-12. By assigning 0000 as the initial state, we can use the CLEAR' input for initialization and to do branches to the initial state. We can use the LOAD' input to do a branch to any other state by simply applying the binary code for that state to the DATA inputs.

Instead of writing equations for flip-flop J and K inputs, the problem then becomes a matter of writing equations for

1. The main count sequence enable (COUNT ENABLE)
2. Branch enable (LOAD')
3. The DATA inputs to control where to branch to
4. THE CLEAR input for initialization and branches to 0

Since we have "don't cares" for all states that do not involve branching, the DATA input logic is usually quite simple. We thus concentrate most of the counter logic in the COUNT and LOAD functions. However, as we showed in Chapter 4, complicated logic functions can be handled quite nicely by MSI multiplexer circuits. We can thus generate the COUNT and LOAD (branch) functions, at worst, with a digital multiplexer for each. These multiplexers would thus *select the proper control signal for each state.*

5.11 AN MSI REDESIGN OF THE CONTROL-STATE COUNTER EXAMPLE

We shall now redesign the data transmitter control-state that we just designed with J-K flip-flops, using MSI. First, the binary state assignments on

---

* Notice that the lettering on MSI counter data sheets is in *reverse* order (i.e., Q_A = 2^0 and Q_D = 2^3).

Figure 5-9 are no longer appropriate, since *we must have a binary count sequence for our main line of counting.* Figure 5-13 shows the same state diagram with sequential binary state assignments top to bottom. Note that if our state diagram did not have an obvious linear count, as this one does, we could arbitrarily define one.

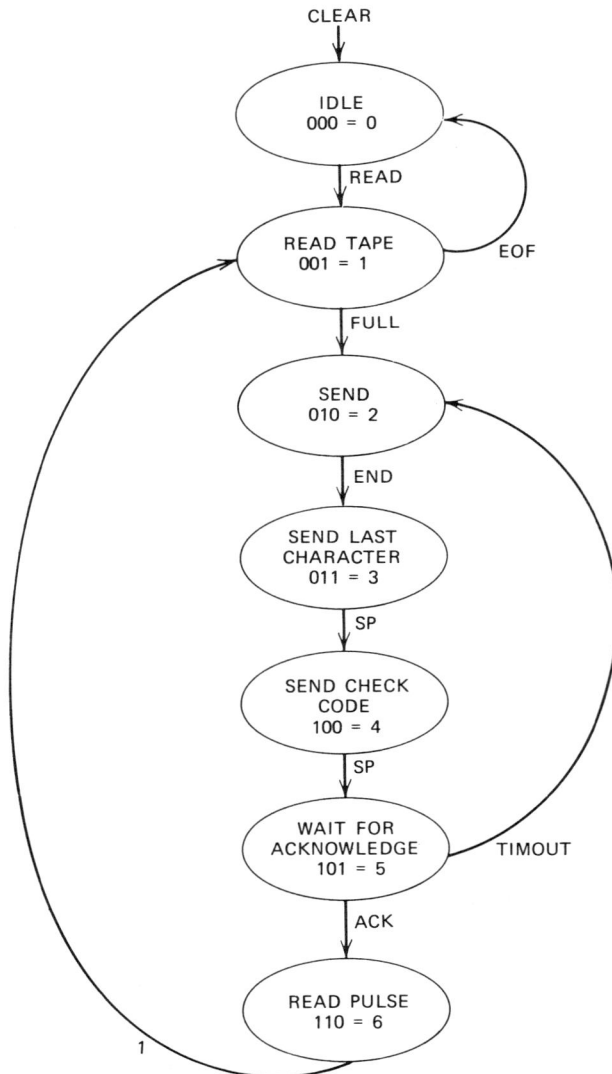

Fig. 5-13.   Data transmitter states reassigned for MSI. Binary numbers = Q_C, Q_B, Q_A.

Figure 5-14a shows a straightforward mechanization of the state diagram in Figure 5-13. Note that the "count multiplexer" simply selects the appropriate control signal for leaving each state via the main binary line of count. *Each multiplexer input corresponds to an "arrow" along the main line of count on the state diagram. Each input on the "branch multiplexer," on the other hand, corresponds to a branch arrow on the state diagram.* Since this multiplexer selects control signals for leaving a state by branching, a 0 (ground) is connected to inputs corresponding to the states that have no branch.

Since the count and branch multiplexer inputs can be read directly from the state diagram, the only real problem in designing an MSI control-state counter is finding the DATA inputs to generate the proper branches. Figure 5-14b shows a Veitch diagram of the various state names, for reference. Figures 5-14c, d, and e represent the DATA inputs to the counter in diagram form. *We can define these functions, one branch at a time, by simply entering the code for the desired destination in the position corresponding to the state from which the branch occurs.*

For example, the branch from the READ PULSE state requires $D_C = 0$, $D_B = 0$, and $D_A = 1$ (to go to the 001 state). We therefore enter 0, 0, and 1 in the upper left-hand corner of the D_C, D_B, and D_A diagrams, respectively. The branch from the WAIT FOR ACKNOWLEDGE state is to state 010, so we enter 0, 1, and 0 in the WAIT FOR ACKNOWLEDGE position on the diagrams for D_C, D_B, and D_A, respectively. The branch from the READ TAPE state is to the 000 state, so we enter 0 in that position on all three diagrams. Since we branch only from these three states, we "don't care" what the DATA inputs are for the other states, so we put X's in those positions on all three diagrams. All that remains is to write simplified logic equations for the three DATA inputs and mechanize the logic as shown on Figure 5-14a.

Since a two-input AND gate (or NAND and inverter) is required to generate D_B, we might consider utilizing the fact mentioned earlier that TIMOUT goes true only during the WAIT FOR ACKNOWLEDGE state. We could thus use the TIMOUT signal to drive D_B, since we set Q_B only for the branch initiated by TIMOUT.

Since there are only three branches, the use of a multiplexer to generate the LOAD signal is only barely justified in this example. We could design conventional logic to control the branching by making a Veitch diagram as shown in Figure 5-15. This function can be mechanized with $1\frac{1}{12}$ gate packages. If we can assume that TIMOUT AND EOF will occur only during the proper states, this equation can be reduced to

$$\text{LOAD} = Q_B \cdot Q_C + \text{TIMOUT} + \text{EOF} \qquad (5\text{-}1)$$

In this case we would require fewer packages without a branch multiplexer. The multiplexer might still be justified however, just because it is easier to understand.

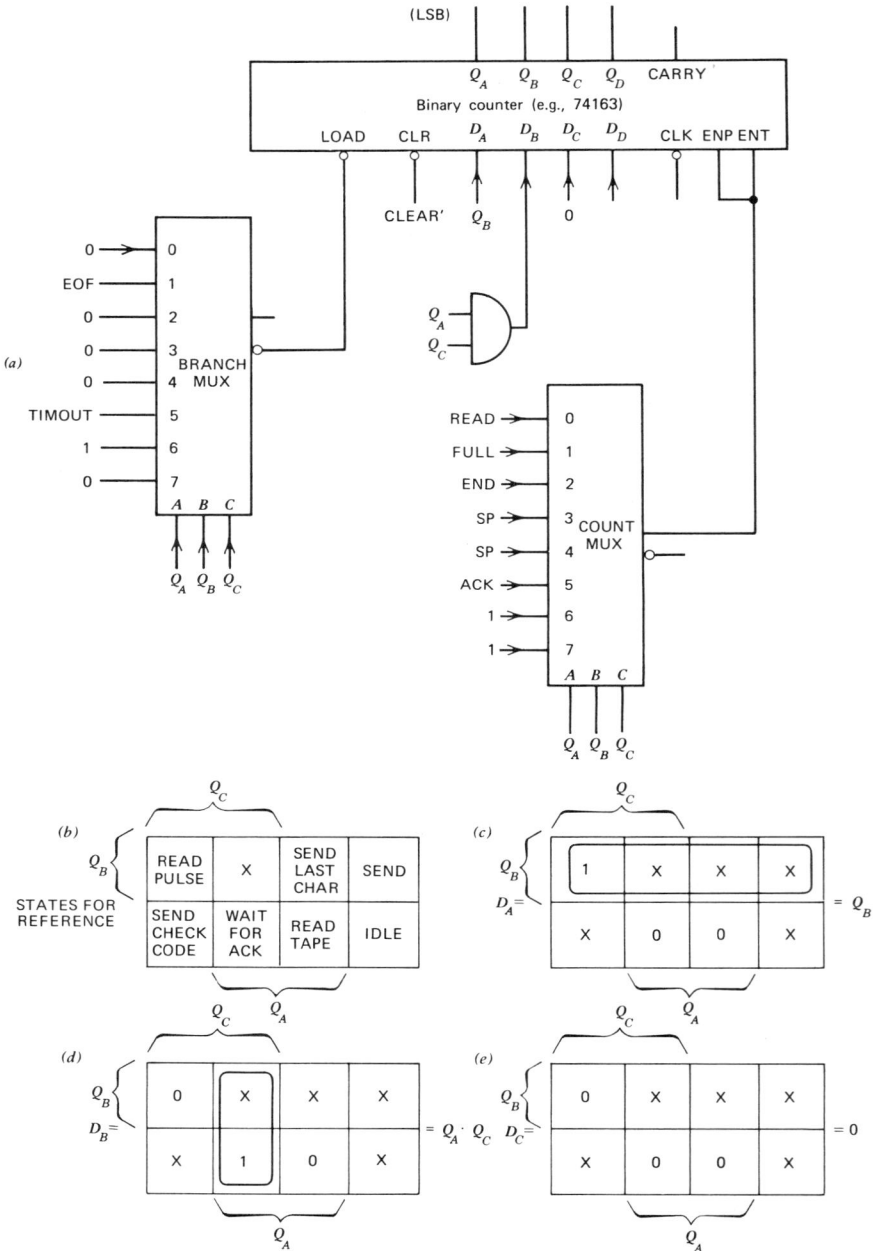

Fig. 5-14. MSI data transmitter control-state counter: (a) logic diagram; (b) states for reference.

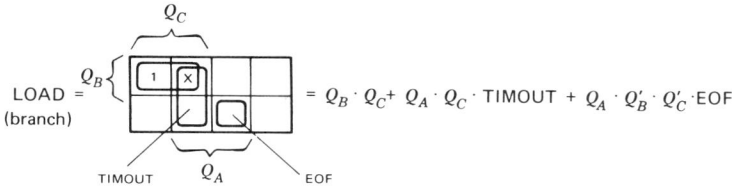

Fig. 5-15

Another possibility is to use the CLEAR input to the counter to generate branches to state 0. We could then omit the EOF term from the LOAD equation and put an X in the READ TAPE state position of the diagrams for D_A, D_B, and D_C. This would allow us to eliminate the gate used to generate D_B by using $D_B = Q_A$.

It is interesting that the MSI mechanization of this counter results in a cost savings even though we are using only three of the four flip-flops available and only three inputs on an eight-input multiplexer. This is an excellent example of using "bargain components" inefficiently and still getting a net savings.

5.12 OTHER MSI SEQUENTIAL CIRCUITS

MSI counter circuits, similar to the one shown in Figure 5-12a, are also available as Up/Down counters, decimal counters, decimal Up/Down counters, and even $\div 12$ counters. However, it is best to try to standardize on one general counter type where possible.

Another standard MSI circuit containing memory elements is the *storage register*. These circuits have common CLOCK and CLEAR inputs and are available with six D flip-flops (e.g., 74174) to a package (with only the Q output) or with four D flip-flops to a package (with both Q and Q' outputs from each flip-flop). Of course, the counter circuit in Figure 5-12 could be used as a storage register by using the LOAD and DATA inputs. Other four-bit storage registers are available with built-in tristate outputs (e.g., 74173) and with built-in two-input multiplexers on the flip-flop D inputs (e.g., 74298).

An *addressable latch* is equivalent to an eight-output demultiplexer with a storage element on each output. An active low enable and eight address bit inputs select one of the eight latches to be loaded with data. By using an addressable latch as the final stage of a demultiplexer tree, as shown in Figure 4-11, the result is equivalent to a tree with a flip-flop on each output. It is thus possible to set or reset any addressed tree output. When a circuit like this is used to demultiplex many signals sent one at a time over a single path, as

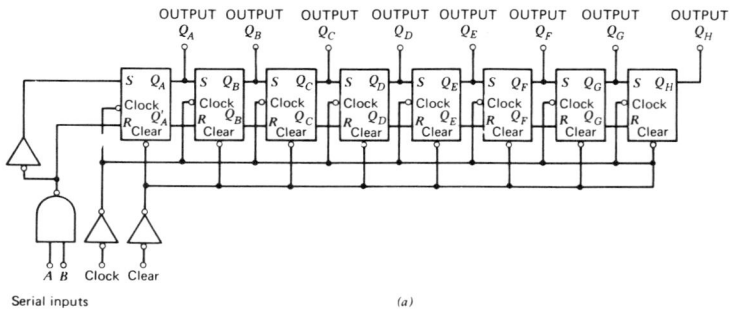

Fig. 5-16.   (a) MSI shift registers: serial-in parallel-out (e.g., 74164).

shown in Figure 4-10*b*, the outputs are *continuous,* sampled replicas of the signals at the multiplexer inputs (e.g., 9334).

Shift registers are simply flip-flops connected internally such that data are loaded into the adjacent flip-flop (and thus "shifted") with each clock. Figure 5-16 shows the three basic types of MSI shift registers:

   1.   Eight-bit serial-in parallel-out (SIPO), where the data are *read in one bit at a time.*

   2.   Eight-bit parallel-in serial-out (PISO), where the data are *read out one bit at a time.*

   3.   Four-bit parallel-in parallel-out (PIPO), where the data are read in *and* out in parallel *or* one bit at a time. The one shown is capable of shifting in either direction *and* loading in parallel.

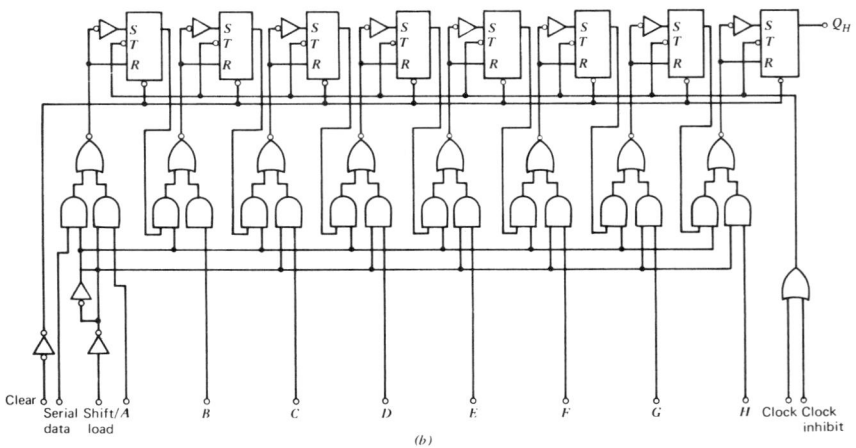

Fig. 5-16.   (*b*) MSI shift registers: parallel-in serial-out (e.g., 74166).

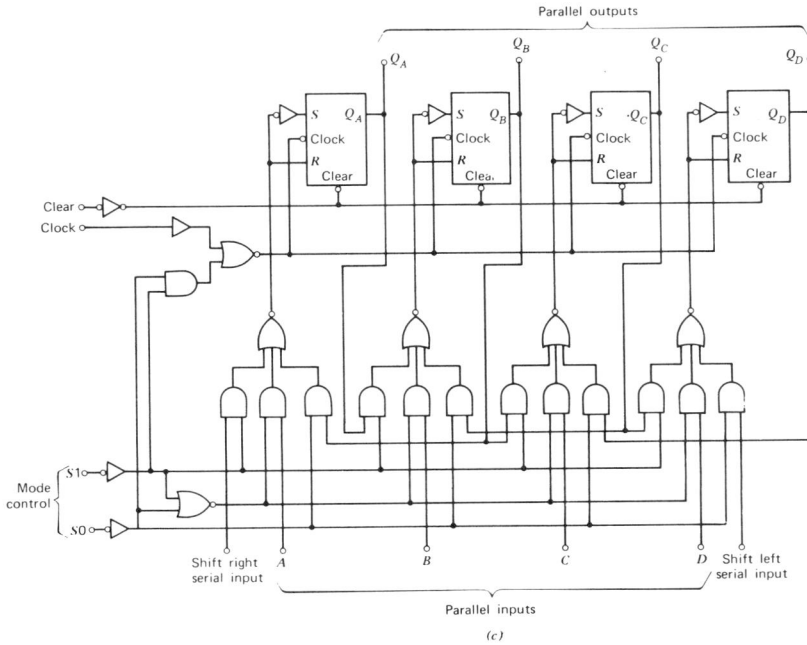

Fig. 5-16. (c) MSI shift registers: parallel-in parallel-out—reversible (e.g., 74194).

Of course shifting can be done by any presettable register (including the counter of Fig. 5-12) by simply connecting the data inputs to adjacent flip-flop outputs and LOADing.

A fourth shift register organization, serial-in serial-out (SISO), is quite popular as an LSI circuit. Shift register lengths of thousands of bits are possible in this configuration since only one input and one output are needed. *Recirculation gates* are often included to connect the shift register output to its own input on command. This makes it possible to keep shifting yet retain the existing data. Multiple shift registers of this type are available with up to six shift registers to a package with common recirculate control or up to four registers with individual recirculate control inputs.

EXERCISES

1. (a) Make truth tables defining each flip-flop type on Figure 5-1, but with Q^n as an input variable. (The Q^{n+1} output column will then be only 1's or 0's.)

    (b) Make an equivalent Veitch diagram.

2. Design gating similar to that in Figures 5-2 and 5-3 to
    (a) Make an *R–S* flip-flop from a *D* flip-flop.
    (b) Make a *J–K* flip-flop from a *D* flip-flop.
    (c) Make a *D* flip-flop from a *T* flip-flop.
    (d) Make a *J–K* flip-flop from a *T* flip-flop.

3. (a) Write the count sequence for a *four-bit* reflected Gray code counter similar to
         Figure 5-6. (Note that the eight-count sequence shown is produced by adding
         bit $A = 1$, then counting B and C back down to zero. Extend this method for
         16 counts.)
    (b) Design a *J–K* flip-flop mechanization.
    (c) Design a *D* flip-flop mechanization.

4. Design a four-bit binary 13 counter (0–12):
    (a) With *D* flip-flops.
    (b) With *J–K* flip-flops.
    (c) With *T* flip-flops.

5. Design a four-bit binary 11 counter that counts *backwards* through binary states 15
    through 5:
    (a) With *J–K* flip-flops.
    (b) With *D* flip-flops.
    (c) With *T* flip-flops.

6. (a) Design a five-bit moebius count sequence.
    (b) Mechanize it with *D* flip-flops.
    (c) Decode the following *single* states: 14, 15, 0, 7, 31.
    (d) Decode two states, three states, and four consecutive states starting with state
         0.
    (e) Do the same starting with state 15.
    (f) Do the same starting with state 3.

7. (a) Make a state diagram for a sequential circuit that will give a true output if
         input *A* goes false for more than 15 clock times (a digital "holdover" circuit).
         Design the logic circuit using
    (b) *J–K* flip-flops and NAND gates.
    (c) *D* flip-flops and NAND gates.
    (d) The MSI counter shown on Figure 5-12 and NAND gates.

8. (a) Make a state diagram for a sequential circuit that will give a true output for 13
         clock pulses after input *A* goes true for one clock time. If *A* goes true again
         during the 13 clocks, it is ignored (a digital "one-shot"). Design the required
         logic circuit using
    (b) *J–K* flip-flops and NAND gates.
    (c) *D* flip-flops and NAND gates.
    (d) The MSI counter shown on Figure 5-12 and NAND gates.

9. Design a control-state counter for a receiver to receive data from the transmitter of
    Figure 5-9. At the *beginning* of each character received CHAR goes true for one
    clock time. At the *end* of the check code character CHECK goes true if there is no
    error. At the end of the sending of ACKnowledge, SCH goes true for one

clock time. As outputs we need a STORE signal to indicate that data are being received and SACK to indicate that an ACKnowledge should be sent.

(a)   Make a state diagram.

(b)   Design a control-state counter using J–K flip-flops and NAND gates.

10.   Mechanize the state diagram of Figure 5-13 with J–K flip-flops and NAND gates.

11.   Redesign and redraw Figure 5-11 taking advantage of the fact that TIMOUT, EOF, and ACKnowledge occur only during the desired state.

12.   Redesign and redraw Figure 5-14 using NAND gates instead of a multiplexer to generate LOAD, taking advantage of the fact that TIMOUT, EOF, and ACKnowledge occur only during the desired state. Also, use the CLEAR input to do the EOF branch.

13.   Make a table of D_A, D_B, D_C, and D_D input required for all count cycle lengths from 1 to 16 using the circuit in Figure 5-12b.

14.   Design a $\div 27$ counter using two MSI counters connected similarly to Figure 5-12b.

15.   (a)   Make a Veitch diagram for the COUNT ENABLE input in Figure 5-14a.

(b)   Write a logic equation in simplest form.

(c)   Mechanize the function using NAND gates.

16.   Mechanize a shift register using the counter circuit of Figure 5-12a.

17.   Connect the circuit of Figure 5-16c so that it will do one of three operations depending on the mode input code S1, S0:

(a)   Shift right

(b)   Shift left

(c)   Reverse the order of the bits: $ABCD - DCBA$.

18.   Design a digital 24-hour clock using a 64-kHz crystal oscillator, eight MSI counters (Fig. 5-12a), and four 7-segment display drivers.

BIBLIOGRAPHY

Booth, Taylor, *Digital Networks and Computer Systems,* Wiley, New York, 1971, pp. 159–223. (Forming state transition diagrams.)

Brock, *Designing With MSI,* Vol. I, *Counters and Shift Registers,* Signetics Corp., Sunnyvale, Calif., 1970, Appendix 2, "Maximum Length $(2^n - 1)$ Shift Counter Sequences."

Hamer, H., "Johnson Counter Decoder," *Digital Design,* November 1973, pp. 56–57.

Phister, M., *Logical Design of Digital Computers,* Wiley, New York, 1958, Chapter 5, "Memory Element Input Equations," and Chapter 6, "Huffman–Mealy Method of Minimizing Number of States."

Su, Stephen Y. H., "Logic Design and its Recent Development, Part 3: Design of Sequential Networks," *Computer Design,* November 1973, pp. 85–92.

CHAPTER SIX

NASTY REALITIES I
Race Conditions
and Hangup States

6.1 INTRODUCTION

In Chapter 5 we developed sequential logic design techniques as though we lived in a purely mathematical, ideal world. This was a useful simplification while we developed our design techniques. But now we must return to the real world, where things do not always work the way we would like. This chapter concerns the "facts of life," which must be known if we are to make reliable, sequential logic designs.

Though seldom admitted, many failures, particularly in new systems, result from design errors. Since most of the problems discussed in this chapter result in *marginal operation,* design problems are *not necessarily found in initial checkout.* With the normal variation in component characteristics, marginal problems occasionally arise in later production. Replacing a component often solves the problem—even though the component is actually within the specification. Because these problems are so insidious, we must be very careful to avoid them in the initial design.

6.2 UNCLOCKED LOGIC AND RACE CONDITIONS

Virtually all flip-flops and MSI circuits containing flip-flops have asynchronous (unclocked) *CLEAR* and/or PRESET inputs. Although these inputs can be quite useful for certain operations, they can also be very dangerous if used improperly. With clocked logic, all flip-flops in the system are "clocked" at the same instant. Each flip-flop comes to a new state based on conditions existing *just before* the clock transition. While the flip-flops are changing, all kinds of *false outputs are produced* due to variations in delay of the various flip-flops and gates.

Figure 6-1 shows an example where $A \cdot B$ is false before and after the clock, yet the function $A \cdot B$ is *momentarily true* just after the clock due to a slight difference in timing of the two signals. With clocked logic this narrow pulse, or "*glitch*," would have no effect, because it is gone long before the next clock. If, however, the $A \cdot B$ signal is connected to the unclocked CLEAR input of a flip-flop, the flip-flop *may be reset* by this signal—even though we only intend to have it reset when both A and B are true as a real system state.

The duration of the "glitch" is so short, however, that it may not reset the flip-flop while we are checking out the system. Later, when a system is built with a faster A flip-flop, a slower B flip-flop, or a faster flip-flop asynchronous CLEAR, the system will fail *sometimes*. Often just a change in temperature is enough to bring on the trouble. The trouble may go away when the cabinet is opened or when an oscilloscope probe is put on the signal.

This kind of problem must be avoided by being very careful in the design phase. The best plan is always to use clocked logic and *avoid the temptation to use unclocked PRESET or CLEAR inputs for all but the simplest CLEAR or PRESET functions*. Often a fair amount of logic can be saved by using asynchronous inputs, but, since logic is cheap and fiascos are expensive, we

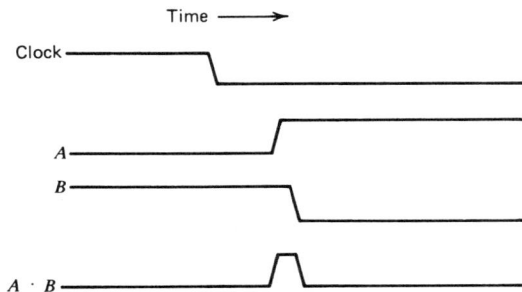

Fig. 6-1. Temporary incorrect output due to "slicing."

must be very careful. It is normally safe to use asynchronous inputs to initialize the system, where there is a good, clean (not decoded) *"initialize"* signal available. It is also usually safe to use unclocked PRESETs to set data from switches, and so on, into flip-flops, providing the PRESET ENABLE signal is carefully generated and distributed.

Notice that the delay difference in Figure 6-1 always occurs if A and B are discrete flip-flops because, when a flip-flop sets or resets, the flip-flop action starts on one output and appears on the other output only after one gate delay. On a TTL flip-flop, for example, one output (e.g., Q) goes *high* first, then one gate delay later the other output (e.g., Q') goes *low*. On MSI circuits the negated outputs are usually not brought out, so any function having a negated signal input requires an inverter. The inverted signals will thus change *after* the uninverted signals. Even though output delays from the same MSI package are usually matched, delay differences can result from unequal loading or even abnormal circuit variations that are still within the specification. Read-only memories can also be depended upon to produce gliches, even after single-bit input changes, because of the increased delay of the internal inverted address lines.

The waveforms in Figures 6-1, 6-2, and 6-3 are examples of a very useful design tool called a *timing diagram*. By drawing signals together as they would be seen on an oscilloscope with many traces, it is often much easier to see the operation of a sequential system. By *exaggerating possible timing differences and glitches* that may occur just after the clock, we can often see potential timing problems.

Figure 6-2a shows a timing diagram of a $\div 4$ binary counter and decoder outputs for the four states. Glitches are shown on all four outputs even though, in reality, all four glitches would probably never occur. The glitch on the State 0 decode, for example, would occur if A' goes true before B' goes false. The timing required for some of the other glitches may contradict this order, but the point is that any of these glitches *might* occur. By showing *all the possible glitches* on the timing diagram we avoid problems by making all of the possibilities clear.

Sometimes we need a clean, decoded counter signal for use as a multiphase clock. As Figure 6-2b shows, a moebius counter gives us this because *only one bit changes at a time*. For longer count sequences a reflected Gray code sequence, such as that shown in Figure 5-6, can be used.

Another common problem when using unclocked logic occurs as follows: if we try to CLEAR a flip-flop with a logic function *that is true only when that flip-flop is set,* the CLEAR signal will disappear as soon as the flip-flop starts resetting. Normally, gate circuit delays will make the CLEAR signal remain long enough to complete the reset, but it is possible for the circuit to go into oscillation. Even if the reset is successful, the CLEAR signal will be a very narrow

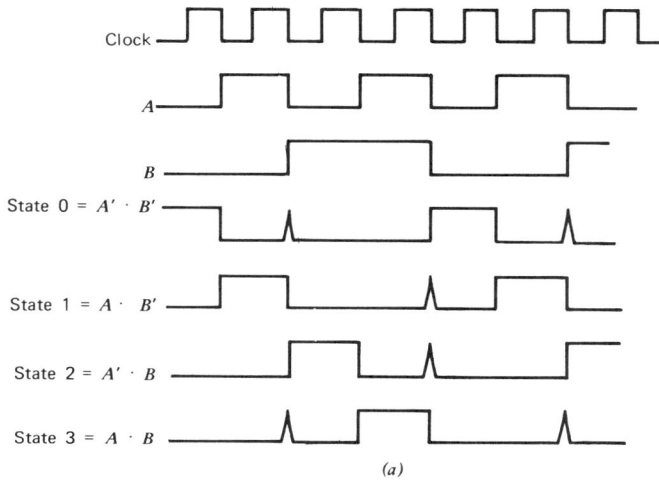

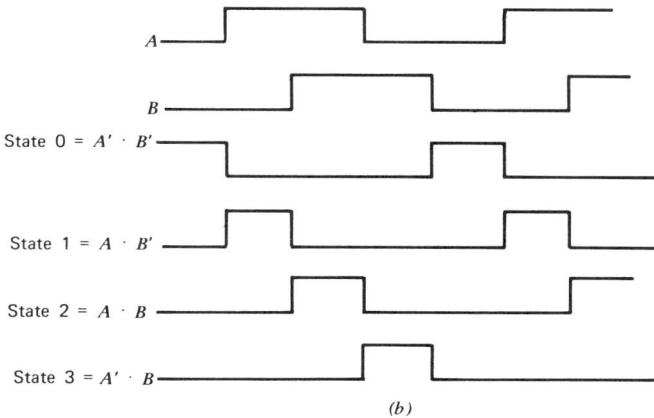

Fig. 6-2. (a) Binary counter decode glitches. (b) moebius counter decodes (no glitches).

"glitch." If *other* flip-flops must be reset by the same signal and they are *slower* than the flip-flop that stops the CLEAR signal, they may not reset at all.

Before fully clocked, presettable counters were available inexpensively, it was common practice to make "*n* state" counters by decoding the $n + 1$ state and connecting that signal to the *unclocked CLEAR* input of the counter.

Since the flip-flops were all on the same chip, they were usually pretty well matched in their reset speed, so the counter normally worked properly. There is still some danger of unmatched, but within specification, flip-flops or mismatching due to *unequal loading of outputs,* so this trick is not recommended. Today, we can buy fully synchronous, presettable counters (like the one in Fig. 5-12) for essentially the same price, so there is just no need to take the risk.

Actually, techniques have been developed (see Ref. 1) for designing reliable *asynchronous* circuits, but they are so difficult to use that their use is limited mainly to *internal* design of IC flip-flops, and so on. For normal digital systems design *clocked logic should be used for all but the simplest functions because it greatly reduces the chances for design error.*

6.3   ASYNCHRONOUS INPUTS TO CLOCKED LOGIC

Even though we use fully clocked logic, we are still stuck with system *inputs* that can change at any time. There is a very brief period of time, just before the clock transition of a flip-flop, during which the inputs to the flip-flop must be stable for reliable operation. This is called the "**set-up time**" and is equal to about one gate delay. If an input changes during this time, the final state of the flip-flop is indeterminate and depends partly on how fast the flip-flop actually is. If we have a system that changes state when an input goes true, and *two or more* flip-flops must change to make that state change, it is possible that *only* the faster flip-flops would change if the input changes just before the clock. The state change would thus be *to the wrong state.*

Figure 6-3 shows a timing diagram of an improper state change example: when the control input goes true, *A* and *B* are supposed to set. The *setup time* of flip-flop *A* is a little faster than that of *B,* so when the control input goes true just before the clock, *A* sets but *B* does not. The state change is thus to state 10 instead of 11. The system will thus do entirely the wrong thing. Again, the problem is insidious because the probability of its happening de-

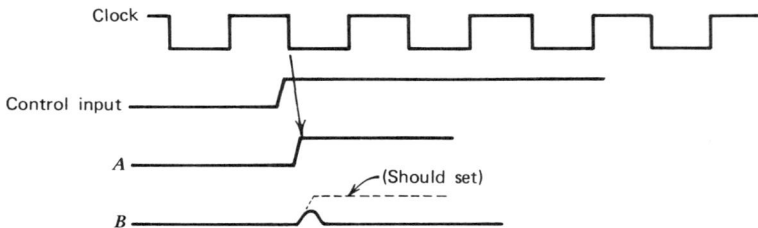

Fig. 6-3.   Improper (2FF) state change due to asynchronous control input.

pends on the difference in setup times of the two flip-flops. If the difference in setup times is only 2 **nanoseconds** (nsec) and the clock transition comes every 500 nsec, *the problem will occur only once every (500/2) = 250 times the control input goes true* (statistically). If the flip-flops on the prototype are very well matched in speed, the problem may not be found until production begins. the best solution is *NEVER TO CHANGE MORE THAN ONE FLIP-FLOP IN RESPONSE TO AN ASYNCHRONOUS INPUT SIGNAL.* If it is impossible to assign control states so that all externally caused state changes are single-bit transitions, then an extra resynchronizing flip-flop should be added.

The little "bump" that appears on the *B* output in Figure 6-3 points up another potential problem with asynchronous inputs. When a flip-flop *almost* sets, its output may go to the edge of being true for a moment and then drop back. The probability of this causing a problem is quite small unless the flip-flop is in a system that is running at close to the maximum clock rate. In this case it is possible that other state changes in the system may (perhaps partially) occur on the next clock. When the original flip-flop drops back to false, an improper operating mode may have been set up. In cases where this could be a problem, it is only necessary to add an additional flip-flop, or (single-bit change) state, to delay any action till the second clock.

Sometimes it is convenient to use several different clocks in a system. For example, the system in Figure 6-4 has a subsystem that uses a slow clock, which clocks once for each data bit, to receive serial data one bit at a time. When a complete word is assembled, it is transferred to a data processing subsystem that has a high-speed clock. The *strobe* input to this subsystem is treated as an asynchronous input. The serial data, however, are synchronous to the clock in the receive-logic subsystem.

Often system inputs come from mechanical devices, such as switches, which may have nasty characteristics. Typically, *switch contacts will bounce for from 5 to 50 msec after they are closed* (see Fig. 6-5a). This can play havoc with a system if it is not dealt with properly. If the raw switch signal is used to gate on a fast system operation, then the operation will start and stop as the switch bounces. This can be avoided by letting the switch closure set the system (via a

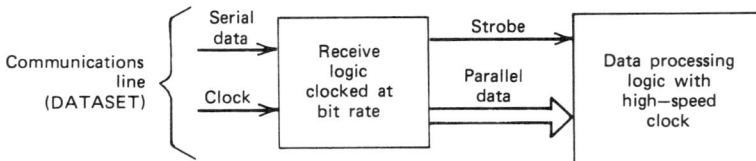

Fig. 6-4.   System with multiple clocks.

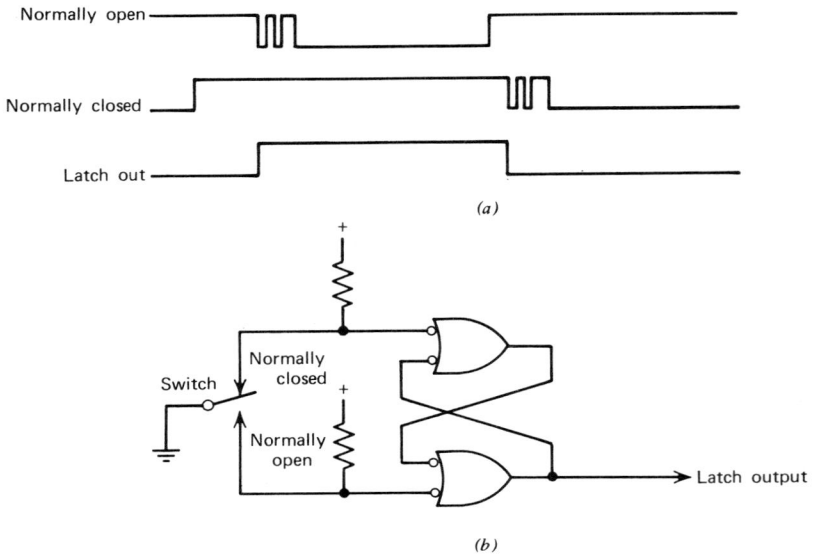

Fig. 6-5.   Switch bounce elimination.

clocked, one-flip-flop change) into a new control state. If something must be done *once* for each time the switch closes, then another (clocked, one-flip-flop change) system state is required *to remember that the switch has been released.* If the operation is finished in less than the switch bounce time, a normally closed contact will have to be used to detect that the switch is released, or the operation will *repeat for each bounce.* Often it is simplest just to use a latch circuit (Fig. 6-5*b*) made from two NAND gates to remove the bounce and make a more ideal control signal. Unless a latch circuit is used to "clean up" switch signals, they should always be drawn on timing diagrams with bounce shown (as in Fig. 6-5*a*). Also, since contacts sometimes open momentarily due to vibration, they should not be depended upon to *stay* closed, even after bounce.

6.4   CLOCK SKEW

The basic principle behind clocked logic is that the next state of all storage elements in the system is determined *at the same instant.* Several practical problems make it difficult to reach this goal. In a large system several clock drivers may be required to drive all the clock inputs. If each one has a different time delay, the clocks will arrive at different times. This timing error is called *clock skew*, and it can cause erratic operation if excessive.

Figure 6-6*a* illustrates an incorrect data transfer between flip-flops due to excessive clock skew. The circuit is essentially a shift register with a 0 input to FF1 and an initial state of 11. We would normally expect the first clock to bring us to the 01 state and the next to the 00 state. Because of the clock skew the first clock, in Figure 6-6*a*, brings us directly to the 00 state. The problem is that clock 1 is *earlier* than clock 2 so the *new* state of FF1 gets clocked into FF2.

Actually, a small amount of skew can be tolerated because of the **propagation delay** of FF1 (delay from clock to change of output). For proper operation, clock skew must be less than the minimum propagation delay minus hold time. Most circuits are guaranteed to work with zero *hold time* (time after clock before data may change), so *maximum allowable skew is usually simply equal to minimum propagation delay.* Unfortunately, IC specifications often do not guarantee a *minimum* propagation delay. This makes true "worst-case" design possible only if all clocks are physically connected to the same driver.

As Figure 6-6*b* shows, clock skew is a problem in only one direction. Data *can* be properly transferred *in one direction* with skewed clocks. As long as the

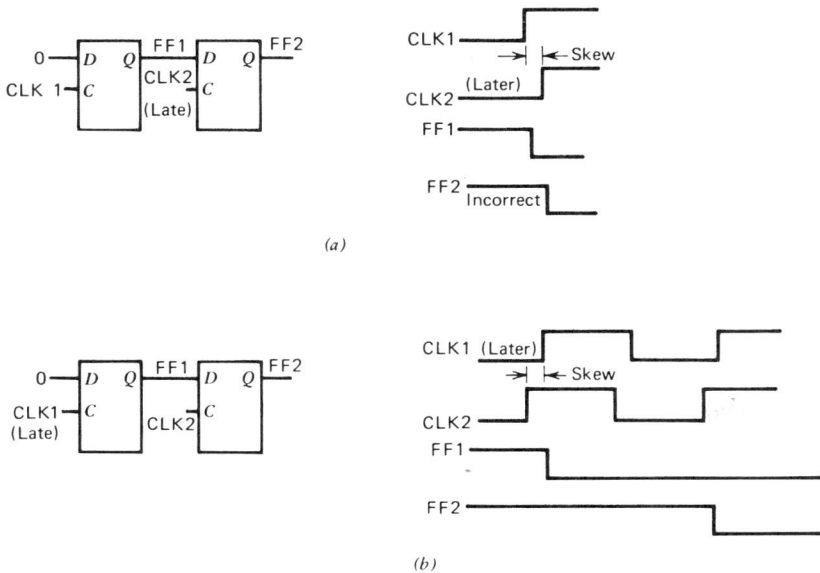

(a)

(b)

Fig. 6-6.   (a) Unsuccessful transfer to later clock. (b) Correct transfer *from* later clock.

transfer is *from* the later clock, there is no problem. In most cases, however, it is necessary to transfer data in both directions. The only solution, then, is to keep the clock skew very low by using a single, high-powered clock driver or very fast matched drivers.

One thing that can make skew a big problem with *TTL* logic is the fact that, *while most MSI circuits are clocked by the* (+) (*low-to-high*) *edge of the clock, the J-K flip-flops are clocked by the* (−) *edge.* This makes it impossible to use a single clock line for all circuits (unless all transfers are in one direction only—e.g., *from* the *J-K's to MSI*). *One solution to this problem is to use only flip-flops with a minimum* propagation time specification and generate the clock for the other circuits by inverting the *J-K* clock with an inverter that is guaranteed faster than the flip-flops (e.g., 74112 *J-K* flip-flops can be used with a 74S series inverter or driver). A better solution is to use only + edge triggered *J-K'* flip-flops such as the 74109.

With extremely high-speed logic (such as ECL or 74S **Schottky TTL**), *wiring delay* can cause clock skew problems. Figure 6-7a shows an example of improper clock distribution. Since the clock line doubles back, the second flip-flop clock is delayed by the 1.5 to 2 nsec/ft it takes for the signal to travel down the wire. This could mean an improper transfer of data *with only a 1-ft loop* with a 74S112 flip-flop (minimum propagation delay = 2 nsec) or even less with an ECL flip-flop. As Figure 6-7b shows, clocked logic can be operated properly with different length clock connections as long as they follow the same path as the logic flow. The delay in the logic connection cancels the delay in the clock line.

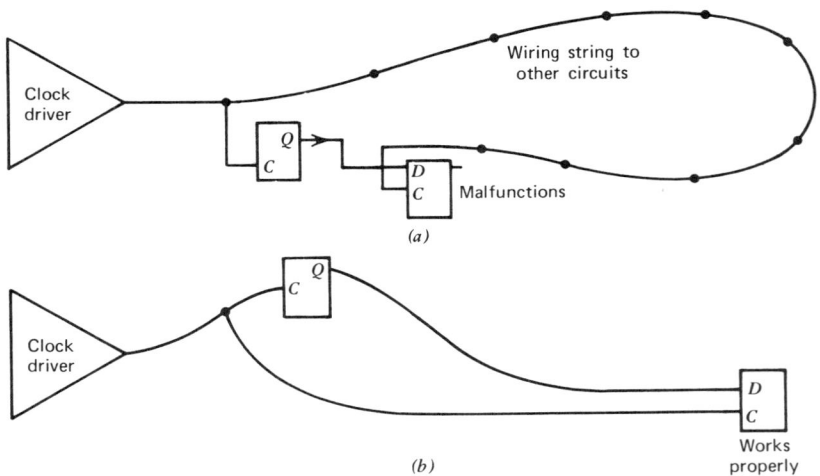

Fig. 6-7. Clock skew due to wiring. (a) Looping back clock line causes skew problem. (b) With proper routing delays cancel.

Note that when there is *gating* between the two storage elements, signal delay is increased so *more skew is allowable.* Clock routing on very high-speed systems can thus usually be optimized for the data paths that have no gates in series.

6.5 MAXIMUM CLOCK RATE

Most flip-flops and MSI circuits have a maximum clock rate specification on the data sheet that, while technically true, is not really practical. First of all "typical" maximum clock rates are often quoted, which simply means "average." In other words, *half the circuits will work at that speed and half will not.* Since we want *all* our systems to work properly, we must use the *worst-case* specifications, such as guaranteed *minimum* operating speed.

Even the worst-case minimum specified maximum clock rate is typically two or three times the PRACTICAL maximum clock rate in a SYSTEM. The reason is that most useful systems require *gating* in the signal path. Since gates add a delay to the propagation delay of flip-flops, the time after the clock (before all flip-flop inputs are stabilized) is the *sum* of these delays. The maximum allowable clock rate is the rate at which the clock comes just after the flip-flop inputs have been stable for the minimum setup time, or: *minimum clock period = flip-flop delay + gating delay + setup time.*

Figure 6-8 shows a system with only two levels of logic, the minimum

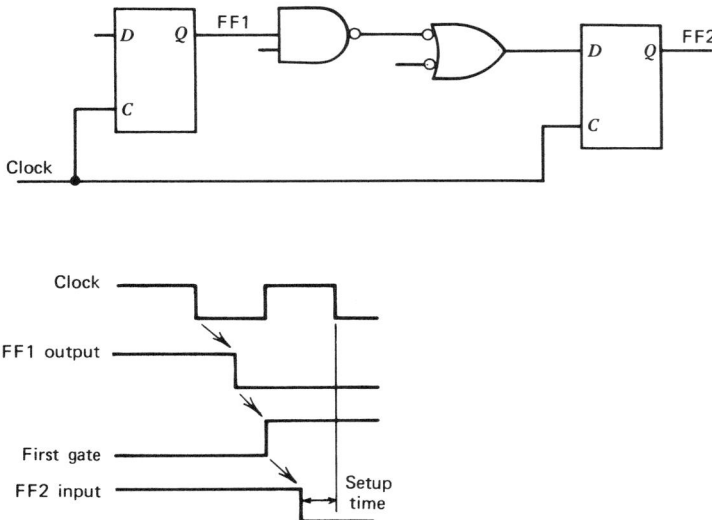

Fig. 6-8. Maximum clock rate.

possible for a practical system. If we take, as an example, a 74S74 flip-flop and 74S00 gates, the maximum clock period is 9 nsec (FF delay) + 5 nsec (gate 1 delay) + 4.5 nsec (gate 2 delay) + 3 nsec (setup time) = 21.5 nsec. This gives a maximum clock rate of $1/21.5$ = *46 MHz*. The mimimum clock rate specified on the 74S74 *data sheet* is *75 MHz*. If we allowed a maximum of four gating levels, we could undoubtedly save some logic by factoring, predecoding, and reducing variables with multiplexers. The maximum clock period would then be 9 + 5 + 4.5 + 5 + 4.5 + 3 = 31 nsec, so our maximum clock rate would be $1/31$ = *32 MHz*, or less than half the specified rate. We thus have a tradeoff between logic cost and speed.

Notice that we used 5 nsec for the delay of one gate level and 4.5 nsec for the other. This is because the specified maximum delay varies, depending on the direction of the transition. Since the logic inverts at each level, we can assume that a signal propagating through the system will have one low-to-high transition (4.5 nsec) and one high-to-low transition (5 nsec).

Some of the *older J-K* flip-flop IC designs (e.g., 7470 through 7473, 7476, 7478, 74107) are *not* edge triggered, which means that they essentially have a preset time that includes *the entire time the clock is high*. Maximum clock rate with these flip-flops is even further from the specification, unless a very narrow clock pulse is used, since *all inputs must settle before the clock goes high*. These flip-flops will change state when the clock falls if the inputs have indicated a change of state at any time while the clock was high. The best solution to this problem is simply to use the newer types of *J-K* flip-flops (e.g., 74112, 74109 or 74S112).

In some systems high-speed flow of data is wanted yet it is impractical to use only two levels of logic. In this case a technique called *pipelining* can be used. For example, Figure 6-9 shows a system that performs four levels of logic on data at a rate that would normally be possible with only two logic levels. The secret is to *resynchronize the data* after the second level by using an extra level of flip-flops. Of course this adds an additional one-clock delay to the appearance of the output data, but the data flows "through the pipeline" at a *rate* normally possible with only two levels of logic. Of course, any number of logic levels can be used between registers, and any number of resynchronization registers can be used in a pipeline system.

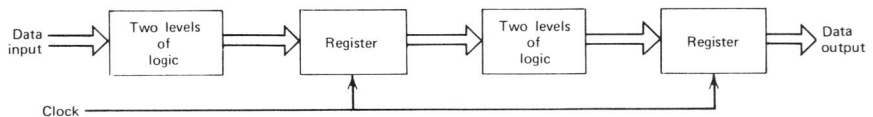

Fig. 6-9. Pipelining: four logic levels at two-level rate.

6.6 HANGUP STATES AND SELF-CLEARING LOGIC

In any system containing storage elements, we normally provide an initialization signal, when power is first turned on, which sets everything to some initial state. This *master clear,* or *initialize,* signal is usually generated automatically by a delay circuit that senses that the power was recently below a certain level. Without such a clear, the system would not start at the correct state and would therefore not work properly the first time. This is fairly minor compared to another problem that results from extra *undefined states* from which there is *no* exit. If a system gets in a state like this, it may continue operating improperly "indefinitely" (or until the system is *reinitialized*).

The moebius counter described in Section 5.5 is a perfect example of this. In Figure 5-7d we show a six-state count sequence for a three-bit moebius counter. Since three bits can define eight states, we might ask, "What happened to the other two states?" If we assume that the counter turns on in the 101 state, the next state will be 010, which is the other undefined state, and the next state will be 101 again (see Fig. 6-10b). The counter will stay locked in a two-state sequence "indefinitely." Of course, while the counter is locked in this sequence, the system appears to have failed. As a matter of fact, *all moebius counters of more than two bits* and many other types of n flip-flop counters with less than 2^n states, *have extra, undesirable count sequences.* That is not to say that all counters with extra states have this problem. The $\div 5$ counter shown in Figure 6-11, for example, is *self-clearing* because *any state* leads back to the main count sequence. State diagrams for specially designed counters should always be drawn with *all possible states accounted for* (as in Figs. 6-10 and 6-11), to be sure there are no hidden states that will hang up the system.

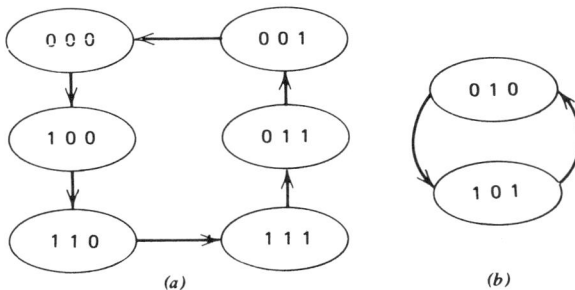

Fig. 6-10. Complete state diagram for a three-bit moebius counter: (a) desired sequence; (b) undesired sequence.

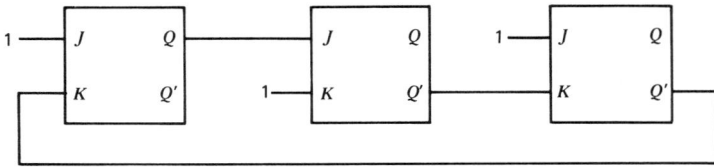

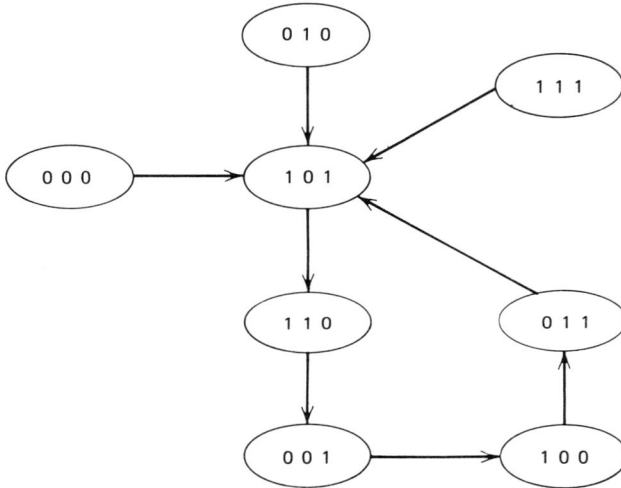

Fig. 6-11.   A self-clearing ÷ 5 counter.

Although it is sometimes possible to depend on the master clear signal to set the system into the proper states, *it is bad design practice.* Flip-flops can be accidentally set or reset by test probes, noise pulses, or unplugging circuit modules during servicing. While this is bound to cause a *temporary* malfunction, *it should be impossible to put the system in a state from which it will not clear ITSELF.* When MSI binary counters are used, this is automatic, since each binary state is part of the defined sequence, whether we use it or not. If we are designing an MSI control-state counter, we simply have to be sure the count control signal goes true on undefined states, as we did on Figure 5-14a by connecting "1" to the unused count multiplexer inputs. In our discrete flip-flop state counter example of Figure 5-9, we provided for the extra state when we used it as a "don't care" (Figure 5-10b) in the equation for J_A. We will thus set A from the 010 state, when READ goes true. We really should have drawn in the 010 state on Figure 5-9, with an arrow marked "READ" pointing to the 100 state.

This principle of making systems self-clearing goes beyond just counter design. For example, it has been found by experience in telephone switching systems that *information in memory will become corrupt in time* if there is no natural process by which incorrect data are "cleaned" out. For example, call information is stored until the call is terminated. If some malfunction causes the termination of a call to go undetected, the call record will stay there forever. Ultimately more and more nonexistent calls would accumulate until the system would overload. By adding "housekeeping logic" to verify occasionally that each call really exists, the system is made *self-clearing*. Many computers that must run continuously are designed to reload all programs and "start from scratch" whenever serious problems are detected. (This event is called a "system crash".)

EXERCISES

1. Make a timing diagram of the $\div 5$ counter in Figure 6-11, including the decodes for the five states. Show all possible "glitches."

2. Design a five-state counter, with five decodes, that can be used as five *clock* phases.

3. Which of the "glitches" in Figure 6-2a would disappear if A could be guaranteed to change before B (as it could if the output of A *clocks* B)?

4. Which states on Figure 5-9 can be decoded without danger of "glitches"?

5. Which states on Figure 5-13 can be decoded without danger of "glitches"?

6. (a)   Assuming none of the control signals for the control-state counter of Figure 5-9 is synchronized to the system clock, is there any possibility of improper state changes?
    (b)   How about with the state assignment of Figure 5-13?

7. If the READ signal in Figure 5-9 is directly from a switch that bounces for 20 msec, is there any problem? (It takes 50 msec to READ TAPE.)

8. Would we have any problem in Figure 5-9 if, instead of using J_C, we directly set C with the PRESET' input, using the signal $(A \cdot B \cdot C' \cdot \text{END})'$?

9. (a)   Would we have any problems in Figure 5-9 if we dropped the $C' \cdot \text{READ}$ term from J_A (Fig. 5-10) and, instead, directly preset C with the signal $(A' \cdot B' \cdot C' \cdot \text{READ})'$?
    (b)   If we used $(B' \cdot C' \cdot \text{READ})'$ instead, what would happen if the READ button was still pressed after the last character was sent?

10. Draw Figure 6-3 as it would look if the setup time of flip-flop B was faster than A.

11. (a)   What is the probability of improper operation per Figure 6-3 if the difference in setup times is 1 nsec and the clock rate is 10 μsec?
    (b)   If the control signal went true once a second, how often would improper operation be expected to occur?

12. Is there any problem with the state counter of Figure 5-13 if the ACK signal is not synchronized to the clock?

13. Alter the state diagram of Figure 5-13 so both the READ signal and the ACK signal can be asynchronous.

14. Draw the other count sequence(s) for the four-bit Moebius counter of Figure 5-7a.

15. Modify the design of the four-bit moebius counter so it is "self-clearing" to the sequence shown on Figure 5-7a.
    (a) With the D flip-flop version.
    (b) With the J-K flip-flop version.
    (c) Using the asynchronous CLEAR input(s) with no race problems.

16. Do exercise 15 for the three-bit moebius counter.

17. Design a system that will generate a single pulse one clock period long each time a push button is pressed.

18. Design a system using an MSI counter, that will produce a signal that goes true for 13 clock times each time a switch is pressed.

19. Design a circuit that will generate one clock pulse each time a *STEP* switch is pushed and will cleanly turn the clocks on and off when a *RUN* switch is turned on and off. No marginal clock outputs are allowed.

20. Could data be passed reliably from one flip-flop to the other in the opposite of the direction shown in Figure 6-7a?

21. (a) Draw a schematic of a two-bit moebius counter using a J-K flip-flop that is triggered on the − edge and a D flip-flop that triggers on the + edge.
    (b) What is the maximum allowable clock inverter delay?

22. Draw a timing diagram, similar to Figure 6-8, but with a *non*edge-triggered, J-K flip-flop as FF2. The circuit should be operating at maximum speed with a clock with a 50% duty cycle.

23. How long could the loop in Figure 6-7a be if there were two levels of 74S00 gates between the flip-flops (74S112's)? (Assume 1.54 nsec/ft.)

24. How would you organize a system, using 74S logic to add one pair of **bytes** (eight-bit characters) every 35 nsec, where eight levels of logic are required to propagate the carry.

25. (a) Determine the count sequences for the counter shown in Figure 6-12.
    (b) Alter the logic so the counter is "self-clearing."

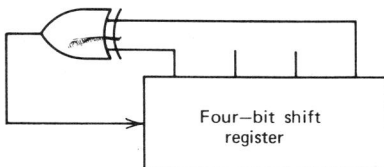

Fig. 6-12

REFERENCE

1. G. A. Maley and J. Earle, *The Logic Design of Transistor Digital Computers,* Prentice-Hall, Englewood Cliffs, N.J., 1963.

BIBLIOGRAPHY

Maley, G. A. and Earle, J., *The Logic Design of Transistor Digital Computers,* Prentice-Hall, Englewood Cliffs, N.J., 1963. (Chapters 8 and 9 develop asynchronous design and analysis techniques.)

Peatman, John, *The Design of Digital Systems,* McGraw-Hill, New York, 1972, Sections 5–6 through 5–12.

Liu, Bede and Gallagher, "On the Metastable Region of Flip Flop Circuits," *Proceedings of the IEEE,* April 1977, pp. 581–583.

CHAPTER SEVEN

PROGRAMMED LOGIC I
Microcomputers

7.1 A UNIVERSAL LOGIC CIRCUIT

In the logic design techniques we have discussed so far, the logical function is defined by *interconnection*. Gate and flip-flop elements are interconnected in an arrangement that produces the required system characteristics. With MSI circuits certain common patterns of interconnection are done within the IC but the system function is still defined by the configuration of interconnecting wires.

Since LSI chips contain so many circuit elements, it has been necessary to develop an entirely different way of defining system function. Instead of a maze of wires, LSI logic characteristics are defined by a *program* that can be stored in an orderly array of bits on the chip. This makes it possible for a single standard LSI part to serve the different needs of thousands of customers. Just as a standard digital computer can be programmed for anything from payroll to rocket guidance, programming gives LSI the flexibility to make *standardized* high-volume components which can be used in a wide variety of applications.

Single chip microcomputers are indeed the ultimate in standardized components because they are essentially an LSI implementation of the general "black box" digital subsystem (Fig. 2-3). A microcomputer has input and output lines which interact in a way which is defined by an internally stored program. The designer's task is thus not concerned with gates and flip-flops, but rather with writing programs.

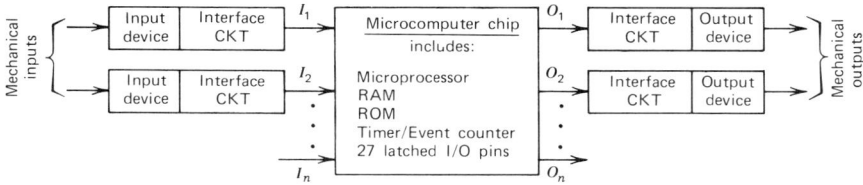

Fig. 7-1. Single chip microcomputer system. A complete "black box" digital logic subsystem on a single chip. For larger systems separate memory, microprocessor, and input/output chips may be used.

The job that was previously done by individual flip-flops is now done by an array of RAM bits that are included on the microcomputer chip. The program itself defines the logical functions—just as the interconnection of gates and flip-flops did in traditional logic design. This program is stored in a ROM that is also included on the chip. Timers on the chip are used by the program to handle any time sensitive operation. The program includes instructions for looking at the input lines and changing the state of the output latches as required by the application.

Since all of the basic logic functions are included in the microcomputer instruction set, *it can perform any logical function* if it is given enough *time* and *memory*. The more complex the logical task is, the more program steps are required to execute it. Since each program step takes time to execute, a point can be reached where the microcomputer is not fast enough to do the required job.

This speed limitation is the fundamental reason we have had to spend four chapters on traditional logic design techniques. Many things in the real world simply happen too fast for programmed logic. When a logical process is broken up into a sequence of small steps it is bound to take longer than if it is handled in a single step by specially designed logic.

With the traditional logic design techniques we have considered so far, every counter and register in the system could change on *each* clock cycle. With the programmed approach the most we can do on each clock cycle is to read *or* write one memory word (usually 8 or 16 bits wide). In addition to this bottleneck more than half of the memory cycles are used to read program instructions. Thus, *with any given IC technology, the programmed approach will be considerably slower than conventional logic design.*

While many jobs can be handled entirely by a single microcomputer, faster moving tasks must often be handled by the hard-wired logic techniques we have been discussing. Often the best solution is a marriage: conventional logic to handle the portions of the job requiring speed, and one or more microcomputers to handle the rest of the job.

Even where speed is not important, conventional logic is often used to tie the system together and perform simple jobs. In the next chapter we discuss some examples of this, but first let us look more closely at programming concepts.

7.2   PROGRAMMING

The basic idea behind all computers and microcomputers is the concept of a program. Instead of designing hardware to fulfill the function required by a specific application, the hardware is designed to perform certain *elementary operations* as directed by a *program*. The program consists of a sequence of these elementary operations or instructions that will perform the function required. The basic hardware can thus be the same for all applications. Programs are then written for the specific application and stored in a memory. The required instructions are stored in consecutive addresses in the memory and are executed automatically, in order, by addressing consecutive memory locations, as determined by a *program counter*. Some instructions cause program *branches* by putting a new address in the program counter. Usually these branches are *conditional branches* in that the branch will or will not occur depending on the result of some previous comparison or mathematical operation.

These program branches (sometimes called *jumps*) are similar to the branches shown on the data transmitter state diagram (Fig. 5-9). In fact, the main counting sequence of that state counter corresponds to the normal, consecutive, program counter sequence of instructions executed by a computer. Each state on the state diagram corresponds to one or more instructions of the program if we mechanize the same function using the programmed approach. For example, the READ TAPE state on Figure 5-9 corresponds to an actual sequence of instructions for reading data from the tape into memory. Some of the instructions in this sequence compare the data read from the tape with a special EOF code. If EOF is detected, a conditional branch instruction sets the program counter back to the IDLE portion of the program. If EOF is not detected, the READ TAPE program simply continues until the last character is stored in memory, then proceeds to the send program, which follows the READ TAPE program in the normal program counter sequence.

7.3   PROGRAM LOOPS

Branches back to previous parts of the program, like the TIMOUT branch in Figure 5-9, are called program *loops* because the program counter repeats a sequence more than once in a *loop*. Often we want to go around a loop a fixed number of times, so we include instructions in the loop to *increment a count and test the result to see if the loop has been executed the required number of times.* The loop is caused by a conditional branch based on this test. When the count reaches the correct number, the conditional branch does not execute, so the *program counter simply proceeds* to the next part of the program.

Program loops can greatly reduce the number of instructions required to perform a function. For example, a program to read 120 characters from a magnetic tape and store them in memory must execute an INPUT instruction and a STORE instruction for each character. Without looping, the program would require 240 instructions as follows:

INPUT #1

STORE #1

INPUT #2

STORE #2

•

•

•

INPUT #120

STORE #120

By using a programmed loop, we can reduce the program *from 240 to only five* instructions as follows:

INITIALIZE INDEX COUNT TO $n = 1$
→INPUT #n
  STORE #n
  INCREMENT INDEX COUNT $(n + 1)$
↳BRANCH IF INDEX COUNT < 120

Programmed loops are so important that computers often have special *index registers* to allow faster looping operations. Also, many computers have a single "increment index and branch if result zero" instruction that does the job of the last two instructions above.

7.4 THE PROGRAM-LOGIC TRADEOFF

Because of the equivalence between programs and logic, we must decide just how far to go in using the programming technique. Programming reduces the cost of the system at the expense of speed. The very simplest microcomputer can be programmed to do *any* data processing job if speed is unimportant. In reality, we always have some speed requirement for doing a job. By using more logic circuits to execute the programs, we can do the job faster and with less programming.

A full range of processor capability is available, from $1 microcomputers to supercomputers costing millions of dollars. In general, the cost is related to the amount and speed of logic hardware provided. With more hardware we can do more actual work in each program step. More hardware also allows us to encode the instruction more efficiently, thereby reducing programming effort and memory requirements for storage. In general, higher speed logic technologies cost more than low-speed logic. But even with the same speed logic, there is a tremendous variation in the amount of work that can be done in a given amount of time depending on the *amount* of logic used in the computer.

Basically, three characteristics affect the *throughput,* or work per unit of time, that a processor is capable of:

1.   Cycle time
2.   Word size
3.   Instruction set and addressing modes.

Cycle time is the time it takes to execute a basic instruction. On larger computers it often equals the memory cycle time, but on small microcomputers it can be much longer. Cycle time depends partly on the type of logic technology used and partly on how much of the instruction execution is done in parallel. Instructions are normally executed as a sequence of smaller, internal processor states. More steps mean less processor hardware but also slower cycle time.

Word size affects throughput directly in several ways. Since only one word in the memory can be changed for each cycle time, the word size directly determines how many bits can be changed per cycle. Although this puts an upper limit on processing speed, we cannot always utilize this limit. For example, if we are handling eight-bit characters (bytes), a 16-bit word will not double our throughput. A 16-bit word will, however, greatly increase our throughput, even when handling 8-bit bytes, because *each 16-bit instruction executed does much more in a single instruction cycle than an 8-bit instruction.* Operations that take one instruction on a 16-bit machine often take two or three on a eight-bit machine.

The *instruction set* of a processor is directly related to word size since the larger the word size, the more complex the operations that can be defined in one cycle. There is quite a variation, however, within processors of the same word size. Many 8 bit processors, for example, have instruction words 8, 16, or 24 bits long, read in 1, 2, or 3 memory cycles. Generally, it takes more logic to make efficient use of the bits of the instruction word. The payoff, however, is threefold: (1) efficient encoding of the instruction bits means that more can be done in each step for a given word size, throughput is therefore increased; (2) with more done per step, fewer instruction words are required to do a given job; memory is thus saved, offsetting the increased logic hardware cost; (3) with fewer programming steps required for a job, the effort to write the program is reduced.

7.5 INSTRUCTION SETS AND ADDRESSING MODES

An instruction set can be thought of as having two dimensions: (1) the type of operation and (2) the addressing mode. A powerful instruction set defines many types of operations and, perhaps more important, many addressing modes. *The weakest instruction set can still perform any operation,* just as we can perform any logical function with a simple NAND gate type. Just as it takes many NAND gates connected together to perform a complex operation, a sequence of many simple instructions does the job of one complex instruction. A processor can do multiplication, for example, with no multiply instruction by simply shifting and adding for each bit of the multiplicand.

Most processors have one or more internal registers called accumulators. *All arithmetic and logical operations are usually performed into an accumulator.* A logical AND instruction, for example, takes an operand from memory, or from another register, logically ANDs it with the contents of the accumulator, and places the results in the accumulator. For example:

Operand from memory or other register:	10101010
Accumulator contents *before* instruction execution:	00001111
Accumulator contents *after* instruction execution:	00001010

Note that each bit of the operand is ANDed with the corresponding bit in the accumulator.

An instruction word is normally divided into several *fields,* with each field encoding a different dimension of the instruction. For example, a 16-bit instruction code might be defined as shown in Figure 7-2.

The operation code field defines the type of operation to be performed. For example, AND, OR, Exclusive OR, and ADD AND SUBTRACT are performed as just demonstrated for AND. LOAD simply places the operand in the accumulator. STORE puts the accumulator contents into the operand address. BRANCH puts the operand address in the program counter, thereby transferring program control to that location. BRANCH ON CONDITION transfers control to the operand address only if previous logical or mathematical operations have set the *condition code flip-flops* to the proper state.

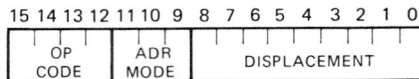

```
15 14 13 12 11 10 9  8  7  6  5  4  3  2  1  0
┌──────────┬──────┬──────────────────────────┐
│    OP    │ ADR  │                          │
│   CODE   │ MODE │      DISPLACEMENT        │
└──────────┴──────┴──────────────────────────┘
```

Fig. 7-2

The operand address (memory address where the operand can be found) is determined from the displacement (D) field by various rules, depending on the *addressing mode* indicated by bits 9, 10, and 11. Some typical modes are the following:

1. *Direct addressing* simply uses the contents of the D field as the memory address. This means we can address memory locations 0 through 512 directly in this example. These locations are therefore used to store important constants. This is also called *base page* or page zero addressing.

2. *Relative addressing* treats the D field as a signed number and adds it to the current program counter contents. We can thus address any word within ± 256 words of the present location. If the present program is reasonably small, this may be quite adequate. References to words used by many, widely scattered programs are made by another mode such as *direct* addressing.

3. *Indexed addressing* adds the D field to the contents of a register, called an *index register,* and uses the result as the operand address. This not only makes it possible to address any location in memory, but it also makes it much easier to perform repeated or *loop* operations. Special instructions often make it possible to increment and test the index register *and* define a branch address, all in a single instruction.

4. *Indirect addressing* is used in combination with one of the other addressing modes. The *contents* of the operand address defined by the other mode is used to indicate the *address* of the operand. The memory contents can thus be thought of as an *address pointer.* For example, we can BRANCH to another program beginning at location 5000 by doing an *indirect* BRANCH to location 100, in the base page, if 5000 (the address "pointed" to) is previously stored in location 100. Indirect addressing is sometimes used in conjunction with indexing; in a mode called *post-indexing* the index register determines the entry in the table whose starting address is stored as a pointer address. Because indirect addressing is used in conjunction with other addressing modes, there is often a separate *indirect address bit* in the instruction format.

5. *Immediate addressing* simply uses the contents of the D field as the operand itself. For example, if we want to add "6" to the contents of the accumulator, we simply use an ADD instruction, in the *IMMEDIATE* mode, with "6" in the D field.

Instruction formats can take many forms that bear no resemblance to the example of Figure 7-2. Designers of different processors produce different

tradeoffs. For example, we could define twice as many operation codes if we set-
tled for only an eight-bit displacement field. Also, it is very desirable to have
more than one accumulator, so we could add a field to define one of four or eight
accumulators by sacrificing displacement field size. It is also useful to have more
than one index register so an index register field can be added by sacrificing some
other field. Actually, there are usually several formats, each optimized for a different type of instruction.

7.6 AN ACTUAL INSTRUCTION SET

As an example of a simple, traditional instruction set, let us look at the instruction formats in Figure 7-3. This instruction set was first used on National Semiconductors multiple chip IMP-16 in 1973. It was later used on the single chip
PACE microprocessor, then on the faster INS-8900. The processor has four
accumulator registers (addressed by the REG field). Two of these can also be
used as index registers. The field labeled INDEX REG/MODE defines four
addressing modes as follows:

TABLE 7-1   INDEX REG/MODE

Reg/mode Field	Addressing Mode	Range
00	Base page (direct)	0 – 255
01	Relative to program counter	PC ± 127
10	Relative to accumulator 2	AC2 ± 127
11	Relative to accumulator 3	AC3 ± 127

We can thus define the memory address in the most suitable way. The first
step of instruction execution by the processor consists of computing the *effective
address* by one of the four rules shown in Table 7-1. If the instruction is an
indirect type, an extra memory cycle is required to get the effective address from
the location defined by the instruction. The actual instruction execution then
uses the effective address.
　Referring to Figure 7-3 briefly observe the available instructions. The LOAD
or STORE instructions cause one of the four accumulators to be *LOADED* or
STORED in the memory location defined by the effective address. If the instruction is JUMP, the effective address is loaded into the program counter. The
ADD instruction adds the contents of the memory location defined by the effective address to the contents of the indicated accumulator and puts the result in

(a) LOAD*, STORE*, ADD, SUBTRACT, AND, OR, SKIP IF GREATER, SKIP IF NOT EQUAL, SKIP IF AND IS ZERO.

(b) JUMP*, JUMP TO SUBROUTINE*, INCREMENT AND SKIP IF ZERO, DECREMENT AND SKIP IF ZERO.

(c) BRANCH-ON CONDITION.

(d) LOAD IMMEDIATE, COMPLIMENT AND ADD IMMEDIATE, ADD IMMEDIATE AND SKIP IF ZERO, PUSH ON TO STACK REGISTER, PULL FROM STACK, EXCHANGE REGISTER AND TOP OF STACK, SHIFT LEFT, SHIFT RIGHT, ROTATE LEFT, ROTATE RIGHT.

(e) REGISTER ADD, REGISTER AND, REGISTER EXCLUSIVE OR, EXCHANGE REGISTERS, REGISTER COPY.

(f) REGISTER IN, REGISTER OUT, HALT, SET FLAG, PULSE FLAG, PUSH FLAGS ON STACK, PULL FLAGS FROM STACK, JUMP TO SUBROUTINE IMPLIED, RETURN FROM SUBROUTINE, RETURN FROM INTERRUPT.

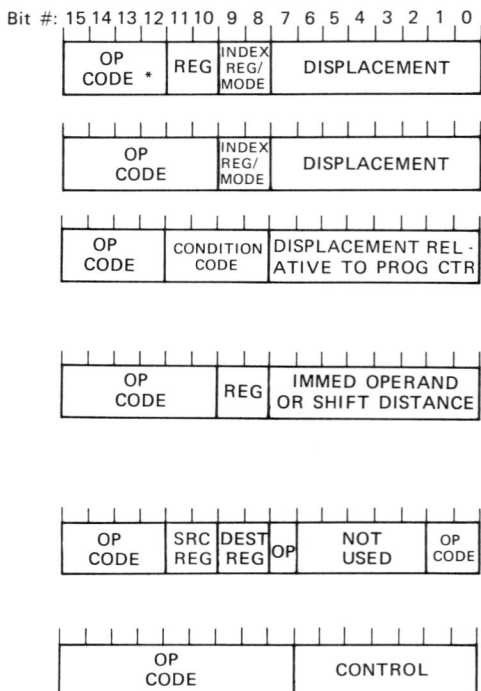

Fig. 7-3. 8900 Instruction formats. The asterisk indicates that two versions of these instructions are available—for direct and indirect addressing.

that accumulator. The SUBTRACT, AND, and OR instructions work in the same way except a different operation is performed.

The SKIP IF NOT EQUAL instruction compares the contents of the effective memory address with that of the specified register. No change is made in the data in the register or memory, but if comparison shows the two words to be unequal, the program counter is incremented by two instead of one. If the next instruction was, for example, a BRANCH, the branch would occur only if there was no skip. Skips are thus used to conditionally execute the single instruction following the SKIP instruction.

The *condition code,* used by the BRANCH ON CONDITION instruction, represents the state of flip-flops in the processor that indicate various system conditions; whether the contents of accumulator 0 is *zero, plus, odd,* or *even;* and whether the most recent operation resulted in a *carry* or *overflow.* More typically, the condition codes indicate whether the most recently executed arithmetic or logical operation produced a result that was zero, plus, had a carry, or overflow, and so on.

The IMMEDIATE instructions use the last eight bits of the instruction word as the operand itself. For example, an ADD IMMEDIATE adds the immediate operand to one of the four accumulators and puts the result in that accumulator.

The STACK instructions shown on Figure 7-3d are used for temporarily storing register contents, that will be used again later, in a *push-down stack*. We can *push* a register into the in stack or *pull* (sometimes called "pop"), previously pushed data out of the stack into a register. No addressing is required because data are pulled out of the stack on a *last in/first out* basis. To interrupt a program, we just push the register contents into the stack in any order, do the other job, then *pull* the register contents back out in reverse order to resume the original program. Even the program counter contents can be stored in this way. The JUMP TO SUBROUTINE instruction (Fig. 7-3f) causes a new address to be loaded into the program counter (a branch) and also pushes the present program counter contents into the stack. A RETURN FROM SUBROUTINE instruction pulls the previous program location from the stack, automatically adding a displacement to it. *Subroutines* are used for doing jobs such as multiplication, which may be needed in several parts of the main program. Instead of repeating the same sequence of instructions each time, we simply jump to the subroutine (in another part of memory) and return when finished.

Returning to Figure 7-3d, we see both SHIFT and ROTATE instructions. In the SHIFT instruction the contents of the specified register are moved to the left or right the number of positions defined by the displacement field. Although bits are simply lost when they shift out the end of the register in the SHIFT instruction, they are shifted into the other end of the register with the ROTATE instructions. If a SET FLAG or PULSE FLAG instruction has been previously executed, the register is effectively a 17-bit register with the *link* flip-flop added at the most significant end.

The REGISTER ADD instruction (Fig. 7-3e) adds the contents of one of the four accumulators (the source register) to the contents of the destination register and puts the result in the destination register. *Exchange registers* simply exchanges the contents of the two specified accumulators. REGISTER COPY loads the contents of the source register into the destination register. REGISTER IN and REGISTER OUT are used for input and output of data from and to external devices.

7.7 THE GRANDDADDY OF MICROPROCESSORS

The first really successful microprocessor was introduced by Intel in 1971. Though it was slow by today's standards (20-μsec cycle time) the 8008 was fast and effective enough to handle a wide range of practical applications. In 1974, just as competition started to appear, Intel announced an upgraded version called the 8080 (Fig. 7-4), which helped them to continue their almost IBM-like

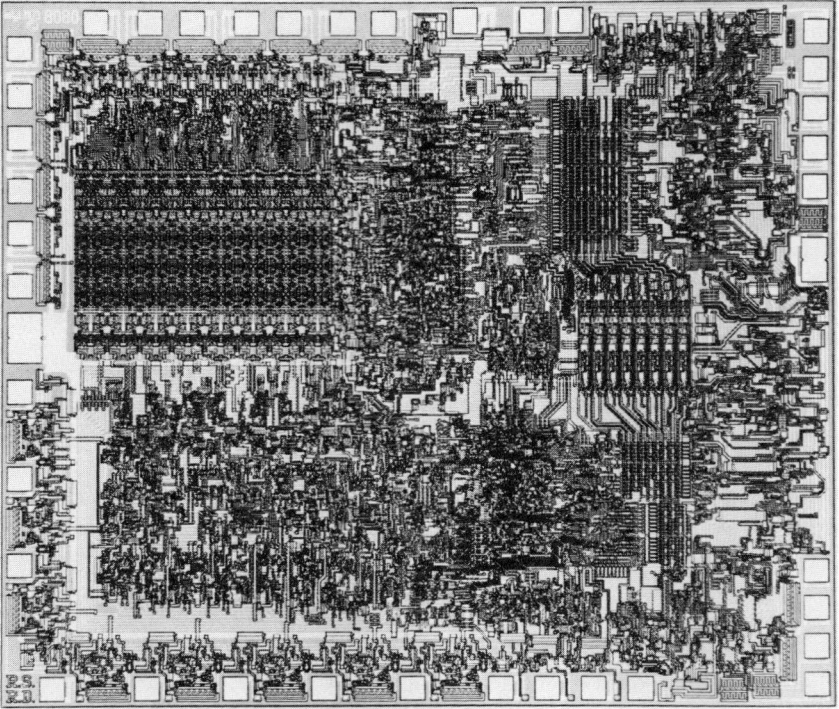

Fig. 7–4.   8080 photomicrograph with pin designations (actual size 0.124 × 0.173 in.).

domination of the industry. Though the instruction sets of these products are less straightforward than many more modern counterparts, they have proven quite effective. Their domination of the industry makes them impossible to ignore.

Though the 8008 is obsolete today its instruction set was the basis of the later 8080 and 8085. We thus start by describing the 8008 instructions and build from there.

The 8008 has eight 8-bit registers. One of these (M) is reserved for memory operations. Whenever the M register is addressed, data comes from, or goes to, the memory location addressed by two other registers $(H$ and $L)$. There are, therefore, only five registers available for general use (called A, B, C, D, and $E)$. As the format in Figure 7-5a shows, the contents of any register can be *moved* into any other register by a one-byte instruction. By specifying the M register as a source or destination register, we can load or store *memory* data to or from any of the other registers. Before doing this, however, we must put the desired memory address in the H and L registers.

These registers together constitute a memory address *pointer* with the high-order address bits in H and the low-order bits in L. We can do something

similar to direct addressing by using two 2-byte MOVE IMMEDIATE instructions (Fig. 7-5*d*) to load the desired address into registers *H* and *L* before executing a memory-referencing instruction. If we are accessing sequential bytes of
memory, we can increment *L* with a single-byte INCREMENT instruction.
This is *similar* to having *indexed addressing* available. Also, we can do something *similar to indirect addressing* by loading the *L* register from memory
before executing a memory referencing instruction.

We do have *true direct addressing* available on the JUMP (same as
BRANCH) and CALL SUBROUTINE instructions (see Fig. 7-5*e*). These
instructions occupy three successive bytes in program memory and can directly

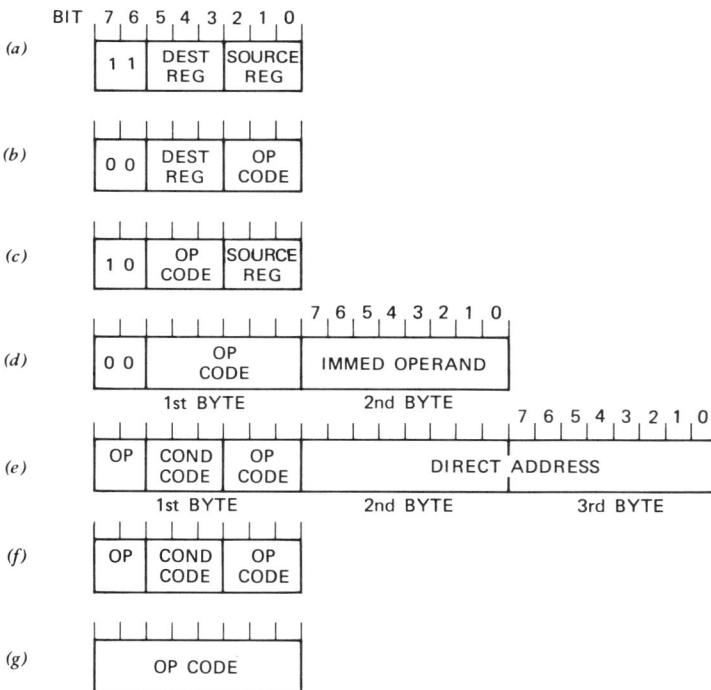

Fig. 7-5. Intel 8008 instruction set and formats. (a) MOVE REGISTER† TO REGISTER†. (b)
INCREMENT OR DECREMENT. (c) REGISTER† TO ACCUM: ADD*, SUBTRACT*, AND,
EOR, OR, COMPARE. (d) IMMEDIATE: MOVE, ADD*, SUBTRACT*, AND, EOR, OR,
COMPARE. (e) JUMP, CALL SUBROUTINE. (f) RETURN FROM SUBROUTINE. (g)
ROTATE A RIGHT*, ROTATE A LEFT*, RESTART, HALT, INPUT, OUTPUT.

* With or without CARRY.
† Note: When the eighth (M) register is addressed, a memory read or write is intiated to the address
specified by the sixth and seventh (H and L) registers. Only the first five registers (A, B, C, D, and E) are
therefore available for general use.

address up to 16,384 bytes of memory. Execution of either of these instructions causes the specified direct address to be loaded into the program counter if the jump conditions are satisfied. The CALL SUBROUTINE instruction also causes the present program counter location to be pushed into the push-down stack. If conditions are satisfied, a RETURN instruction causes the top of the stack to be loaded (popped) back into the program counter, thus returning to the main program sequence. The JUMP, CALL, and RETURN instructions all can be unconditional or can have a *condition code* field that determines the conditions under which the program branch will occur, as shown in Table 7-2. The conditions listed refer to results of the most recent arithmetic or logical operation (not move).

All arithmetic and logical operations use the A register as their destination. We can thus ADD, for example, an immediate operand (Fig. 7-5d), or the contents of any register (Fig. 7-5c), to the contents of the A register and the result will appear in the A register. If we ADD from the M register, we actually add the contents of the memory location, "pointed to" by the H and L register, to the A register. Of course, the other operations, SUBTRACT, AND, Exclusive OR, and OR, work similarly. COMPARE is also similar, but is used only to set the condition codes. A subtraction is performed, and the condition codes are set according to the result, but the A register remains unchanged.

The ROTATE instructions (Fig. 7-5g) perform a single shift left or right on the A register with or without the *carry* flip-flop as a ninth bit. The data shifted out the end of the register are fed into the other end to prevent loss. To shift more than one position, this instruction must be repeated.

The INPUT instruction inputs data to the A register from external devices via the same input pins used for the memory data. Three bits of the instruction define up to eight different input devices (ports). The OUTPUT instruction outputs data to one of 24 external devices (output ports). Four bits in the instruction indicate the device number.

TABLE 7-2   CONDITION CODES

Condition Code	Required Condition
000	No carry
001	Not zero
010	Positive
011	Parity odd
100	Carry
101	Zero
110	Minus
111	Parity even

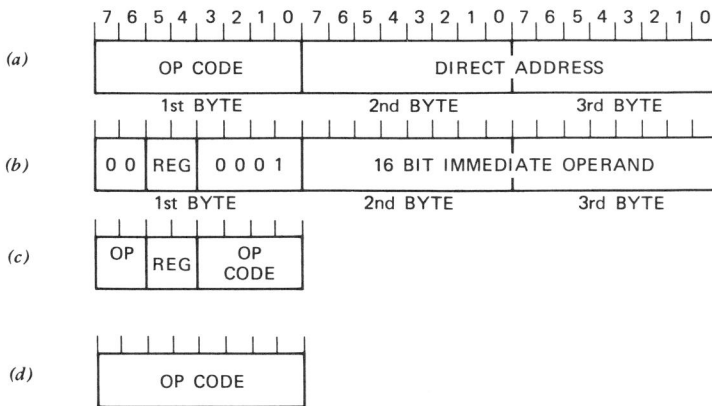

Fig. 7-6. Intel 8080 additional instructions. (a) LOAD A from memory, LOAD HL from memory, STORE A to memory, STORE HL to memory. (b) LOAD EXTENDED IMMED (registers BC, DE, HL, or SP). (c) PUSH (BC, DE, HL, or AF), POP, (BC, DE, HL, or AF), ADD TO HL (BC, DE, HL, or SP), STORE A (memory address in BC or DE), LOAD A (memory address in BC or DE), INCREMENT (BC, DE, HL, SP), DECREMENT (BC, DE, HL, SP). (d) EXCHANGE HL ↔ DE, EXCHANGE HL ↔ STACK, LOAD SP WITH HL, LOAD PC WITH HL, COMPLIMENT A, SET CARRY, COMPLIMENT CARRY, DECIMAL ADJUST A, ENABLE INTERRUPT, DISABLE INTERRUPT, NO OPERATION.

Notes: BC = registers B and C taken as a 16-bit register (likewise for DE, HL, and AF); SP = stack pointer register; PC = program counter; F = flag register (carry, sign, etc.) The 8080 also executes all instructions shown on Figure 6-7.

In addition to all the 8008 instructions just described, the 8080 and 8085 have all of the *additional instructions* shown on Figure 7-6. Although the 8008 has direct addressing only for JUMP instructions, the 8080 has *direct addressing* also for LOAD A, STORE A, LOAD HL, and STORE HL. We can thus load or store a byte, or pair of bytes, from any of 65,536 memory locations with a single (three-byte) instruction.

Many 16-bit instructions are added to allow easy handling of 16-bit words and *addresses*. For example, the LOAD HL instruction (Fig. 7-6a) loads the contents of the directly addressed memory location into the L register and the contents of the next memory location into the H register. *We are thus effectively treating the H and L registers together as a 16-bit register* and effectively loading it from a 16-bit memory location.

As Figure 7-6b shows, we can also load 16-bit immediate operands into register pairs *BC, DE, HL,* or the 16-bit stack pointer (SP) register. These immediate operands are defined by the second and third bytes of a three-byte instruction. We can also PUSH or POP any of four register pairs into and out of the push-down stack. The extended ADD instructions allow us to 16-bit add any of four register pairs to the contents of *HL*. This, of course, is quite useful for computing memory addresses. The INCREMENT and DECREMENT instructions are extended to allow a 16-bit increment or decrement in one single-byte instruction. We can also LOAD A and STORE A using registers *B* and *C* or *D* and *E* as the address pointer instead of just *H* and *L*. This is like having three separate index registers that can be used to address three different lists of operands. We can increment each index with a single-byte instruction and load each operand with a single-byte instruction.

As Figure 7-6*d* shows, it is also possible to compute jump addresses in *HL*, then execute the jump by using LOAD PC WITH HL. Instructions are also added for manipulating the SP (stack pointer register), carry flip-flop, and interrupt enable flip-flop. A DECIMAL ADJUST A instruction makes it much easier to perform decimal addition and subtraction by converting the results of a binary operation on two decimal numbers into two 4-bit BCD digits. The INPUT and OUTPUT instructions are changed to two-byte instructions on the 8080. This makes it possible to address up to 256 different external devices for both input and output.

The 8080 *push-down stack* is mechanized in the main RAM by simply using the 16-bit SP register as a memory address to PUSH and POP data (see Fig.

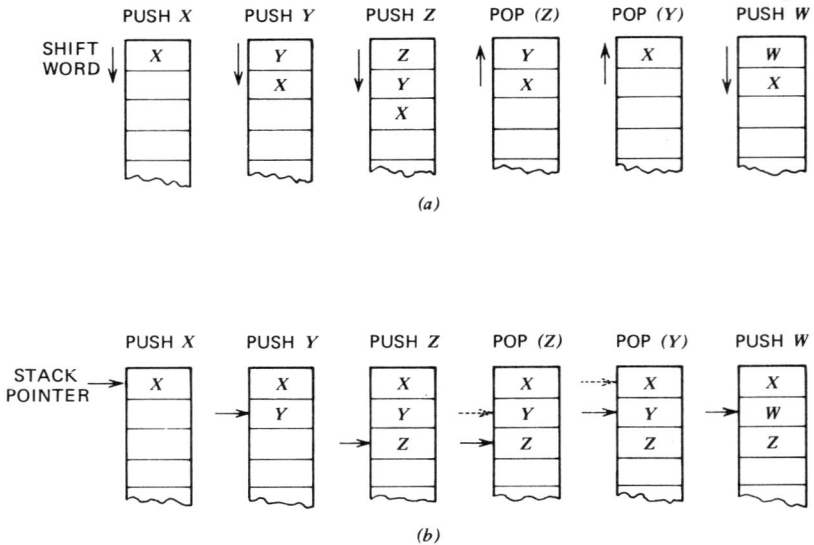

Fig. 7-7.   Push-down stacks. (a) Shift register type (e.g., 8008) (b) Stack pointer type (e.g., 8080). A solid arrow indicates stack pointer state used as memory address; a broken arrow indicates stack pointer state at completion of instruction.

7-7b). A PUSH instruction decrements the SP by two, then stores the data being "pushed" in the memory location pointed to by the SP register. A POP instruction loads 16 bits of data from the memory locations indicated by the SP, then increments the SP by two.

The number of words that can be PUSHed into the stack is thus limited only by the number of words of main RAM we want to reserve for the push-down stack. The push-down stacks in the 8008 actually exist in hardware as 8 reversible shift registers (see Fig. 7-7a). When an address is pushed (by CALL SUBROUTINE), it is simply shifted into the shift register. When it is pulled again (RETURN), it is shifted back out. With this system the size of the stack is limited by the length of the shift registers provided (seven words for the 8008).

7.8   ADVANTAGES OF PROGRAMMED LOGIC

Looking at the instruction formats in Figures 7-3, 7-5, and 7-6, we see that they encode the operations to be performed quite efficiently. Each instruction class has a format well suited for defining it efficiently. The processor requires additional logic to interpret the bits of the instruction differently depending on the operation code bits. The payoff, however, is that less memory is required to store the program. This is the same tradeoff encountered in Section 4.8, where we found that, by adding some logic to encode ROM inputs, we could greatly reduce the size ROM required to perform a logic function. The concept of an instruction word gives us even better ROM utilization because, instead of trying to save ROM bits by encoding functions that resulted from a conventional logic design approach, we *start from the beginning* to define the problem so that it makes the most efficient possible use of the ROM, that is, the program approach.

Of course programs need not be stored in ROMs. In fact, processors are almost always designed so that *any type of memory, with any speed, can be connected to the memory bus.* It is thus possible to mix blocks of ROM, PROM, and RAM devices with various speeds. During program checkout it is most convenient to store the program in RAM so it can be easily changed. Later, the program portions of memory can be replaced by PROM if desired. If the system is a high-volume product, the program can eventually be put on mask-programmed ROMs when it is fully field tested. It thus becomes possible to make instant field changes to add new features by simply plugging in new ROMs containing additional program features. With conventional wired logic, such major changes are impossible.

7.9   FLOW CHARTS: THE PROGRAMMER'S BLOCK DIAGRAM

The procedures for designing programmed and wired logic systems are identical in principle. The *block diagram* used in wired logic design has a direct counterpart in programming called a *flow chart*. Both can define system logic to any degree of detail, and both allow us to represent the overall system and then subdivide it into increasingly detailed representations (see Fig. 2-6) until we reach the detailed "component" level. *In the case of programs, the "components" are the instructions.* We must know our "components" (the instruction set) well to make the proper decisions at all levels of design. Instructions correspond roughly to MSI logic in level of complexity.

Just as we join the boxes in a *block diagram* with lines and arrows indicating *signal flow between components,* we join the boxes in a *flow chart* with lines indicating the *program counter flow* between instructions. A flow chart on the general system level represents many instructions by a single box, just as a general block diagram represents many IC components by each block. At the most general level, for a large system, each box may represent a complete program to be written by a separate person. Programs are also often written in *"modules"* with well-defined inputs and outputs. By using different combinations of general program modules, custom programs can often be assembled easily, just as custom-wired logic systems are assembled by connecting general module types.

The data transmitter state diagram shown in Figure 5-13 is actually almost identical with a programming flow chart at the general system level. Figure 7-8 shows it redrawn in the correct flow chart form. All *major* decision points (conditional branches) are shown as diamond shapes. Since this flow chart is at the general system level, the blocks and diamonds actually represent entire programs. Though the *routines* (small programs) represented by the blocks and diamonds, may contain many conditional branches, these are internal decisions and need not be shown on the flow chart at this level. They will, however, appear on the detailed flow charts for individual routines.

Note that there are several *loops* in this program as indicated by arrows looping back to previous parts of the program. After the ⅓-sec TIMOUT, for example, the program branches back to the start of the SEND DATA routine. Also, after an ACKNOWLEDGE signal is received, the program branches back to the start of the READ TAPE routine. Once the final program is written, *each arrow actually represents a particular program counter state* (the memory address of the first instruction of the routine to which it points).

In general, there is a consecutive program counter sequence from the top to the bottom of the flow chart, with only departures from that sequence requiring branches. We can label the arrows, just as we identify the signals on a block diagram in wired logic design. We simply invent descriptive mnemonic abbreviations that suggest the function of the instructions starting at that location, just as signal names suggest their functions on a block diagram. On Figure 7-8, for example, we call the first location of the READ TAPE routine RTAPE. *Both arrows pointing to that box on the flow chart thus represent the program counter*

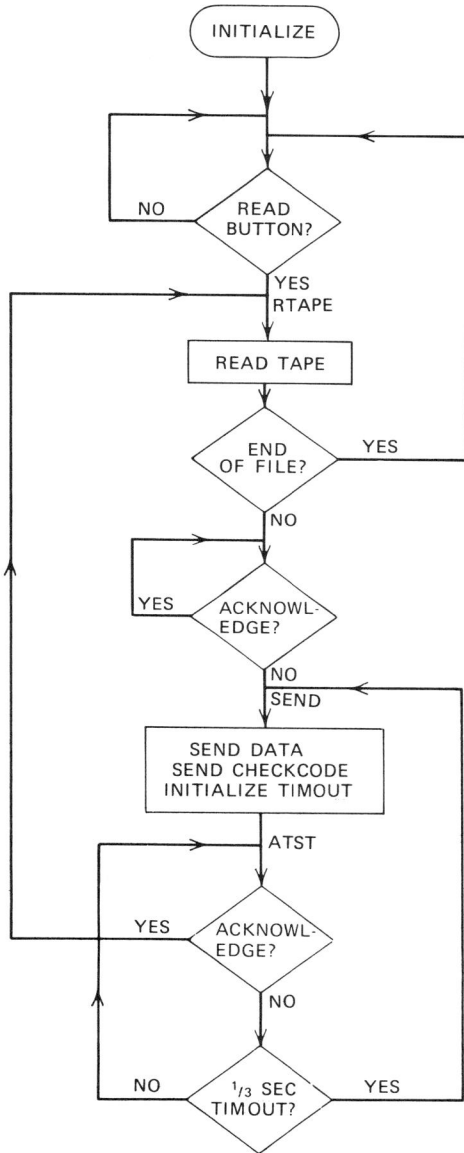

Fig. 7-8. Flow chart of data transmitter (from Fig. 5-13).

entering that state. In one case it is simply the next instruction in sequence (after the READ BUTTON test fails to branch back); the other arrow represents a program branch to the location called RTAPE, when the ACKNOWLEDGE signal is detected.

7.10 THE EQUIVALENCE OF PROGRAMS TO WIRED LOGIC

There is a direct equivalence between the structure of a flow chart and the Boolean logic functions. Figures 7-9a and b, for example, show AND and OR structures as they appear on a flow chart. As with gates, these structures can be combined to make any desired function. Notice that the functions f_1 and f_2 are generated *at separate times* using the *same processor logic. This is the key to the logic savings with programmed logic* and also why it is basically slower. Instead of generating f_1 and f_2 with two separate gates simultaneously, we use the same gates and "remember" the result with a conditional program branch.

Figure 7-10 shows, as an example, a detailed flow chart for generating the logic function:

$$f = A \cdot B' \cdot C' \cdot D' \cdot E \cdot F \cdot G' \cdot H + A' \cdot B \cdot C \cdot D' \cdot E \cdot F \cdot G' \cdot H' \qquad (7\text{-}1)$$

where $A, B, C, D, E, F, G,$ and H are the bits of the A register in an 8008 or 8080 microprocessor. We first do a COMPARE IMMEDIATE 10001101 instruction (Fig. 7-5d). If the first term of the equation is true, the contents of the A register will be 10001101, so the COMPARE instruction will set the condition flip-flops to "remember" and zero result. We next do a conditional JUMP

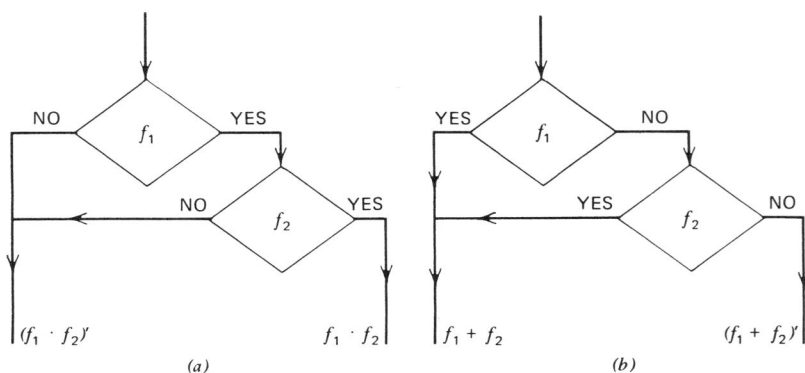

Fig. 7-9.   Flow chart Boolean structures: (a) AND structure; (b) OR structure.

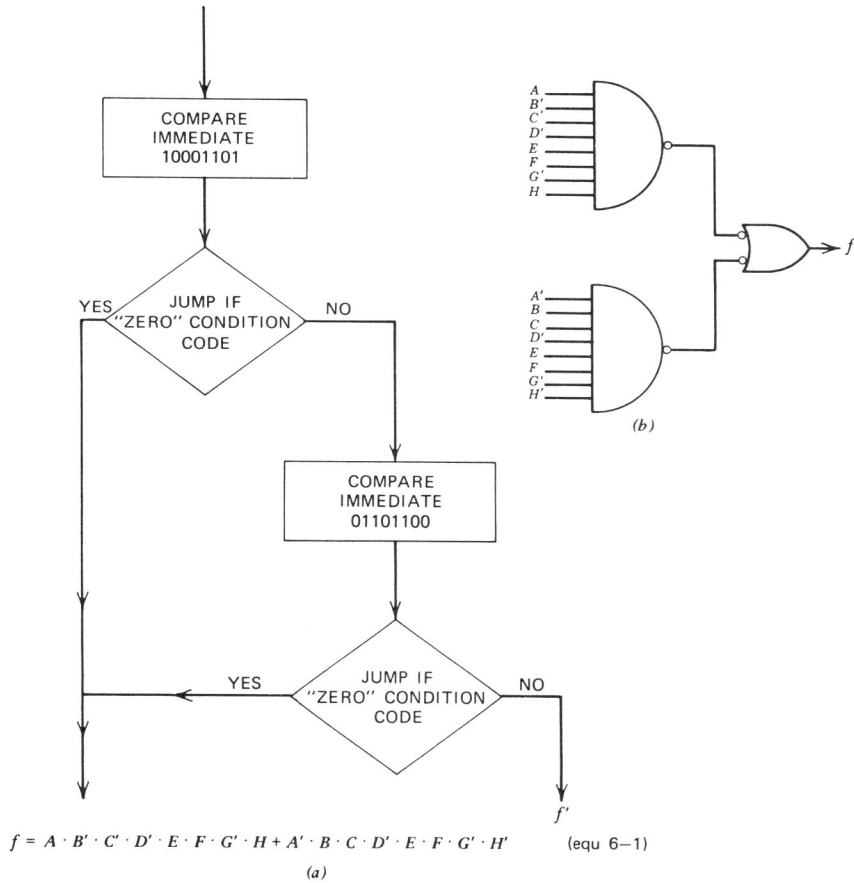

$$f = A \cdot B' \cdot C' \cdot D' \cdot E \cdot F \cdot G' \cdot H + A' \cdot B \cdot C \cdot D' \cdot E \cdot F \cdot G' \cdot H' \qquad \text{(equ 6-1)}$$

(a)

Fig. 7-10. Programmed logic example: (a) programmed microprocessor. (A register contains A, B, C, D, E, F, G, H); (b) NAND gate logic.

(jump if zero set) instruction, which will transfer program control (to the program that does whatever we want to do if function f is true), if the first term is true. We then do the same thing for the second term. If neither term is true, the program counter simply continues to whatever we want to do if function f is not true.

Note that this is the most detailed possible flow chart in that each step actually represents an instruction. We normally do not have to go into this much detail with the flow chart, but would stop with something looking like Figure 7-9b with "A = 1000110?" written in the first diamond and "A = 01101100?" in the second. Just as we seldom make block diagrams down to the gate level, the most detailed flow charts we will make usually require several instructions for each box or diamond.

To define the logic for this function we need 10 bytes of ROM, since the COMPARE IMMEDIATE instructions (Fig. 7-5d) each requires two bytes, and the JUMP instructions (Fig. 7-5e) each requires three bytes. We can thus do the function with 10 bytes, or $10 \times 8 = 80$ bits, of ROM. As Figure 7-10b shows, it would take 2.25 IC packages to do the same job with wired logic. In other words, $80/2.25 = 36$ ROM bits replace one IC gate package. Experience with actual complete systems confirms that this is a good *average* ratio to expect when replacing wired logic with programmed logic. Since *we can buy thousands of ROM bits for the installed cost of a single IC,* the economy of this approach is fantastic. A single 65,536 bit ROM package, for example, can replace $65,536/36 = 1820$ IC gate packages!

As another example of directly equivalent programmed logic, we can mechanize the seven-segment display decoder of Figure 3-2 as a programmed *table look up.* We first generate a 10-word table, in memory, that looks exactly like the truth table of Figure 3-2d. This table can be anywhere in memory, but let us assume we put it in the 10 locations beginning with 11110000 (F0 in **hexadecimal**) as shown in Table 7-3.

Now, if we have a digit code in the A register (of an 8008 or 8080), instead of outputting it to external decoder circuitry, we can convert it to a seven-segment code before output by simply executing the following program sequence (assuming H already contains the correct upper half of the correct address):

IMMEDIATE OR TO A   11110000

MOVE TO L FROM A

MOVE TO A FROM M

We thus convert the digit code into a table address, put that address in the L register, then load the proper seven-segment code from the memory table (memory location H,L) to the A register. The A register will then have a seven-segment code in bits A_6 through A_0 that can be output directly to the display drivers.

As a specific example, if we have 0011 ("3") in the A register, we will have 11110011 after the IMMEDIATE OR. Moving this to the L register and loading the contents of that memory location to the A register, we end up with 01111001 in the A register. The seven right-hand bits of this code represent the seven display segments "a through f."

TABLE 7-3 SEVEN SEGMENT
DISPLAY TABLE IN MEMORY

Memory Location	Contents
11110000	01111110
11110001	00110000
11110010	01101101
11110011	01111001
11110100	00110011
11110101	01011011
11110110	00011111
11110111	01110000
11111000	01111111
11111001	01110011

Our memory requirements for the seven-segment decoder are 10 bytes for the table and four bytes for the program, or a total of $14 \times 8 = 112$ bits. (Note that the program could be only two bytes if the table started at location 0 since the first instruction could be omitted.) We previously determined (Fig. 3-8) that the seven-segment decoder would take 5.25 packages to mechanize with discrete gates. Our bits/gate package ratio in this case is thus $112/5.25 = 21.4$ bits/gate package, which is even better than our previous figure of 36. Of course, we can buy a single MSI package that does the same job and includes drivers. Though the MSI package may be a better choice in this case, the point is that any *logic* can be done more economically in the program. Our "package count" for this logic in a 65K bit ROM is $112/65,536 = 1/585th$ *of a package!*

Of course, real systems may still need wired logic ICs for the really high-speed portions of the logic, and things such as interface driver circuits must still be done with discrete logic ICs. For moderate-speed, complex logic, however, the programmed approach is unbeatable.

In Chapter 4 we discussed the use of ROMs to generate Boolean functions. If we had mechanized equation 7-1 in this way, we would have needed a one-bit $\times$ $2^8 = 256$-word ROM. We would have thus used 256 bits to do what could be done by only 80 bits. Even more important, our 80 bits can be *a small part of any size ROM*, while the 256 bits require a certain organization. The lesson is that the programmed approach is the way to make optimum use of ROM for logic. We normally *never use logic equations in program design*. The functions are defined naturally by making flow charts of the desired functions, then writing a sequence of instructions to implement the flow chart.

7.11   COMPARING INSTRUCTION SETS

To illustrate how the instruction set affects computing power, program storage requirements, and programming ease, let us take a single 8900 instruction and try to perform the same function on the 8008 and the 8080. Just to be fair, we will assume that we are working with an eight-bit operand. The 8900's advantage would be much greater with a 16-bit operand, of course. The ADD instruction on the 8900 can add the contents of any memory location to any of four accumulators and put the result in that accumulator. Let us assume that we must keep the AC0 (first accumulator) contents for future use, so we will ADD into AC1 on the 8900. To do the same job on the 8008, we must first save the contents of the A register, then load the B register into A. This is necessary because we can do arithmetic only into the A register. We must then set up the H and L address and ADD. We thus need the following instruction steps:

MOVE C FROM A   (save A)

MOVE A FROM B   (put operand in A)

MOVE H IMMEDIATE   (half of address)

MOVE L IMMEDIATE   (half of address)

ADD MEMORY TO A

Since the third and fourth instructions each require two bytes (Format 7-5d), we need seven bytes of instruction memory on the 8008 to duplicate the two bytes required to store the ADD instruction in the 8900. Since memory for program storage and programming labor cost money, *the more expensive processor may give a cheaper overall system. In general, it is false economy to use a processor with a crude instruction set if the memory cost will be several times greater than the processor cost.*

Let us now try the same thing with the extended instruction set of the 8080/8085. We can now load A directly from memory, then add from B:

MOVE C FROM A   (save A)

LOAD A   (direct addressed from memory)

ADD B TO A

Our program is now simpler, but since the second instruction requires three bytes (Format 7-6b), it still occupies five bytes of memory versus two for the 8900.

Of course, we stacked the cards somewhat by starting with an 8900 instruction (though we could have stacked them much more by using a 16-bit operand). Actually, the variable-length instruction format of the 8008 and 8080 is occasionally more efficient. For example, adding the B register into the A register can be done with a single-byte instruction on the 8008 and 8080, but it requires a two-byte instruction on the 8900. Also, we quite often access memory from consecutive locations, so the L pointer register can often be simply incremented to get the next operand address. This can be done by a single byte instruction on the 8080 but it requires two bytes on the 8900. Other processors, such as the Texas Instruments 9900 series, can actually add two numbers separately indexed from memory *and* increment both indexes all with a single instruction.

Powerful instruction sets can save considerable memory in large systems, but it is difficult to optimize instructions for both large and small systems. The problem is that the flexibility that improves efficiency in large systems often causes instruction bits to be wasted on small systems. This fact has been recognized in the newer microcomputer products: Intel now has a new instruction set optimized for small systems on its 8048/9 microcomputer chip. Another instruction set optimized for very large systems is used on the 16-bit 8086 microprocessor. Neither of these is directly compatible with the 8080 instructions but both are similar enough to be easily learned by 8080 users.

A surprisingly large part of any real program is concerned with calculating, saving, and setting up address pointers. For this reason 16-bit arithmetic capability is very important even in applications where 8-bit data are being processed (except in very small systems). Often the data handling itself turns out to be a relatively small part of the program.

Another kind of programming "overhead" that can use up surprising amounts of programming effort and execution time is *context changing*. Most microcomputers must handle several different tasks which are each done by separate programs. In real microcomputer systems program control is continually jumping about between different programs and subroutines. Modern instruction sets can make these context changes quickly and effortlessly. The Texas Instruments 9900 series, for example, has a single instruction which does a complete change of context by simply selecting a new set of 16 working "registers." Since the registers are actually RAM memory locations, any number of different contexts is possible.

The choice of instruction set is an extremely important decision because there is a tendency for the entire company to become "locked in." Although it is easy to upgrade to newer microcomputer chips with the same instruction set, changing instruction sets means that old programs must be rewritten, people must be retrained, development systems must be changed, and so on. The difficulty in changing instruction sets makes it important to look into the future in making a choice. The additional cost of a more sophisticated instruction set tends to disappear with time. Company needs also tend to grow with time towards more complex systems. Microprocessor choice should thus strongly consider your own future needs and the future commitment and leadership prospects of the vendor.

EXERCISES

1. (a) For the two program approaches at the end of Section 7.4, which is capable of the higher speed?
   (b) If each instruction takes 2 μsec, find the time required to store all 120 characters with each approach.
   (ci How much time would it take if the cycle time is 1 μsec?
   (d) How much time would the looping approach take if an "Increment index and branch if result is zero" instruction is available?
   (e) Make a detailed flow chart of the looping approach.
2. The 4 accumulators on an 8900 contain the following: AC0 = 5, AC1 = 27, AC2 = 3, AC3 = 255.
   (a) What will they contain after the following sequence of stack instructions?

   PUSH   AC0
   PUSH   AC1
   PUSH   AC2
   PULL TO   AC1
   PULL TO   AC0
   PUSH   AC3
   PULL TO   AC2
   PULL TO   AC3

   (b) What will they contain after executing the following instructions?

   REGISTER ADD AC0 TO AC1
   REGISTER AND AC2 TO AC0
   REGISTER EXCLUSIVE OR AC2 TO AC3

3. What will the A register of an 8008 microprocessor contain after the following sequence of instructions:

   MOVE IMMEDIATE 11001000
   MOVE TO B FROM A
   INCREMENT B
   ADD B TO ACCUMULATOR

4. Make a flow chart for a program that branches to one of eight subroutines, depending on the three least significant bits of the code in the A register.

5. Make a flow chart for a program that branches to one of four subroutines depending on the binary value of the contents of the A register as follows:

Value	Go to
0–60	SR1
61–130	SR2
131–138	SR3
139–255	SR4

6. Write a program for the 8080 that will
   (a) Make the A register count (forever) as a binary counter.
   (b) Make the A register count (forever) as an eight-bit Moebius counter.

7. (a) Make a flow chart of a program that will mechanize equation 4-7.
   (b) Write the program for the 8008 microprocessor.
   (c) Calculate the number of gate packages required to mechanize the function with wired logic and use this number to compute program bits/gates package.

8. Make a flow chart of a program that will branch to one of three subroutines depending on whether the contents of the A, B, or C register is greater.

9. What will the A register contain after executing "Exclusive Or to A from A"?

10. Make a flow chart of a program that will branch if the contents of the A register satisfies
    (a) Equation 4-5.
    (b) Equation 4-6.

11. How many ROM bits are required to implement equation 4-5 and 4-6 in this way?

12. Flow chart a looping program which will branch if the contents of the A register satisfies equation 4-7. Use a general subroutine from problem 10 to evaluate each minterm.

13. How many bits of program ROM are equivalent to the 1024-bit ROM which implemented equations 4-4 through 4-7?

BIBLIOGRAPHY

Microprocessors

Intel, *MCS-85 Users Manual*, Intel Corp., Santa Clara, Calif., 1978, Pub. #9800366D.

Katz, Jeffery, Morse, Stephen P., Pohlman, William B., Ravenel, Bruce W., "8086 Microcomputer Bridges the Gap Between 8 and 16 Bit Designs," *Electronics*, February 16, 1978, pp 99–104.

Lofthus, Alan and Ogden, Deene, "16-Bit Processor Performs Like Minicomputer," *Electronics*, May 27, 1976, pp. 99–105.

Morse, Stephen, Pohlman, William B., and Ravenel, Bruce W., "The Intel 8086 Microprocessor: A 16-Bit Evolution of the 8080," *Computer*, June 1978, pp. 18–27.

Scrupski, Stephen, "Minis, Look Out: Here Comes a Powerful Family of 16-Bit μC Chips," *Electronic Design 12*, June 7, 1978, pp. 54–58.

Texas Instruments, *990 Computer Family Systems Handbook*, Texas Instruments, Austin, Texas, 1975, Manual #945250-9701.

Microcomputers

Bryant, John D. and Longley, Rick, "16-Bit Microcomputer Is Seeking a Big Bite of Low-Cost Controller Tasks," *Electronics*, June 23, 1977, pp. 118–123. (TI 9940)

Intel, *Application Techniques for the MCS-48 Family*, Intel Corp., Santa Clara, Calif., 1977, Pub. #98-413B.

Intel, *MCS-48 Users Manual*, Intel Corp., Santa Clara, Calif., 1978, Pub. #98-270C.

Instruction Sets

Osborne, Adam, *An Introduction to Microcomputers*, Osborne, Berkeley, Calif., 1975 (Two volumes. The second volume is a large collection of data sheets.)

I/O and Interrupt Structures of Specific Microprocessors

Klingman, Edwin E., *Microprocessor Systems Design*, Prentice-Hall, Englewood Cliffs, N.J., 1977.

Programming Concepts

Booth, Taylor, *Digital Networks and Computer Systems*, Wiley, New York, 1971. (Much information on programming concepts, assemblers, compilers, algorithms.)

Computer Automation, *The Value of Power*, Doc. #89A00064A-C, General Automation, Anaheim, Calif., 1972. (An excellent general coverage of instruction sets and addressing modes.)

Mini and Larger Computers

Bell and Newell, *Computer Structures: Readings and Examples*, McGraw-Hill, New York, 1971. (Descriptions of 40 actual computers and their structures and design philosophies.)

CHAPTER EIGHT

PROGRAMMED LOGIC II
Computer-Aided
Programming

8.1   ASSEMBLERS

Although programs are sometimes used to generate wire lists and otherwise aid the *wired logic* designer, the lack of standardization has kept computer-aided wired-logic design from really catching on. Programmed logic, on the other hand, is perfectly suited for computer assistance. In fact, *programs are virtually always automatically assembled by computer programs called assemblers.*

An assembler automatically converts a convenient "people language" into binary "computer language." The programmer simply writes out the program in the people-oriented format. This program, called **source program**, is then the input to a computer that is programmed with the assembler program. The assembler program converts the source program into a binary coded **object program**, which is executable on the microcomputer. A side-by-side printout, or listing, is also produced showing the binary object code and the easily readable source program. Each line of binary machine code can thus be understood by reading the source program **statement** that produced it, which is printed on the same line to the right.

The assembly program also checks the format of the source program for errors and handles all the details of address assignments. To make the program more readable (and therefore self-documenting) the assembler allows the programmer to make *comments,* which have no effect on the program. These comments are simply printed on the right of the program listing to make it more readable.

For each desired instruction, the programmer (engineer) writes a line of source program. Instead of writing out the full name of the instruction, one simply writes a *mnemonic* abbreviation. Mnemonics are usually three or four key letters from the instruction name. This makes it easy to remember yet quick to write. Table 8-1 shows the mnemonics recognized by the 8008 (MCS8) assembler and the binary operation codes they produce. For example, if we want a "Rotate *A* register Right thru carry" operation, we simply write RAR and the assembler automatically produces 00011010, which is the correct operation code, for the object program.

Most instructions require additional information from the programmer to fill in variable fields such as source or destination register, address, or immediate operand (see Fig. 7-5). This information is provided after the instruction mnemonic, with one or more spaces separating it from the operation mnemonic. For example, to "Increment the C register," we write INR C and the assembler translates it to 00010000, which is the code for the increment instruction with 010 filled into the destination register field to indicate the *C* register.

Some operation codes require two additional pieces of information, separated by a comma. For example, to move data to the *B* register from the *C* register, we simply write MOV B,C. The assembler converts this to 11001010, where 001 indicates the *B* register and 010 the *C* register.

The *data* for immediate operands and addresses (*adr*) can be defined as a decimal number, or as a hexadecimal number followed by *H*. For example;

ADI 5	is translated to	10000111
		00000101
MVI M,5AH	is translated to	00111110
		01011010
MVI B,1	is translated to	00001110
		00000001
JNZ 127	is translated to	01001000
		01111111

Notice that in the last example JNZ stands for "Jump if Not Zero condition" (jump if accumulator not zero). The assembler translates NZ to condition code

TABLE 8-1 INTEL 8008 INSTRUCTION MNEMONICS[a]

Description of Operation	Assembler		Oper. Code Bits		
	Mnemonic	Operand	76	543	210
Move data to reg/mem from reg/mem	MOV	dst, src	11	DDD	SSS
Move immediate data to reg/mem	MVI	dst, data	00	DDD	110
			BB	BBB	BBB
Increment register contents	INR	dst	00	DDD	000
Decrement register contents	DCR	dst	00	DDD	001
Add reg/mem to A register	ADD	src	10	000	SSS
Add immediate to A register	ADI	data	00	000	100
			BB	BBB	BBB
Add reg/mem to A register with carry	ADC	src	10	001	SSS
Add immediate data to A register with carry	ACI	data	00	001	100
			BB	BBB	BBB
Subtract reg/mem from A register	SUB	src	10	010	SSS
Subtract immediate data from A register	SUI	data	00	010	100
			BB	BBB	BBB
Subtract reg/mem from A register with borrow	SBB	src	10	011	SSS
Subtract immediate data from A register with borrow	SBI	data	00	011	100
			BB	BBB	BBB
Logical AND reg/mem to A register	ANA	src	10	100	SSS
Logical AND immediate data to register	ANI	data	00	100	100
			BB	BBB	BBB
Exclusive OR reg/mem to A A register	XRA	src	10	101	SSS
Exclusive OR immediate data to A register	XRI	data	00	101	100
			BB	BBB	BBB
OR reg/mem to A register	ORA	src	10	110	SSS
OR immediate data to A register	ORI	data	00	110	100
			BB	BBB	BBB
Compare reg/mem to A register	CMP	src	10	111	SSS
Compare immediate data to A register. No change to A.	CPI	data	00	111	100
			BB	BBB	BBB
Rotate A reg. Left (A_7 to A_0 and Carry)	RLC	—	00	000	010
Rotate A reg. Right (A_0 to A_7 and Carry)	RRC	—	00	001	010

TABLE 8-1   Continued

Description of Operation	Assembler		Oper. Code Bits		
	Mnemonic	Operand	76	543	210
Rotate A reg. Left thru Carry	RAL	—	00	010	010
Rotate A reg. Right thru Carry	RAR	—	00	011	010
Unconditional Jump to specified address	JMP	adr	01 BB XX	XXX BBB BBB	100 BBB BBB
Jump to specified address if condition satisfied: [a] (JNC, JNZ, JP, JPO, JC, JZ, JM, JPE)	J[a]	adr	01 BB XX	CCC BBB BBB	000 BBB BBB
Unconditional Call subroutine at specified address, push current address into stack for use by Return	CALL	adr	01 BB XX	XXX BBB BBB	110 BBB BBB
Call subroutine at specified address and push current address if condition: [a] (CNC, CNZ, CP, CPO, CC, CZ, CM, or CPE)	C[a]	adr	01 BB XX	CCC BBB BBB	010 BBB BBB
Unconditional Return from Subroutine	RET	—	00	XXX	111
Return if: [a] (RNC, RNZ, RP, RPO, RC, RZ, RM, or RPE)	R[a]	—	00	CCC	011
Restart at mem adr AAA000, push prog counter	RST	adr	00	AAA	101
Input port 0–7 to A register	IN	port	01	00M	MM1
Output port 0–24 from A register	OUT	port	01	RRM	MM1
Stop program until interrupted	HALT	—	00	000	00X
Stop program until interrupted	HALT	—	11	111	111

[a] Note 1: Remaining letters of conditional branch mnemonics, meaning, and corresponding condition codes as follows: NC (No Carry) 000, NZ (Not Zero) 001, P (Positive) 010, PO (Parity Odd) 011, C (Carry) 100, Z (Zero) 101, M (Minus) 110, PE (Parity Even) 111.

Note 2: src and dst indicate source and destination of data as follows: A reg(000), B reg(001), C reg(010), D reg(011), E reg(100), H reg(101), L reg(110), Memory address H, L (111).

Note 3: X = "don't care," additional bytes are designated BB   BBB   BBB.

[a] See also Figure 7-5.

001. The JUMP, CALL, and RETURN instruction mnemonics each have eight possible forms, as indicated by Note 1 on Table 8-1.

Though we sometimes want to specify an address as a number, as was done in the last example above, this is often inconvenient. Usually we want to jump to the beginning of some other part of the program whose address we do not know as we write the program. For this reason all assemblers allow *symbolic addressing, which allows us to assign an arbitrary name to any address, then refer to that address by name.* For example, we can *label* the first instruction of the READ TAPE routine TAPE1. To branch to that routine, we simply write JMP TAPE1, and the assembler will produce the correct three-byte binary instruction code. The label (name) is assigned by preceding the first instruction of the READ TAPE routine with it. For example, if the first instruction of that routine is an INPUT from port #3, we would write:

TAPE 1: IN 3   ;READ TAPE ROUTINE

This line will generate only a single-byte instruction (01000111), but the assembler will "remember" the RTAPE1 symbol and its location for future reference. READ TAPE ROUTINE is *totally ignored* by the assembler because it occurs after a semicolon following the operand (3). These words are added strictly to make the program more readable and are called *comments.* Actually, *each line of the program can be thought of as having four "fields" as follows:*

Label	Operation	Operand	Comment
TAPE 1:	IN	3	;READ TAPE ROUTINE

As far as the assembler is concerned, only one space character is needed to separate the operation field from the operand field. The *label* field (sometimes called *name* or *location*) always starts with the first character on the left of the line, and is always terminated with a colon. Only lines that we want to refer back to are given labels, so *most lines will start with one or more spaces to indicate no label field.*

Next is the *operation* field, which mnemonically defines the instruction. If the instruction is a type that requires an operand, one or more spaces after the instruction mnemonic indicate the start of the *operand* field. For operations requiring no operand (e.g., RLC), a semicolon after the operand field starts the comments field.

The operand field may indicate source or destination (*src* or *dst*) as "$A,B,C,D,E,H,$ or L" registers or as "M" for memory operations. Immediate operand *data*, addresses (*adr*), and input or output *ports* can be defined in any of the following ways:

1.   Decimal constants, for example, 1, 15, 255, or 2586.

2.   Hexadecimal constants followed by *H*, for example, 1H, FH, FFH, or A1AH.

3.   ASCII characters enclosed by quotation marks, for example, "M", "N", "2", "&".

4.   Address symbols defined as labels elsewhere, for example, RTAPE1.

5.   Arithmetic expressions containing labels, for example, RTAPE1+4.

6.   Arithmetic expressions containing the present location, for example, $-3.

This last type is necessary because, since the assembler keeps track of locations, we do not know when we are writing the program where the present instruction will end up in memory. If, for example, we want to branch back to three bytes before the beginning of the current instruction, we can simply write $-3 in the operand field and the assembler will compute the proper address. Note the $ indicates "present address" here. If we want to go to the fourth word (byte) after the address we have labeled RTAPE1, we simply write RTAPE1+4 and the assembler will compute the required address during assembly.

Notice that arithmetic expressions used in assemblers are converted to actual binary codes *during assembly* so they *must include only data fully defined during assembly*. The *comments* field is strictly optional and is ignored by the assembler. It is used simply to make the program more readable. With proper use of comments, a program can be self-documenting in that anyone can follow its operation without any other writeup. If a line of the program starts with a *colon* in the leftmost character position, the assembler will ignore the entire line, so the entire line can be used as a comment. It is thus possible to include a "writeup" of the program within the body of the program.

8.1.1   A Programming Example

Before proceeding with more details of assembly programs, let us try using some of the basic principles we have just covered by writing an actual program in assembly language. We will write a program for receiving serial data of the type sent by a Teletype machine equipped to communicate over the Telex or the Western Union network. This is similar to the way computer data are transmitted, but it is slower and each character is only five bits. Figure 8-1a illustrates the signal format in general and shows, as an example, an *R* (or 4 in Figures mode) code. Bits are sent one at a time with a duration of 22 msec per bit. When no data are being sent, the line is continually in the 1 state (stop code). When data are sent, a 0 "start bit" is always sent first. Synchronization is obtained by measuring time from this 1 to 0 "start transition" to the center of each bit. It is thus 1.5 bit times, or 33 msec, from the start transition to the center of the first bit, and 22 msec between each bit thereafter.

Most teleprinters that use this code use a mechanical commutator, released by the start bit and rotating at 420 rpm, to convert this serial signal into five parallel bits latched mechanically. Crude as this mechanical approach sounds, it is still by far the most common way of receiving this type of signal today. Of course, we can design an electronic circuit to do the same thing using a shift register, a bit counter, a clock generator, and some logic. There is also a special LSI circuit, called a Universal Asynchronous Receiver Transmitter (UART), that will do the job with a relay or **optical isolator** to interface it to the line, and some logic to interface it to the processor input bus.

We accomplish the same logic by programming and simply interface a *single* processor input bit to the line with an optical isolator. This is thus a direct example of replacing wired logic with programming. In chapter 11 we attack the same problem from yet another approach.

Figure 8-1b shows a fairly detailed flow chart for the required program. Notice that we have two blocks representing time delays. Figure 8-2 shows a detailed flow chart of the "11 msec DELAY" routine. This routine simply uses the D register to count around a loop a certain number of times calculated to produce the required delay. With the Intel 8008 the INCREMENT instruction takes 20 μsec and the JUMP takes 44, so the total delay around the loop is $20 + 44 = 64$ μsec. We thus need to go around the loop $11,000/64 = 172$ times to generate an 11 msec delay. Our delay program in symbolic form is thus

(label)	(operation)	(operand)	(comment)
LOOP1:	MVI	D,172	;11 MS DELAY
	DCR	D	
	JNZ	LOOP1	

Here we can see the usefulness of symbolic addresses, since at this stage we have no idea where this part of the program will be in memory, yet we must refer back to a previous address with the JUMP instruction. By simply making up an arbitrary name, LOOP1, we can refer to it without worrying about addresses and let the assembly program take care of the details. Since this is such a simple Jump, we might prefer to define it in terms of the present address ($) and not define a symbol at all:

```
MVI   D,172  ;11 MS DELAY
DCR   D
JNZ   $-1
```

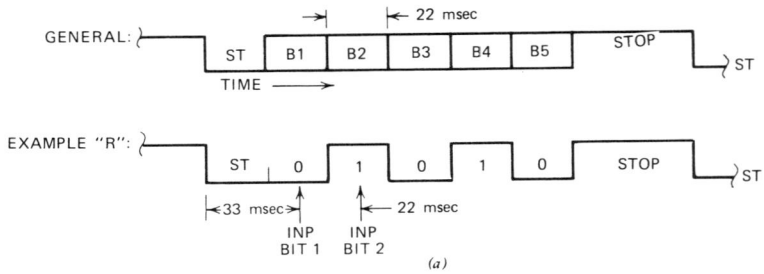

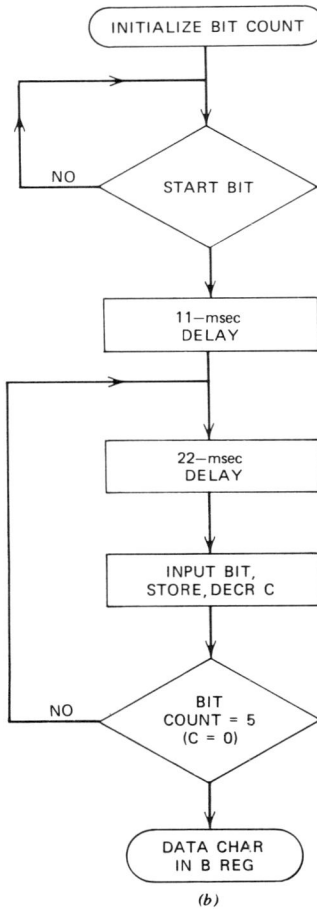

Fig. 8-1. Serial receive data (Baudot code): (a) serial receive signal waveform; (b) program flow chart.

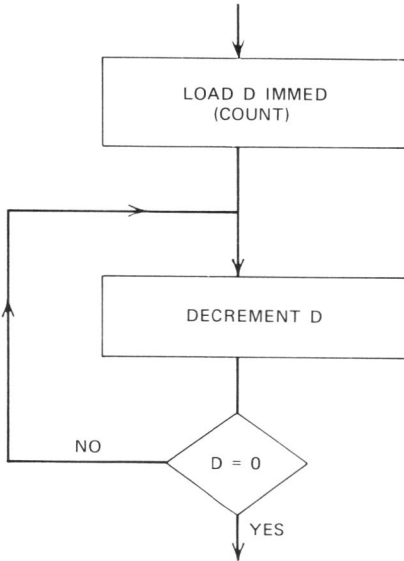

(a)

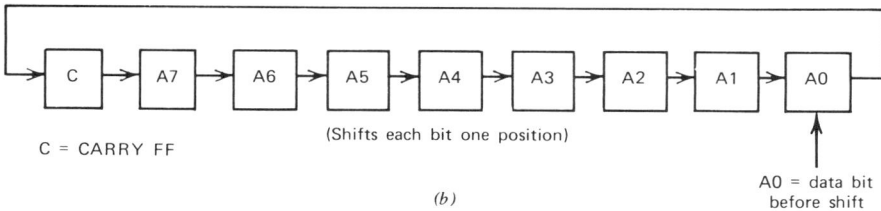

C = CARRY FF

(Shifts each bit one position)

(b)

A0 = data bit
before shift

Fig. 8-2. (a) Routine for 11-msec delay (block on Figure 6-12b). (b) ROTATE A RIGHT instruction.

Either way, the resulting object program, produced by the assembler, will be identical. The second method is quite dangerous in a variable word length machine due to the possibility of making errors in counting bytes.

The "22-msec delay" is a slightly larger problem because the maximum delay we can generate with the simple loop would be $256 \times 64 = 16,700 = 16.7$ msec. The usual way of generating longer delays is to put *a loop within a loop*. This method could give a delay as long as $16.7 \times 256 = 4.27$ sec. We can gain enough additional delay for our purposes, however, by simply adding another instruction in the loop. If we add, for example, "Compare the A register to memory" (CMP M), no harm is done; but we add another 32 μsec to our loop time, thereby increasing it to $64 + 32 = 96$ μsec. The required count is thus $22{,}000/96 = 230$. Our "22-msec DELAY" routine is thus

```
        MVI   D,230     ;22MS DELAY
LOOP2   CMP   M
        DCR   D
        JNZ   LOOP2
```

At the end of each 22-msec delay it will be the center of a new bit time, so we will input a bit to bit 0 of the A register. We will then put that bit in the carry flip-flop by doing a ROTATE A RIGHT (see Fig. 8-2a), MOVE any previously accumulated bits into A from the B register, then ROTATE A LEFT to shift the new bit back from the carry into A_0 and move the others over one place. We next MOVE the accumulated data from B to A, then increment and test the bit count. The symbolic coding (with comments) is as follows:

```
IN    O        ;INPUT NEW BIT TO A₀
RAR            ;PUT NEW BIT IN CARRY
MOV   A,B      ;PUT PREVIOUS BITS IN A
RAL            ;SHIFT ALL BITS LEFT
MOV   B,A      ;HOLD IN B
DCR   C        ;DECREMENT BIT COUNT
JNZ   LOOP2    ;LOOP IF NOT 5TH BIT
```

The only thing remaining is the beginning part of the flow chart of Figure 8-1b. We must simply initialize the bit count (C register) to five and execute a HALT instruction. The processor will thus stop until an interrupt is received. Since the DATA' line is connected to the INTERRUPT input of the processor as well as bit 0 of input port 0, the *start transition* will cause the program to start up again with the first instruction of the 11-msec delay. Another way to wait for START would be to loop on an input test as follows:

```
IN    O
CPI   O
JNZ   $-3
```

As long as the input is 1 (stop), the program stays in this loop. The START TRANSITION causes the *zero* condition to go true, allowing the program to proceed to the 11-msec delay. Since the above method requires three extra bytes of program, we will use the interrupt to detect the start bit. The complete program can thus be written by simply putting together the parts we have discussed. It is shown in Table 8-2 as printed out (listed) by an assembler.

TABLE 8-2 ASSEMBLY LISTING

	;SERIAL DATA RECEIVER ROUTINE 45 BAUD TELEX	
0000 1605	MVI C, 5	;INITIALIZE BIT COUNT
0002 00	HLT	;WAIT FOR START INTERRUPT
0003 1EAC	MVI D, 172	;INITIALIZE 11 MS DELAY
0005 19	DCR D	;DECREMENT D
0006 480500	JNZ $-1	;LOOP UNTIL D = 0 (11MS)
0009 1EE6	LOOP2:MVI D, 230	;INITIALIZE 22MS DELAY
000B BF	LOOP1:CMP M	;ADDL TIME DELAY
000C 19	DCR D	;DECREMENT D
000D 480B00	JNZ LOOP1	;LOOP UNTIL D = 0 (22MS)
0010 41	IN 0	;INPUT NEW BIT
0011 1A	RAR	;PUT BIT IN CARRY
0012 C1	MOV A, B	;PUT PREV BITS IN A
0013 12	RAL	;SHIFT ALL BITS LEFT
0014 C8	MOV B, A	;HOLD IN B
0015 11	DCR C	;DECREMENT BIT COUNT
0016 480900	JNZ LOOP2	;LOOP IF NOT 5TH BIT

Notice that the assembler listing reprints the source program statements in the same (label, operation, operand, comment) format they were written in. To the left of the source statements it prints the assigned memory locations and instruction codes in *hexadecimal* form. This printout is in four columns. The left-hand column shows the hexadecimal memory location of the first byte of the instruction. Next, the instruction code is printed with two hex digits for each byte of the instruction (left to right).

The first line of the printout shows that the first instruction (MVI C,5) is stored in memory bytes 0 and 1. The first byte is hexadecimal 16, or 00 010 110, where the 010 represents register C. The second byte is the immediate operand decimal 5 or 000 001 01. The next instruction (HLT) is stored in location 2 and is simply all zeros. The assembler automatically generates the required operation codes and assigns consecutive addresses. Notice that the second byte of the JNZ -1 is computed by the assembler by subtracting 1 from the present location ($6 - 1 = 5$). Notice also that the assembler automatically looks up the address labeled LOOP2 and fills in "9" as the address field for the source statement JNZ LOOP2.

The resulting program requires 25 bytes of instruction memory. Since it performs the same function as a wired-logic serial data receiver, or the LSI UART receiver circuit, it is interesting to compare the cost of the program bits with that of conventional logic. In early 1978 the cost of ROM bits was .05¢ per bit if purchased as part of a 65 k bit MOS ROM in 100 quantities. The incremental cost of the programmed serial data receiver is thus 8×25 bytes $\times$ .05 = 10¢! This is another example of how unbeatable programmed logic is.

Looking again at the program itself, it is clearly self-documenting in that anyone trained in programming can follow the listing. Also, modifications to the receive format are extremely easy. For example, we can convert it into an eight-bit receiver by just changing the first instruction to MVI C,8. To change the timing, we simply change the operands of the MVI instructions in the two delay loops. In fact, if these parameters are stored in read/write memory, we can set them before entering the program and thus receive several different (start-stop, serial) formats and data rates with the same program.

Now that we have seen what we can do with an assembler, let us look at other things they can do for us.

8.1.2   Assembler Instructions (Pseudo-Ops)

Up to now, all our statements were simply converted by the assembler to instruction codes. Assembler instructions, or *pseudo-ops,* are *operation* field mnemonics that do not convert directly to machine instructions but are used to control the assembly process, to define symbols, to generate data, or to allow the generation of often used sequences of instructions with a single statement.

The assembler program automatically assigns consecutive memory addresses to each line of program in order. It starts with address zero unless told to do otherwise by an *Origin (ORG)* pseudo-op. If, for example, we want our program to start from location 1000, we simply write the following, before the first line of the source program:

$$\text{ORG} \qquad 1000$$

We can use ORG again in the program if we want to start another part of the program at another location. Note that no machine instructions are produced by an ORG statement.

We can define symbols (labels) as having any value we want by using the *Equate (EQU)* pseudo-op. For example,

$$\text{TADR} \quad \text{EQU} \quad 2$$

defines the TADR symbol as having a value of 2. No machine instructions are produced in the final program by this statement, but once defined the TADR symbol can be used in the operand field and the result will be the same as if we wrote a 2. This can be useful when some parameter (such as tape unit address) is used repeatedly in the program. The parameter can then be redefined, and the program reassembled, by changing just one EQU statement instead of going through and changing the parameter each time it appears.

We can *define constants and data tables* by using the *Define Byte* (DB) pseudo-op. For example,

MESS1   DB   "PRINTOUT"

defines eight consecutive memory locations containing character codes for P, R, I, N, T, O, U, and T. The address of the first letter can be referred to symbolically as MESS1. Note that we indicated alphanumeric (ASCII) data by using quotation marks before and after the data. We can also define data tables. For example,

MASKS   DB   1,0FH,240,TADR+2

defines four consecutive words: 00000001, 00001111, 11110000, and 00000100. The symbolic address of the first word is MASKS, so we can, for example, use the third mask by simply putting MASKS+2 in an operand field. Notice that the fourth word was computed by the assembler from our previous definition of TADR.

We could generate the seven-segment display data table (Table 7-3), used by the example in Chapter 7, as follows:

SEG7   DB   7EH,3OH,6DH,79H,33H,5BH,1FH,3OH,7FH,73H

The program to convert BCD codes in the *A* register to seven-segment codes could then simply use this symbolic name as follows:

ORI   SEG7   ;GENERATE ADDRESS (SEG7+A CONTENTS)

MOV   L,A   ;PUT ADDRESS IN L REG

MOV   A,M   ;LOAD 7 SEG CODE

Tables of 16-bit addresses can be generated by using the *Define Word* (DW) pseudo-op. For example,

LINK   DW   SUB1,SUB2,SUB3

will make a six-byte table with the binary address of SUB1 in the first two bytes (least significant first), followed by SUB2 and SUB3. The DW pseudo-op is needed because, while symbolic addresses are easily used in the assembler, they end up in the assembled program *as locations containing instructions or in address field of instructions, only.* By defining an address as data, we allow the *assembled object program* itself to use the address (e.g., to compute addresses at execution time). *It is important to be aware of this distinction between what is available during execution of the assembler program and what is available during execution of the assembled object program.*

Sometimes it is desirable to assemble several versions of a program by adding or deleting certain sections of the program. For example, if a system is sold with different I/O equipment options, we may want to delete those portions of the program for which the correct equipment is unavailable. Of course, this could be done by writing the program in separate modules of source program tape or cards and physically putting together the right combination of tapes or cards. A much cleaner way to do the same thing is to use the *conditional assembly* pseudo-op, *IF*. Each module of the program is preceded by an *IF* statement and followed by an *ENDIF* statement. If the expression in the *operand* field of the *IF* statement is zero, the entire block of source program will be ignored by the assembler. A READ TAPE program, for example, could be conditionally assembled as follows:

```
IF    TAPE

•
 (the tape program)
•

•

ENDIF
```

By simply adding a single control statement:

```
TAPE    EQU    1
```

or

```
TAPE    EQU    0
```

we can effectively include or delete the entire tape program.

We sometimes find ourselves repeatedly using a similar sequence of instructions in our programs. The *MACRO* pseudo-op allows us to give a name to a sequence of instructions, then *let the assembler produce those instructions whenever we put the name of the MACRO in the operation field of our source program.* For example, on the 8008 we often want to move 14-bit addresses into both the *H* and the *L* registers to set up an address prior to a memory operation. Instead of using two instructions to do this, we can define a MACRO called LXI (Load eXtended Immediate) as follows:

```
LXI    MACRO    ADR

       MVI      H,ADR/256

       MVI      L,ADR

       ENDM
```

Once this MACRO is defined, we can simply write

```
       LXI   3597
```

and it will have the same effect as if we had written

```
       MVI   H,3597/256

       MVI   L,3597
```

The assembler will automatically produce the correct four-byte sequence from the single statement. The division by 256 gives us the most significant byte of the address, while *L* is loaded with the eight least significant bits of the address.

Notice that a MACRO is *not subroutine* because the entire sequence is inserted in the program each time the name is used. We could, for example, define a DIVIDE MACRO, then use it repeatedly in a program, but the result would be less efficient than jumping to the same DIVIDE subroutine each time. Assemblers often include several, already defined MACRO's. For example, LXI is included in the MCS-8 assembler. This essentially gives the programmer an expanded instruction set to work with.

The *END* pseudo-op indicates the end of the program to the assembler. It must *always* be included after the last statement of the source program.

An important feature of assemblers is that *they check for errors in the source program.* The *program listing* will have an *error flag* (letter) on each line where an error is detected. Sometimes a separate printout can be requested that shows only lines with errors and the error messages. The following are examples of error messages and the kinds of errors that cause them:

ILLEGAL CHARACTER: The special character (such as $,/.,) appears in the statement (not in the comment), or perhaps a required operand field is missing.

MULTIPLY DEFINED SYMBOL: The symbol has been defined more than one time.

UNDEFINED SYMBOL: The symbol has been used but never defined.

ILLEGAL NUMERIC CONTAINS CHARACTER: A hexadecimal number includes an illegal digit (such as G or N), or the numeric contains nonnumeric characters.

ILLEGAL OPCODE: The operation code is not one of the acceptable mnemonics.

MISSING OPERAND FIELD: No operand found for an operation code that requires one.

ILLEGAL VALUE: The numeric value of a hexadecimal or decimal number of an expression has overflowed its limit.

ILLEGAL SYMBOL: A location field contains a symbol that has more than five characters or that does not start with an alphabetic.

MISSING LABEL: The label, which is required by the EQU pseudo-op, is missing.

SYMBOL TABLE OVERFLOW: Too many symbols in source program to fit into allocated symbol table.

LINE OVERFLOW, MAXIMUM: Input line exceeds 48 characters; or missing carriage return.

ERRONEOUS LABEL: Opcodes END and ORG may not have a label.

ILLEGAL ORIGIN: Value of new origin is less than correct program count.

ILLEGAL OPERAND: DAD opcode requires symbolic operand.

In addition to producing a combined *listing* of the source and object programs, the assembler can output *punched cards or paper tape* containing the *binary* codes of the object program. This tape is normally in a format for loading into the microprocessor read/write memory using a *binary loader program* stored in the microprocessor. Sometimes the tape starts with a *bootstrap loader* program, making it possible to load the tape into a processor that contains only the first few instructions of the loader. The bootstrap program then takes over and controls its own loading, then the loading of the object program itself.

TABLE 8-3 8900 ASSEMBLER MNEMONICS AND OPERANDS

Instruction	Mnemonic	Operand[a]	Format (Fig. 7-4)
Load	LD	r, disp(Xr)	a
Load Indirect	LD	r, @disp(Xr)	a
Store	ST	r, disp(Xr)	a
Store Indirect	ST	r, @disp(Xr)	a
Add	ADD	r, disp(Xr)	a
Subtract	SUB	r, disp(Xr)	a
Skip if AND is Zero	SKAZ	r, disp(Xr)	a
Skip if Greater	SKG	r, disp(Xr)	a
Skip if Not Equal	SKNE	r, disp(Xr)	a
And	AND	r, disp(Xr)	a
Or	OR	r, disp(Xr)	a
Jump	JMP	disp(Xr)	b
Jump Indirect	JMP	@disp(Xr)	b
Jump to Subroutine	JSR	disp(Xr)	b
Jump to Subroutine Indirect	JSR	@disp(Xr)	b
Increment and Skip if Zero	ISZ	disp(Xr)	b
Decrement and Skip if Zero	DSZ	disp(Xr)	b
Push on to Stack Register	PUSH	r	d
Pull from Stack	PULL	r	d
Add Immediate, Skip if Zero	AISZ	r, data	d
Load Immediate	LI	r, data	d
Complement and Add Immediate	CAI	r, data	d
Shift Left	SHL	r, #	d
Shift Right	SHR	r, #	d

A *relocatable loader* in the microprocessor makes it possible to load the program starting at any desired memory location. It is thus possible to load several program modules into different parts of the memory. The object program tape must be in a *relocatable format* so the loader program can *compute the correct addresses of all program fields representing memory references during loading.*

Yet another punched tape format is required if the program tape is going to be used to program a PROM on a PROM programmer. These tapes use *one character for each* bit to be programmed. The BNPF format uses *B* to define the start of a word, *N*'s or *P*'s to indicate 0's and 1's, and *F* to indicate the end of a word.

TABLE 8-3   Continued

Instruction	Mnemonic	Operand[a]	Format (Fig. 7-4)
Rotate Left	ROL	r, #	d
Rotal Right	ROR	r, #	d
Exchange Registers and Top of Stack	XCHRS	r	d
Register Copy	RCPY	sr, dr	e
Exchange Registers	RXCH	sr, dr	e
Register And	RAND	sr, dr	e
Register Exclusive OR	RXOR	sr, dr	e
Register Add	RADD	sr, dr	e
Branch-on Condition	BOC	cc, disp	c
Return from Subroutine	RTS	ctl	f
Return from Interrupt	RTI	ctl	f
Jump to Subroutine Implied	JSRI	ctl	f
Set Flag	SFLG	fc	f
Pulse Flag	PFLG	fc	f
Push Flags on Stack	PUSHF		f
Pull Flags from Stack	PULLF		f
Register in	RIN	port	f
Register out	ROUT	port	f
Halt	HALT		

[a] r = register number; Xr = index register number (3 or 4) (relative or direct used if none specified); disp = displacement; fc = flag code: L, OV, CY or GFO-12 for Link, Overflow, Carry, or General Flags; ctl = control (added to return address); port = input or output port (added to AC3 contents to determine port number).

8.1.3   The 8900 Assembler

The preceding section mainly concerned the MCS-8 assembler (for the Intel 8008). Although all assemblers are based on common principles, they differ in details such as control characters used, pseudo-op mnemonic abbreviations, and acceptable data formats. Although the MCS-8 assembler is typical in many ways, its limited addressing modes make it simpler than most assemblers.

The 8900 is more typical in some respects because the instruction set includes a full range of addressing modes. Because of this, the *operand* field of the source program is somewhat more complex. Table 8-3 shows the mnemonics and operand format used by the 8900. For example,

LD   0,7(3)

is written for: "Load register 0 from memory address in index/register 3, increased by 7." This is translated by the assembler to 1000 00 11 00000111 (see Fig. 7-3a). *Indirect* addressing is indicated by preceding the address with @. For example,

LD   0,@7(3)

produces a LOAD INDIRECT instruction, causing the data to be loaded into register 0 from the memory address *pointed to* by the *contents* of memory location 7, *plus* the contents index register 3.

If no index register is specified, the assembler automatically computes the displacement and uses the *program relative* addressing mode if possible. For example,

LD   1,TABLE2

causes register 1 to be loaded with the contents of the location labeled TABLE2. The assembler subtracts the current address from the address labeled TABLE2. If the result is in the allowable range ±127, the resulting object code specifies *program relative* addressing mode with the proper displacement in the displacement field. If the result is outside this range, a *LOAD INDIRECT* instruction is produced with *base page* mode addressing to a *pointer word* that is also automatically produced by the assembler. This pointer word contains the 16-bit direct address of TABLE2.

There are other minor differences between the 8900 and the MCS-8 assemblers. For example, the symbol for *present address* is "." instead of "$." Binary operand data are represented in hexadecimal form *preceded* by instead of followed by *H*. The 8900 object code printout is also different in that the *location* is printed in decimal form in the first column and also in hexadecimal form in the second column. The operation code is printed only as a hexadecimal number in column 3.

These minor differences are found with each different assembler, since there is no real standardization. The purpose here is simply to introduce the general concepts. The manuals provided by the manufacturers provide the required information on specific details of the particular assembler being used. In fact, a manufacturer often has several different assembler versions available since a more elaborate assembler requires a more elaborate computer system (e.g., with more memory or peripheral devices) to run it.

8.2 COMPILERS

Compilers are programs similar in purpose to assemblers, but they push the level of design automation back further towards the flow chart or problem statement level. The source program on a compiler bears no resemblance to the "machine language" of the instruction set. The languages of compilers are *problem oriented*: they are designed to state the problem in the most efficient way. The compiler program then translates this statement of the problem into *machine language,* producing a binary object code for the processor for which it was written.

Some programming languages have become so standardized that compilers for them are available for virtually every computer on the market. FORTRAN is an example of such a language; programs written in the FORTRAN language can be compiled and run on virtually any computer (except today's microcomputers). This fact is put to good use by the microcomputer manufacturers. By writing their programs that require a full-scale computer system, in FORTRAN, these programs can be compiled and run on virtually any computer type. Intel, for example, has a *cross assembler* (for assembling microcomputer programs), a *simulator* (to allow checkout of microcomputer programs on a large machine by using programs to execute the microcomputer instructions), and even a compiler written in the FORTRAN language.

Another advantage of compilers is that the programs can survive long after the computer is obsolete. FORTRAN, for example, was originally introduced by IBM for use on their 704 and 709 computers. Now those computers are obsolete, and even current IBM computers use a completely different instruction set. Programs written in FORTRAN, however, can still be compiled not only on new IBM computers, but on virtually any other computer from a minicomputer up.

A disadvantage of compilers is that they generally program like an unimaginative, but accurate, programmer. Since they basically just plug in subroutines to do the job described by the source program, they often use many instructions to do things that a clever programmer could do in fewer instructions. Of course some compilers are more efficient than others, but, in general, compilers produce programs that use more memory and/or run slower than programs written in assembly language. Time, however, is on the side of the compiler; as programmers' and engineers' salaries are rising, memory costs are going down and processor speeds are going up. Even now, the programming of systems that will *not* be produced in high volume is best done in *high-level languages* (as compiler languages are called). The added memory and processor cost is more than offset by reduced programming cost.

Another problem is that in general today's standard programming languages are oriented toward business (COBOL, RPG) or scientific (FORTRAN, ALGOL, BASIC) problem solving. Scientific problems require very high accuracy, so the scientific languages like FORTRAN handle data primarily in floating-point notation. For simple digital systems this can be quite wasteful. Also, the scientific languages are optimized for inputing formatted data (e.g., from punched cards), performing complex algebraic operations on them, and outputting formatted results (e.g., printouts). While BASIC is more down-to-earth, it significantly sacrifices speed and storage efficiency to allow quick interactive development of small terminal-oriented programs.

8.2.1   PL/M

Probably the most widely used compiler for microcomputers is called PL/M, MPL, PL/I, and so on, depending on the manufacturer. It is actually a subset of IBM's PL1 language, but it also resembles PASCAL and their common ancestor ALGOL.

A PL/M program is written in a much freer form than an assembly program. Instead of just one statement per line, *statements* are made anywhere in the line. *A semicolon after the statement* indicates the end of the statement. *Labels are always followed by a colon,* and comments are always enclosed by /* and */ on each side (e.g., /*COMMENT*/). Symbolic *identifiers* can be any length and use any alphanumeric characters as long as they begin with a letter and have no spaces. Instead of spaces, dollar signs can be used to improve the readability of identifiers (e.g., INPUT$COUNT). Special *reserved words,* which are part of the language, cannot be used as identifiers (e.g., OR, PLUS, DO, IF, etc.).

Generally the programmer can forget about addresses. The problem is defined by simply defining the variables, then making certain statements about the variables. These *statements* can be made in very readable form, such as

$$WIDTH = AREA/HEIGHT;$$

Though the Intel microprocessor has no *divide* instruction, this simple statement is sufficient to define what we want done. The compiler automatically puts a *DIVIDE* subroutine in the object program to perform the operation and keeps track of the variables by their symbolic *identifiers.*

Before an identifier can be used in a statement, it must be defined with a *DECLARE* statement. The declaration defines whether the variable should be handled as a *BYTE* (8-bit) or as an ADDRESS (16-bit) variable. It also sometimes defines an *initial value. Data,* in declarations and statements, can be stated in any of the following forms:

212	*Decimal* (can use *D*)
11010100B	*Binary*
324Q	*Octal* (can use *O*)
D4H	*Hexidecimal*
'T'	*ASCII Character*

Note that all five examples above represent the *same* binary data in a different way.

To save space in describing PL/M we now present the allowable declarations and statements in an outline form. We use *A, B, C, D,* and *E* to represent any *identifiers*. We define the acceptable forms by example. Some examples of variations of the general form are indented below the general form, and explanations appear to the right where necessary.

First the *Declarations:*

DECLARE (A,B,C) BYTE;	*Identifiers A, B,* and *C* are eight-bit variables
DECLARE A BYTE;	*A* is an eight-bit variable
DECLARE (A,B) ADDRESS;	*A* and *B* are 16-bit variables
DECLARE A(100) BYTE;	Identifier *A* represents a table of 100 subscripted eight-bit, variables A(0), A(1), . . . A(99), which can be referred to as such in statements; *A* is called a *vector*
DECLARE A(50) ADDRESS;	A 100-byte table is formed, which can be referred to as A(0), A(1), . . . A(49)
DECLARE (A,B,C) (100) BYTE;	Defines three 100-byte tables
DECLARE (A,B,C) BYTE INITIAL (1,27,42);	Byte variables *A, B,* and *C* have initial values of decimal 1, 27, and 42
DECLARE A ADDRESS INITIAL 27H;	16-bit variable *A* has an initial value of hexidecimal 27
DECLARE A DATA (1,27,575);	A(0), A(1), and A(2) are stored in ROM area as decimal 1, 27, and 575; variable size is automatically assigned, eight or 16 bits, so A(0) and A(1) will be byte, A(2) will be address.

DECLARE A LITERALLY 2350H;	Whenever *A* is used elsewhere in the program, 2350 hexidecimal will be substituted
DECLARE A LITERALLY 'ERROR STOP';	Any length string of characters, up to 255 bytes acceptable
DECLARE A BYTE, (B,C) ADDRESS, D DATA 27Q;	Several declarations can be listed after one "DECLARE"

That takes care of all of the general types of *declaration*. Note again that the letters *A*, *B*, *C*, and so on, are used here just for simplicity. For example, DECLARE A BYTE; would normally have a more descriptive identifier than *A*, such as DECLARE WIDTH BYTE. Having declared the variable WIDTH as eight-bit data, we can use it freely in *statements*, which we outline next. The compiler automatically computes functions defined by statements to the accuracy of the most accurate variable used. If multiplication or division appears, however, the result is always a 16-bit variable. Generally the statements can use *expressions* that are similar in form to algebraic or logical expressions. However, since the expressions must be typed as a sequence of characters, there may be some confusion unless parentheses are used. Unless indicated otherwise by parentheses, the PL/M compiler always performs operations in a fixed order of precedence. *The listing below shows all the operators allowed in PL/M expressions in order of precedence:*

$$/,*,MOD$$

$$+,-,PLUS,MINUS$$

$$<,<=,<>,=,>=,>$$

$$NOT$$

$$AND$$

$$OR,XOR$$

Most of these symbols have obvious meanings, but since they must be made from a character set that does not include some of the standard mathematical symbols, some explanation may be in order: * indicates multiply, MOD indicates the *remainder after division*, $<=$ indicates *less than* or equal, $<>$ indicates not equal, $>=$ indicates greater than or *equal*, NOT *complements* all bits in the variable, and XOR is simply *EXclusive OR*.

Expressions are normally evaluated by PL/M by performing the operations on the top line of the above list first, then working down in order. Operators that appear on the same line have equal precedence and are evaluated *from left to right* as they appear in the expression. Parentheses are thus sometimes required to correctly define the desired expression. For example,

$$AVERAGE\$VALUE = (SAMPLE1 + SAMPLE2)/2$$

must have parentheses or SAMPLE2/2 will be evaluated first, then added to SAMPLE1.

Let us continue now with our outline form, showing examples of possible *statements:*

$A = B + C*D;$	Simple statement
$A,E = (B + C)*D;$	Defines both A and E as equal to $(B + C) \bullet D$
$A = A + 1;$	Increment A
$A = B(6) - 3;$	A = the seventh entry in table B, minus 3($B(0)$ is the first entry)
$A = B(C + 27) + 5;$	The subscript can be an expression itself
$A = B$ AND $15;$	Masks the 4 lsb of B into A
IF $A > B$ THEN $A = B;$	A conditional statement $A = B$ applies only if $A > B$
IF $A > B$ THEN $A = B$ ELSE $A = 0;$	A more complex conditional statement; if A is not greater than B, $A = 0$ applies
IF $A = 1$ THEN $B = C/D;$	
IF $A = B/C$ THEN $D = 1;$	
GO TO $A;$	Will transfer program control to location A; A is a label defined by "A:" in front of a statement elsewhere in the program. (GO TO statements make a program hard to follow so their use is considered bad form.)
HALT;	Stops machine till interrupt

DO WHILE (A =3; • • • • END	A group statement; a "loop" is formed causing all statements be-tween DO and END to execute repeatedly as long as the expres-sion $A = 3$ is true
DO A =1 TO 10; • • • END;	Variable A will start at an initial value of 1 and increment each time through the loop; when A reaches 10, the program will pro-ceed
DO A =1 TO B +1: • • END;	Both the initial and the final value can be *expressions*
DO A =1 TO 10 BY 2 • • END;	Each time around the loop A will be incremented by 2, so the loop will only be executed five times

IF A =B THEN
DO;
•
•
•
END;
	If $A =B$, will do any number of statements between DO and END, otherwise goes to statement after END; an entire block of program can thus be condition-ally executed during program ex-ecution time

DO CASE A –5; statement 0 statement 1 statement 2 • • • END;	A multiple branch; only one of the statements will be executed depending on the expression $A - 5$, e.g., if $A - 5$ is 0, only statement 0 is executed; if $A - 5$ is 1, only statement 1 is executed; the program proceeds with the statement following END

Each of these statements defines a job that would take many instructions in assembly language to define. The compiler handles all the details for us from our problem-oriented statements of what we want done. Since we can use *expressions* to determine such things as subscripts, addresses, branch condi-tions, and number of loop iterations, a great many steps can often be combined in a single statement. To add even more power to the language, we can define

PROCEDURES (subroutines) much as we defined *MACROS* in the assembly language, but here *we can actually use them in expressions*.

First we must define the procedure:

A: PROCEDURE (B,C) ADDRESS;

    DEFINE (B,C,D) ADDRESS;

    .

    .

    .

    RETURN D;

    END;

The statements between **DEFINE** and **RETURN** define the result to be *returned* (D) as a function of variables B and C. We can now simply use $(A(E,F)$ as a variable in any expression. E and F are used by the subroutine, as variables B and C respectively, to compute D during program execution. The result (D) is then automatically substituted in the expression in place of $A(E,F)$.

In addition to being able to write our own *PROCEDURES,* several of them have been included as part of the PL/M language:

SHL (A,B)   Shift eight- or 16-bit variable A left B places

SHR (A,B)   Shift eight- or 16-bit variable A right B places

ROL (C,B)   Rotate byte variable C left B places

ROR (C,B)   Rotate byte variable C right B places

CARRY ⎫

ZERO   ⎪

         ⎬   =1 if indicated condition true

SIGN   ⎪

PARITY ⎭

INPUT (A)   Represents the byte obtained by inputting
                     from port A

OUTPUT (A)   Causes an output of a byte to output port A

We can use these *PROCEDURE* names in expressions as if they were variable identifiers:

IF CARRY =1 THEN B =SHL(B,1)   Will shift the data represented
                                        by identifier *B* left one place if
                                        the condition flip-flop indi-
                                        cates a carry occurred pre-
                                        viously

OUTPUT (2) =A +1   Will increment *A* and output it
                                        to port 2

OUTPUT (2) =INPUT (3)*INPUT(2) +A;   Will output to output port 2;
                                        the product of the values at
                                        input ports 2 and 3 plus the
                                        value of *A*

Another type of *PROCEDURE* does not return any data to the program, so we simply *CALL* it instead of using it in expressions. One such *PROCEDURES* is defined as part of the PL/M language—a time delay program. We simply CALL it as follows:

CALL TIME(A); A delay of *A* times 0.1 msec will be produced; then
                the program will resume
CALL TIME(25);   Will cause a delay of 2.5
                                        msec.

Address labels can be attached to any statement in the program by preceding the statement with a label followed by a *colon*. For example,

A:B =C +1   Assigns identifier *A* as the statement's address
247:A =B +1   Is similar to an ORG statement in an assembler in that
                 it assigns address 247 to that statement; the compiler
                 will proceed with consecutive addresses from there
B =.A +27   .A means "the address of label *A*", addresses can thus
                 be used as a part of expressions.
EOF   Is required at the end of PL/M programs to indi-
                 cate the end (End Of File).

That concludes our basic outline of the PL/M language. More details should be known to use the language effectively, but these are available in the manufacturer's PL/M manual.

Now to put to use what we have learned about PL/M, let us try reprogramming in PL/M the *Serial Receive Data* program we wrote before in assembly language. Referring to the flow chart of Figure 8-1*b*, we see that it is actually *more detailed* than our PL/M program need be. First, the loop where we count bits and the initialization of the bit count can be handled by simply using a *DO* group. Also, we need not concern ourselves with which registers to use; we just use a symbol *CHAR* to represent the data. We can also forget about the trick with the carry bit, as we can simply shift *CHAR* left and *OR* in the new bit. The PL/M program itself is shown below:

```
/*SERIAL DATA RECEIVER ROUTINE 45 BAUD TELEX*/
    HALT;      /*WAIT FOR START INTERRUPT*/
    DECLARE (CHAR,I) BYTE;
    CALL TIME (110);      /*START DELAY*/
    DO I = 1 TO 5:
    CALL TIME (220);      /*BIT DELAY*/
    CHAR = SHL(CHAR,1) OR INPUT (0);
    END;
```

Basically, it follows the same pattern as the assembly language program. Instead of writing delay loops, we can simply call the *TIME* procedure for the two delays. Notice that it takes only one statement to shift the character left after the previous iteration *and* input and merge in the new bit. The important thing is that where the assembly language program took 16 lines of rather abstract code, this one takes only seven. Moreover, the PL/M program is much more readable and therefore easier to write and harder to make mistakes on. The resulting code, however, undoubtedly takes a few more bytes of storage.

In large system programs PL/M makes it possible to greatly reduce programming time and errors by organizing the program into a *block structure*. The block structure is the programming equivalent of the "black box" subdivision of hardware systems discussed in Section 2-9. Each program block fits on a single page and contains just enough program to be challenging but not overwhelming. The first step in programming a system is thus a one page block of program that defines the general logic of the entire system just as a system block diagram does for hardware. This top level block ignores details and simply defines the interaction of major program blocks which are given descriptive PROCEDURE names. The next level of subdivision consists of one page of code for each procedure. In a simple system these one page blocks may be adequate to completely define the procedure directly. If this is not possible, details are again ignored by simply using descriptive procedure names again. These procedures are later defined in still more detailed blocks of programming.

It is thus possible to use *procedures within procedures within procedures* to subdivide the program into a tree structure with as many levels as necessary to define an entire program of *any* complexity without ever having to deal with more than one page of program code at a time. Since all work is done on "bite sized pieces," confusion is minimized and fewer errors occur. For true structured programming each block must start at the top and end at the bottom with no entry or exit from the middle allowed. For this reason GO TO statements are not allowed.

Figure 8-3 is a sample PL/M program consisting of five short blocks. The purpose of the program is to print a table of square roots of numbers from 1 to 1000. The top level block of the program is on lines 52–69. Line 68 is a perfect example of how a large amount of code can be compactly represented in readable form by ignoring details that are defined in lower level blocks. Line 68 causes the number that is the square root of I to be printed. The actual computation of the number is defined by a procedure called SQUARE$ROOT which is defined in detail by the lower level block on lines 5–12. The details of the printing itself are defined by the PRINT$NUMBER procedure block on lines 37–50. Within this block yet another procedure is referenced on line 49. The PRINT$STRING procedure defines more details of the printing operation but still defers the details of outputting individual characters by referring to the PRINT$CHAR procedure on line 33. The final details of timing out individual bits serially are defined on lines 14–27. This represents the lowest level of subdivision just as a detailed schematic diagram does with wired logic. When making a change or fixing a problem we can work from the top down, following the treelike structure to the block requiring attention. Unstructured programs, on the other hand, are like one giant schematic of the *entire system* and require plowing through mountains of detail to get to the problem.

Since there is no one-to-one relationship between lines of program and the machine code produced, PL/M prints out the object code separately as shown in Figure 8-4. Program checkout with PL/M is done by checking the *results* of the source statement with the same top-down approach used to write the programs. The program flow in a structured program directly follows the listing from top to bottom. If a program reaches a particular point, we know that all code before that point has been executed and all code after that point has not. Details in lower level blocks can be initially ignored. Checkout thus consists of checking for correct results first on the general level and then on a more and more detailed level.

8.3 COMPILERS, OPERATING SYSTEMS, AND VIRTUAL MACHINES

The interchangeability of programs and wired logic causes an interesting philosophical problem in that we almost always work, not with a "real machine,"

A Sample Program in PL/M

block level

PASS-1

```
00001  2      2048: /* IS THE ORIGIN OF THIS PROGRAM */
00002  2    DECLARE TTO LITERALLY '2', CR LITERALLY '15Q', LF LITERALLY '0AH',
00003  2           TRUE LITERALLY '1', FALSE LITERALLY '2';
00004  2
00005  2    SQUARE$ROOT: PROCEDURE(X) BYTE;
00006  3        DECLARE (X,Y,Z) ADDRESS;
00007  3        Y = X; Z = SHR(X+1,1);
00008  3            DO WHILE Y <> Z;
00009  3            Y = Z; Z = SHR(X/Y + Y + 1, 1);
00010  4            END;
00011  3        RETURN Y;
00012  3        END SQUAREROOT;
00013  2
00014  2    PRINT$CHAR: PROCEDURE (CHAR);
00015  3        DECLARE BITSCELL LITERALLY '91',
00016  3           (CHAR,I) BYTE;
00017  3        OUTPUT (TTO) = 0;
00018  3        CALL TIME (BITSCELL);
00019  3            DO I = 0 TO 7;
00020  3            OUTPUT(TTO) = CHAR; /* DATA PULSES */

00021  4            CHAR = ROR(CHAR,1);
00022  4            CALL TIME(BITSCELL);
00023  4            END;
00024  3        OUTPUT (TTO) = 1;
00025  3        CALL TIME (BITSCELL+BITCELL);
00026  3        /* AUTOMATIC RETURN IS GENERATED */
00027  3        END PRINTSCHAR;
00028  2
00029  2    PRINT$STRING: PROCEDURE(NAME,LENGTH);
00030  3        DECLARE NAME ADDRESS,
00031  3           (LENGTH,I,CHAR BASED NAME) BYTE;
00032  3            DO I = 0 TO LENGTH - 1;
00033  3            CALL PRINTSCHAR(CHAR(I));
00034  4            END;
00035  3        END PRINTSSTRING;
00036  2
00037  2    PRINT$NUMBER: PROCEDURE(NUMBER,BASE,CHARS,ZERO$SUPPRESS);
00038  3        DECLARE NUMBER ADDRESS, (BASE,CHARS,ZERO$SUPPRESS,I,J) BYTE;
00039  3        DECLARE TEMP (16) BYTE;
00040  3        IF CHARS > LAST(TEMP) THEN CHARS = LAST(TEMP);
00041  3            DO I = 1 TO CHARS;
00042  3            J = NUMBER MOD BASE + '0';
00043  4            IF J > '9' THEN J = J + 7;
00044  4            IF ZERO$SUPPRESS AND I <> 0 AND NUMBER = 0 THEN
00045  4                J = ' ';
00046  4            TEMP(LENGTH(TEMP)-I) = J;
00047  4            NUMBER = NUMBER / BASE;
00048  4            END;
00049  3        CALL PRINTSSTRING(.TEMP + LENGTH(TEMP) - CHARS, CHARS);
00050  3        END PRINTSNUMBER;
00051  2
00052  2    DECLARE I ADDRESS,
00053  2        CRLF LITERALLY 'CR,LF',
00054  2        HEADING DATA (CRLF,LF,LF,
00055  2        '                    TABLE OF SQUARE ROOTS', CRLF,LF,
00056  2        ' VALUE  ROOT VALUE   ROOT VALUE   ROOT VALUE   ROOT VALUE   ROOT',
00057  2        CRLF,LF);
```

Fig. 8-3.   A sample program in PL/M.

```
00058  2             /* SILENCE TTY AND PRINT COMPUTED VALUES */
00059  2             OUTPUT(TTO) = 1;
00060  2             DO I = 1 TO 1000;
00061  2             IF I MOD 5 = 1 THEN
00062  2                 DO; IF I MOD 250 = 1 THEN
00063  3                     CALL PRINTSTRING(.HEADING,LENGTH(HEADING));
00064  4                 END; ELSE
00065  4             CALL PRINTSTRING(.(CR,LF),2);
00066  3             CALL PRINTSNUMBER(I,10,6,TRUE /* TRUE SUPPRESSES LEADING ZEROES */);
00067  3             CALL PRINTSNUMBER(SQUARESROOT(I), 10,6, TRUE);
00068  3             END;
00069  3
00070  2
00071  2    DECLARE MONITORSUSES (10) BYTE;
00072  2    EOF
NO PROGRAM ERRORS
```

but rather with a programmed abstraction. Somewhere, deep in the bowels of every computer or microprocessor is an instruction set that the user never sees. At this *microinstruction* level the instructions are actually executed by wired logic hardware.

The instruction set that we normally call "machine language" (eg. Figs. 7-3, 7-5, and 7-6) is in reality a *virtual machine* created by microprograms inside the processor. Since machine language is awkward to work with, we create another level of abstraction called assembly language. When we write and check out programs in assembly language, we can essentially ignore the existence of machine language. In a sense we are working with an abstract machine that executes assembly language. With high level languages like PL/M the same thing is true: we write and debug our programs in the language as though we had a machine that executed that language.

With both assemblers and compilers there is an extra step added between writing a program and actually running it: the source code must be assembled or compiled into object code before it will run. This extra step becomes particularly annoying during program checkout because when an error is found it is often necessary to reassemble or recompile just to try the solution. This extra delay can be very time consuming if there are a large number of errors.

Another approach to high-level languages uses a program called an *interpreter* to execute high-level commands directly *at execution time*. The programmer can then simply enter programs or changes and immediately execute them without having to do an assembly or compilation first. This greatly speeds up checkout because it turns into an *interactive* process. BASIC and FORTH are both examples of interpretive languages.

Just as the machine language instruction set is implemented by programs written in microinstructions, the interpreter implements high-level language commands with programs executed in machine language. A higher level virtual machine is thus created which executes programs written in the high-level language directly.

LINE NUMBER - ADDRESS CORRESPONDENCE

```
 2=0800H    6=0803H    7=080AH    8=081DH    9=0838H   10=089DH
11=08A9H   12=08B1H   13=08B2H   16=08B5H   17=08BAH   18=08BCH
19=08C5H   20=08C8H   21=08D2H   22=08D7H   23=08D8H   24=08E8H
25=08EBH   26=08EFH   27=08F7H   28=08F8H   30=08FBH   31=0904H
32=0907H   33=090DH   34=0923H   36=092DH   37=092EH   38=0931H
39=0938H   40=093CH   41=093FH   42=0952H   43=096FH   44=0972H
45=0999H   46=099DH   47=09AFH   48=09CCH   49=09D1H   51=09F5H
52=09F6H
```

```
S00064 02553      115
   2DH 0AH 0AH 0AH 20H 20H 20H 20H 20H 20H 20H 20H 20H 20H 20H 20H 20H
   20H 20H 20H 20H 20H 20H 20H 20H 20H 20H 20H 54H 41H 42H 4CH 45H 20H 4FH
   46H 20H 534 51H 55H 41H 52H 45H 20H 52H 4FH 4FH 54H 53H 0DH 0AH 0AH 20H
   56H 41H 4CH 55H 45H 20H 52H 4FH 4FH 54H 20H 56H 41H 4CH 55H 45H 20H
   20H 52H 4FH 4FH 544 20H 56H 41H 4CH 55H 45H 20H 20H 52H 4FH 4FH 54H 20H
   56H 41H 4CH 55H 45H 20H 20H 52H 4FH 4FH 54H 20H 56H 41H 4CH 55H 45H 20H
   20H 52H 4FH 4FH 54H 0DH 0AH 0AH
   60=0A6CH   61=0A6FH   62=0A78H   63=0AA4H   64=0AC3H
   65=0AC6H   66=0ACFH
S00081 02773        2
   0DH 0AH
   67=0AD7H   68=0AF7H   69=0B09H   70=0B2BH
S00001 00BCCH  S00002 00BCDH  S00003 0ABCEH  S00004 00BCFH
S00005 00C00H  S00021 00BD0H  S00023 00BD2H  S00024 00BD4H
S00029 00BD6H  S00031 00BD7H  S00039 00BD8H  S00040 00BDAH
S00042 00BDBH  S00047 00BDCH  S00048 00BDFH  S00049 00BE0H
S00050 00BE1H  S00052 00BE2H  S00053 00BE3H  S00054 00BE4H
S00063 00BF4H  S00078 00BF6H  S00079 00BCAH  S00080 00BC8H
0000H HLT HLT HLT HLT HLT HLT HLT HLT HLT HLT HLT HLT HLT HLT HLT HLT
```

(a)

GENERATED OBJECT CODE

```
0800H JMP,82H,08H LHI,0BH LLI,D0H LMB INL LMC DCL LBM INL LCM INL LMB
0810H INL LMC LLI,D0H LAM INL LCM ADI,01H LBA LAC ACI,00H ORA RAR LCA
0820H LAB RAR LLI,D4H LMA INL LMC LHI,0BH LLI,D2H LAM INL LCM INL SUM

0830H INL LBA LAC SBM ORB JTZ,A9H,08H DCL LBM INL LCM LLI,D2H LMB INL
0840H LMC DCL LBM INL LCM LLI,C8H LMB INL LMC LLI,D0H LBM INL LCM LLI
0850H,CAH LMB INL LMC JMP,8AH,08H LEM DCL LDM LMI,11H LBI,00H LCB LAD
0860H RAL LDA LAE RAL LEM DCE LME LEA RTZ LAB RAL LBA LAC RAL LCA DCL
0870H DCL LAB SUM LBA INL LAC SBM LCA JFC,83H,08H DCL LAB ADM LBA INL
0880H LAC ACM LCA INL SBA SBI,80H JMP,5FH,08H CAL,57H,08H LAD LLI,D2H
0890H ADM INL LDA LAE ACM LEA LAD ADI,01H LDA LAE ACI,00H ORA RAR LEA
08A0H LAD RAR INL LMA INL LME JMP,27H,08H LHI,09H LLI,D2H LAM INL LCM
08B0H RET RET JMP,F8H,08H LHI,0BH LLI,D6H LMB XRA 010 LBI,5BH DCB JTZ
08C0H,C5H,08H JMP,BEH,08H INL LMI,00H LAI,07H LHI,0BH LLI,D7H SUM JTC
08D0H,E8H,08H DCL LAM 010 LAM RRC LMA LBI,5BH DCB JTZ,E1H,08H JMP,DAH
08E0H,08H INL LBM INB LMB JMP,C8H,08H LAI,01H 010 LAI,5BH ADI,5BH LBA
08F0H DCB JTZ,F7H,08H JMP,F0H,08H RET JMP,2EH,09H LHI,0BH LLI,D8H LMB
0900H INL LMC INL LMD INL LMI,00H LLI,DAH LBM DCB LAB INL SUM
0910H JTC,2DH,09H LAM LLI,D8H ADM INL,LBA LAI,00H ACM LLB LHA LAM LBA
0920H CAL,85H,08H LHI,0BH LLI,DBH LBM INB LMB JMP,07H,09H RET JMP,F6H
0930H,09H LHI,0BH LLI,E0H LMB INL LMD LLI,0FH DCL SUM JFC,41H,09H LMI
0940H,0FH LHI,0BH LLI,E2H LMI,01H LHI,0BH LLI,E0H LAM LLI,E2H SUM JTC
0950H,D9H,09H LLI,DFH LBM LLI,C8H LMB INL LMI,00H LLI,DCH LBM INL LCM
0960H LLI,CAH LMB INL LMC CAL,57H,08H LAB ADI,30H LBA LAC ACI,00H LLI
0970H,E3H LMB LAI,39H SUM JFC,7CH,09H LAM ADI,07H LMA LHI,0BH LLI,E2H
0980H LAM SUI,00H ADI,FFH SBA DCL NDM LLI,DCH LBA LAM INL LDM SUI,00H
0990H LCA LAD SBI,00H ORC SUI,01H SBA NDB RRC JFC,A1H,09H LLI,E3H LMI
09A0H,20H LAI,10H LHI,0BH LLI,E2H SUM LLI,E4H ADL LBA LAH ACI,00H DCL
```

Fig. 8-4. (a) Line number-address correspondence. (b) Generated object code.

```
09B0H LDM LLB LHA LMD LHI,08H LLI,DFH LBM LLI,C8H LMB INL LMI,00H LLI
09C0H,DCH LBM INL LCM LLI,CAH LMB INL LMC CAL,57H,08H LLI,DCH LMD INL
09D0H LME LLI,E2H LBM INB LMB JMP,47H,09H LHI,08H LLI,E4H LCH LAL ADI
09E0H,10H LBA LAC ACI,00H LCA LAB LLI,E0H SUM LBA LAC SBI,00H LLI,E0H
09F0H LDM LCA CAL,FBH,08H RET JMP,6CH,0AH R01 RRC RRC RRC INE INE INE
0A00H INE INE INE INE INE INE INE INE INE INE INE INE INE INE INE
0A12H INE INE INE INE INE JMP,41H,42H JMP,45H,2CH I07 CAL,20H,53H 0PB
0A20H 010 I00 CFS,45H,20H CFS,4FH,4FH JMP,53H,0DH RRC RPC INE CAL,41H
0A30H,4CH 010 I02 INE INF CFS,4FH,4FH JMP,20H,56H I00 JMP,55H,45H LBA
0A40H INE CFS,4FH,4FH JMP,20H,56H I00 JMP,55H,45H INE INE CFS,4FH,4FH
0A50H JMP,20H,56H I00 JMP,55H,45H INE INE CFS,4FH,4FH JMP,20H,56H I00
0A60H JMP,55H,45H INE INE CFS,4FH,4FH JMP,0DH,0AH RRC LAI,01H 010 LHI
0A70H,08H LLI,F4H LMI,01H INL LMI,00H LAI,E0H LCI,03H LHI,08H LLI,F4H
0A80H SUM INL LBA LAC SBM JTC,28H,0BH LLI,C6H LMI,05H INL LMI,00H LLI
0A90H,F4H LBM INL LCM LLI,CAH LMB INL LMC CAL,57H,08H LAB SUI,01H LBA
0AA0H LAC SBI,00H ORB JFZ,D2H,0AH LLI,C8H LMI,FAH INL LMI,00H LLI,F4H
0AB0H LBM INL LCM LLI,CAH LMB INL LMC CAL,57H,08H LAB SUI,01H LBA LAC
0AC0H SBI,00H ORB JFZ,CFH,0AH LBI,F9H LCI,09H LDI,73H CAL,FBH,08H JMP
0AD0H,E0H,0AH JMP,D7H,0AH R01 RRC LBI,D5H LCI,0AH LDI,02H CAL,FBH,08H
0AE0H LHI,0BH LLI,F4H LBM INL LCM LLI,DCH LMB INL LMC LLI,DFH LMI,0AH
0AF0H LBI,06H LDI,01H CAL,31H,09H LHI,08H LLI,F4H LBM INL LCM CAL,03H
0B00H,08H LHI,08H LLI,DCH LMA INL LMI,00H LLI,DFH LMI,0AH LBI,06H LDI
0B10H,01H CAL,31H,09H LHI,08H LLI,F4H LAM INL LCM ADI,01H LBA LAC ACI
0B20H,00H DCL LMB INL LMA JMP,78H,0AH HLT
```

(b)

Since there is no need to reassemble or recompile every time a change is made, program development is much faster with interpretive languages. Their disadvantage is that execution speed is generally much slower since each command must be interpreted *at execution time.* By translating the program in advance, assemblers and compilers avoid the time penalty at execution time.

Another useful programming abstraction is an *operating system.* Operating systems simplify programming by creating a virtual machine which is easier to use than the machine they operate on. For example an operating system can allow the programmer to separately program many different tasks as though each had a processor all to itself. The operating system schedules the running of these tasks in such a way that they each get to use the machine some of the time. Common resources such as input/output equipment are automically scheduled so that each task can use them without worrying about the others. Tasks can also send messages to each other.

At any given moment each task is either *running, ready,* or *waiting.* Each task is given a priority that determines its relative importance in the system. The operating system always executes the highest priority task that is ready. *The operating system thus creates a number of virtual machines from a single real machine.* It also greatly simplifies programming by taking care of all details of routine tasks such as communicating with input/output devices or responding to interrupts.

Again the price that is paid is slower execution time. Each level of programming abstraction is somewhat like a "middleman" who makes things more convenient but takes his profit in execution speed.

TABLE 8-4   LEVELS OF PROGRAM ABSTRACTION

	Program Level	Translated by	Stored in
	System inputs	Application program	Main memory
	High-level programming language	Compiler, interpreter, operating system	Main memory
	Assembly language	Assembler	Main memory
Higher	Machine language	Microcode	Fast inner memory
Level	Microinstructions	Wired logic	Gate, register interconnections

In a sense a microcomputer system that is programmed for a particular application is a virtual machine which executes extremely high level commands. Pressing a key on a microcomputer based CRT terminal, for example, can be thought of as a high level command. As Table 8-4 shows, the application program may be isolated by as many as four levels of abstraction from the wired logic deep in the processor. Just as the user of the product is often unaware of its internal workings, programmers can generally ignore levels below the one they are working on. At each level we effectively work on a virtual machine and ignore all lower levels.

8.4   MICROPROGRAMMING

Since we have shown that programmed logic can be substituted directly for wired logic at a great cost savings, it is not surprising that *the logic of virtually all modern computers is internally mechanized by program.* Of course, we must draw the line somewhere since the more we replace wired logic with programs, the slower the result becomes. Since inexpensive logic circuits can operate at a clock rate about 10 times faster than reasonably priced main memory components, we can make a good tradeoff, with negligible sacrifice in speed, by building just enough wired logic to permit execution of an instruction in about five to 10 steps.

With a microprogrammed approach, conventional logic design techniques are used to design a very fast, logically simple *"inner computer"* that addresses registers only. The *microprogram* for the "inner computer" is in a small (100 to 1000 word), high-speed, ROM *control memory.* The microprogram reads *macroinstructions* out of the large main memory and carries out all the functions required by the (macro)instruction set of the *virtual machine,* or "outer computer." The virtual machine acts in all respects like a computer with a particular (macro) instruction set and addressing modes as determined by the control memory. This instruction set bears no resemblance to the *microinstruction set* that is mechanized by conventional logic. In a sense, we can think of the microinstructions as components, "connected together" by the microprogram, to mechanize the (macro)instruction set.

For a real example let us look at the microinstruction set used by National Semiconductor to implement the 8900 instruction set (Fig. 7-2) in their bit slice version called the GPC/P. A virtual machine with the 8900 instruction set is produced by a particular microprogram on the GPC/P "inner computer." Internally, the GPC/P microinstruction word length is *24 bits* and bears no resemblance to the *macro*instruction format of Figure 7-3. As Figure 8-5 shows, the 24-bit microinstruction word is divided into several encoded fields. Each microprogram step can combine arithmetically or logically the contents of two *source registers* (*A* and *B*) and put the result in a *destination register*. The operation to be performed is determined by a four-bit *Arith Operation* field. Two of the bits of this field select an AND, OR, EXclusive OR, or ADD operation. The other two determine whether source register *A* will be complemented and if there will be a carry input. Another field controls shifting, and the remaining bits control the word format, whether to set the condition code, and so on. Other formats are used for microprogram jumps, shifts, and input/output. *Microcontrollers* use instruction sets like this to do programmed system jobs directly, since this kind of instruction format is actually more efficient for some applications.

Though the 8900 *virtual* computer has only four working registers, the inner computer that executes the microinstructions has eight (see Fig. 8-6). Four of these micro-level registers function as the four registers accessed by the *macro*instructions. The other four are used as program counter, memory address, memory data, and flag register in the virtual machine. Each instruction cycle of the virtual computer starts with a microinstruction that loads the next instruction from the main memory. Next the microprogram uses the operation code bits of the macroinstruction word to *point* to the starting address of the *micro*routine that executes that instruction. During the execution of a macroinstruction the microprogram may branch several times. For

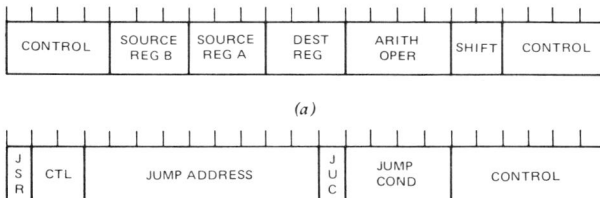

Fig. 8–5. GPC/P microinstruction format: (a) arithmetic instructions; (b) Microprogram jump instructions.

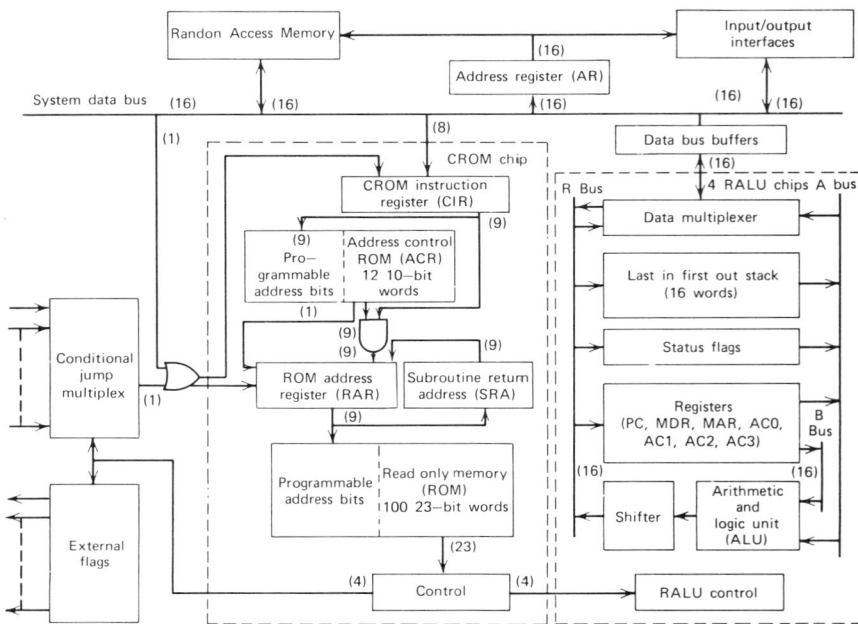

Fig. 8-6.   GPC/P functional block diagram.

example, the same operand address computation routine can be used by several different macroinstructions with the same addressing mode.

We can see how microprogramming significantly reduces the conventional logic requirements if we look at the execution steps for an ADD macroinstruction with relative addressing:

1.   Read macroinstruction from main memory
2.   ADD 1 to the program counter (next macroinstruction address)
3.   Compute the operand address by ADDing the displacement field of the macroinstruction to the program counter or index register
4.   Read the operand from memory and ADD it to the designated register, putting the result in the designated register, set condition codes

Notice that we had to ADD in three of the four steps. Since we do each addition as a separate step, *we can use a single adder for three separate things.* In fact, we are not only using the same adder for several things, we are also using the same register selection logic and control logic. Of course, we can divide the instruction execution into even more steps allowing an even simpler "inner computer" structure, but the result would be a slower macroinstruction cycle time.

Actually, since microprogram branches require a separate microinstruction cycle in the GPC/P, more than four microinstruction steps are required to execute an ADD instruction. Many microprogrammed computers include jump conditions and jump address in the *same* microinstruction format with arithmetic and logical operations. On a large machine, like the IBM 360/50, the microinstruction word is 90 bits long with 29 different fields. Generally, modern minicomputers all seem to use microinstruction words of about 24 bits.

Some minicomputer manufacturers encourage the customer to program his own special instructions, and even offer microassembler programs and have *writable control stores* making it possible to simply *load the microprogram* just as one loads a program. This makes it possible to add special instructions to the normal instruction set for your particular application. Actually, microprogramming is similar in principle to writing *macros* or *procedures* using assembler or compiler programs except that the microprograms instructions execute on the "inner computer" hardware, which operates at a very high speed and is normally limited to register addressing.

The National Semiconductor GPC/P microprocessor has a particularly efficient control store because it uses two programmable logic arrays instead of a ROM to store the microinstructions (see Fig. 8-7). The first control store makes it possible to scatter the operation code bits around in the macroinstruction word as convenient for the various formats illustrated in Figure 7-3. Normally this would complicate the process of jumping to the first microinstruction required to execute a particular macro operation code.

The 8900 instructions, for example, have the operation code determined by nine of the instruction register bits. This would directly point to $2^9 = 512$ *different starting addresses*. To avoid this problem, a PLA with 12 nine-bit words and nine inputs is used to generate a special *mask* for each of 12 instruction formats. This mask is used to gate out "don't care" bits in the macroinstruction word. The format shown in Figure 7-3a, for example, needs only bits 12 to 15 to link to the proper subroutine (if, for example, bits 10 and 11 were included, a different microprogram starting address would result with each of the four possible registers). The format of Figure 7-3b, however, requires bits 10 to 15 to be used as a pointer. Each word of the address control PLA thus represents a different mask for a different macroinstruction format.

Although this masking according to format greatly reduces the number of microprogram starting addresses pointed to by the nine macro operation code bits, it still does not alter the fact that they are scattered over $2^9 = 512$ different locations. By using a PLA for the control memory itself, we can mechanize only those words that contain useful microprogram words. The results of this method are quite impressive in that *only 100 23-bit microprogram words* are needed for the control memory to mechanize the complete 8900 instruction set as shown in Figure 7-3. An instruction set of this complexity would normally require at least 256 words of ROM to do the same job.

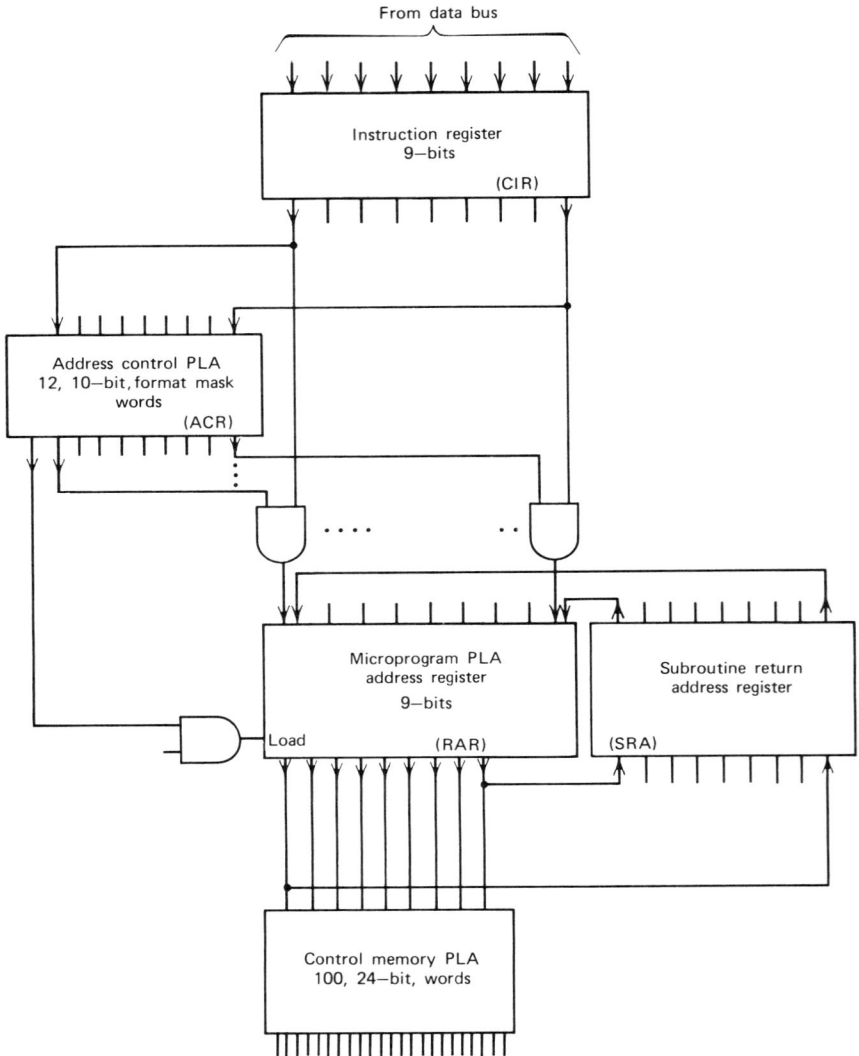

Fig. 8-7.   GPC/P PLA microprogram addressing.

When ROMs are used to store the microprogram, instead of masking the macroinstruction bits, a *starting address generating ROM or PLA* is used to generate the actual starting address for the microprogram from the operation code bit inputs.

Ideally, we can think of each microprogram word as a micro operation in that the pattern of output bits produced causes a certain combination of operations (e.g., increment program counter). Macroinstructions are then mechanized by simply going through a particular sequence of these micro operations.

The microprogram approach is also sometimes used to implement extremely high-speed general purpose logic. *Bit slice* LSI chips (e.g., the 2901) can be used to implement very fast special purpose processors with any desired word size. For example, a 36-bit processor can be made by putting together nine 4-bit wide slices with a control ROM. Microprocessors with instruction sets at the micro level are generally called *microcontrollers*. Although they are less convenient to program, they can achieve much higher speed in executing simple logical tasks because they do not pay the penalty of higher levels of program abstraction.

8.5 INTERRUPT PROGRAMMING

The job of most digital systems is to produce outputs in response to the inputs received. In very simple systems the program can often loop on an input test while waiting for inputs to happen (e.g., for a button to be pressed). Since no useful work can be done while in such a loop, this technique is wasteful. *Interrupts* allow input or output devices to interrupt a program that is doing useful work when the device is ready. The processor can thus *service* I/O devices only when they are ready and spend the rest of the time doing useful work.

An interrupt causes the current program counter contents to be pushed into the stack, then forces the program counter to a specified location. Before this happens, however, the instruction being executed when the interrupt occurs is allowed to complete its cycle normally. The location to which the program counter is forced is called the *interrupt location* and normally contains an unconditional branch to an *interrupt service routine.* This routine performs the desired task (usually an input and/or output and related processing), then returns program control to the main program by executing a RETURN instruction (see Fig. 8-8a). The RETURN instruction pulls the previously stored program counter state out of the stack and back into the program counter. The last instruction of all interrupt service routines is, thus, always a RETURN instruction.

Use of interrupts opens up a *whole world of problems.* The interrupt service routine must be written in such a way that *nothing* is changed when we return to the main program. This normally means that the interrupt service routine must start out by *saving the contents of all registers* it will use, including the states of the flag bits (carry, zero, sign, parity, etc.). At the end of the interrupt routine all registers and flags are restored to their previous state before returning to the original program. The push-down stack is ideal for this job, as the

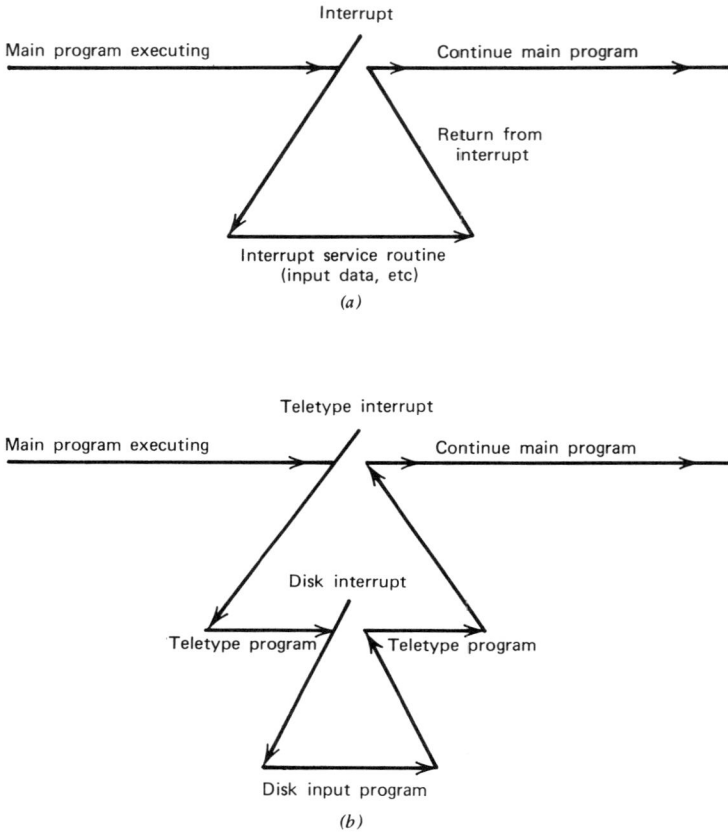

Fig. 8-8. (a) Single program interrupt. (b) Two-level interrupt (interrupting the interrupt routine).

registers and flags can simply be pushed in a fixed order, then popped from the stack, in reverse order at the end.

This saving and restoring of registers can be eliminated if all registers used in the interrupt routine are *reserved* for that purpose and *never* used in other parts of the program. The saving and restoring of flags are still necessary, however; if a program is interrupted between an arithmetic operation and a conditional branch based on condition codes generated by that arithmetic operation, the program will *make a wrong branch* if the flag bits are altered by an intervening interrupt routine. Notice that subroutines can cause the same problems, but since we intentionally call a subroutine, it is easier to avoid problems. An interrupt, on the other hand, may occur from *any* point in the program.

Programming must be done very carefully when interrupts are used, because mistakes may cause trouble only when the interrupt occurs from a certain location and with certain combinations of data. *A program "bug" may thus appear only once a month or even once a year* when that combination occurs. If, for example, we forget to save and restore the flag bits, there will be no problem unless an interrupt occurs *between* an arithmetic operation and a conditional branch based on the results of that operation. There will still be no problem if the last arithmetic operation in the interrupt routine produces the same (carry, zero, etc.) result as the main program did. The problem will thus show up only very occasionally, when the interrupt occurs during certain key parts of the program *and* the condition code results are different. To make matters worse, the symptoms may be different each time the problem occurs. The only solution is to be very careful when writing interrupt programs.

There are many other possible programming mistakes with interrupts. If the interrupt routine changes a memory location that is also used by the main program, there will be problems when the main program resumes. One way that this can happen is if a shared subroutine being used by program *A* is interrupted and subsequently used by program *B*. If the subroutine uses memory locations as counters or temporary storage, the original values will be lost when the interrupt is finished. If subroutines are to be shared by different programs, they must be *reentrant,* which means that there must be *no shared RAM storage.* Registers can be used for temporary storage if they are saved and restored but if memory locations are used there must be separate areas for each user.

Unless a microcomputer has more than a few working registers (e.g., the TI 9900 has 16), it is difficult to write reentrant code. Some instruction sets (e.g., 8086) provide *base registers* which can be used to offset temporary storage locations to different areas for each user or calling program. In machines such as the 8080 the push-down stack may be used for additional temporary without sacrificing reentrancy. Time delay loops, such as the ones used in the serial receiver program of Figure 8-1, cannot be used when interrupts are possible, since the interrupt routine execution time will add to the programmed delay time. When interrupts are used, we can avoid wasting time in delay loops by, for example, providing an interrupt at the end of an external timing circuit period that indicates the end of a bit time.

A **real-time** clock interrupt is often used to control all time-dependent operations in a system. For example, an interrupt can be generated by the 60-Hz line voltage transitions every 16.7 msec. The interrupt routine can keep a real-time clock *count* in a memory location or register than can be used for timing various operations. If the present state of a clock count is stored by a subroutine, the *elapsed time* can be determined at any time later by simply subtracting the *initial* time count from the *current* time count.

Sometimes certain operations must be done periodically, on each clock interrupt or perhaps every 10 clock interrupts. The real-time clock interrupt routine can keep track of these counts and execute the jobs at the proper time. Sometimes absolute time and even date are needed by the system. This can be set initially from the *keyboard* and maintained automatically by the clock interrupt routine. The number of cycles per day on the AC line is carefully controlled at the power station to keep electric clocks correct, so the long-term accuracy of this type of clock is excellent.

Another potential problem from interrupts comes with certain operations that must proceed at high speed once started. If we are writing a record of data on a disk or magnetic tape, for example, an interrupt during output would be disastrous. To prevent this type of problem, an *interrupt enable* flip-flop (which is usually a flag bit) is used. Before doing an operation that cannot be interrupted, the interrupts are disabled by a special instruction. When the operation is complete, interrupts are reenabled.

More complex systems use *priority interrupts* to do the various required jobs on a priority basis. The most time sensitive jobs are given highest priority so they can actually interrupt an interrupt routine of lower priority, as shown in Figure 8-8b. Lower priority interrupts are disabled during execution of interrupt routines of high priority, then reenabled at the end of the routine. Priority interrupts can be used to mechanize input and output *buffers* in the main memory. The main program simply gets its input data from a buffer area in memory. This buffer can be automatically written into by an interrupt routine each time an input interrupt occurs. The two processes—writing into the buffer and taking data from the buffer for processing—can thus proceed independently. A register or memory location can be used to keep track of the next address to write in or read out of the buffer. Output interrupt routines can likewise read out of memory buffers independently of the main program whenever an output device is ready.

The method of handling interrupts varies greatly between microcomputers. Generally an interrupt input causes execution of a microprogrammed instruction which reads an address pointer from a particular ROM location and transfers program control to that location. If the processor has *vectored interrupts,* there is a different address pointer or vector for each interrupt level so control is immediately transferred to the proper interrupt routine. If there are multiple interrupts on a machine without vectored interrupts, the single interrupt routine must test the various input devices to find out which interrupt routine to branch to.

Most microcomputers do more than just branch to the indicated starting location. Generally they disable interrupts and execute the same instruction which is normally used to *call* subroutines. This instruction pushes the program counter onto the stack and often also pushes the program status word and other working registers.

The time required to respond to an interrupt, including saving the working registers, is called *interrupt latency time.* In many applications it can be a significant part of the execution time of the most time-critical parts of the program. Often a small amount of external hardware buffering can free up a tremendous amount of processor time by reducing the frequency of interrupts.

EXERCISES

1. (a) Make a flow chart for a "loop within a loop" delay routine that will provide 3 sec of delay on the Intel 8008.
    (b) Write a program for the loop in assembly language.
    (c) Make an assembly listing of the program, with hexadecimal addresses and object code, as in Table 8-3.
    (d) Write a program in 8900 language. Assume instruction execution times as follows: AISZ without skip = 5.6 μsec, JMP = 4.2 μsec.
    (e) Write a PL/M program for a 3-sec delay.

2. What is the allowable range of the assembler operand field for the INP and OUT instructions:
    (a) For the Intel 8008?
    (b) For the Intel 8080?

3. Write *single* assembler statements that would generate the following 8008 object code sequences (see Table 8-1):
    (a) 00000010
    (b) 11000111
    (c) 00001100
         0001001
    (d) 01001100
         00010110
         00000000

4. (a) Write an MCS-8 assembly language statement that will generate a table of ASCII character codes for printing out a hexadecimal digit. The first entry on this table should be *labeled* HEXTAB.
    (b) Make a flow chart for a program that will output the proper ASCII characters from the table to port 3. The first character should represent bits 7, 6, 5, and 4 of the A register, and the second should represent bits 3, 2, 1, and 0.

(c)   Write the program in MCS-8 assembly language.

(d)   Write the same program in PL/M.

5.   Make an "assembly listing" similar to Table 8-3 for the following program:

;AVERAGING PROGRAM: OUTPUT 0 = AVG OF INP 1&2

START	ORG	1000H
	INP	1000H
	MOV	1
	INP	B,A
	ADD	2
	RRC	B
	OUT	(A + B)/2 =AVERAGE
	END	0

6.   Rewrite the program of Table 8-3 for an eight-bit character with 10-msec-bit duration. Make the program start at location 2000, and call a delay subroutine at location 100 instead of using two separate delay loops in the program. Write the delay subroutine so it will be assembled at the same time.

7.   (a)   Write an MCS-8 assembler program for the program flow charted on Figure 7-11. Assume the data (ABCDEFGH) are on input port 3; if the function is true, we want to branch to a location defined elsewhere as CONTRL. The program should loop until the function is true.

(b)   Write an 8900 program for the same thing.

(c)   Write a PL/M program for the same thing.

8.   (a)   Write an MCS-8 assembler program that will output to port 2 the successive ASCII characters of the message DATA INCORRECT. The program should start at location 1100 and should *Return* at the end.

(b)   Write two PL/M statements to do the same thing.

9.   Define a MACRO, for the MCS-8 assembler, called DELAY. This MACRO should produce a time delay of 1 msec times the value in the operand field, up to a limit of 256 msec. For example, DELAY 22 should produce a delay of 22 msec.

10.   Define a "RARN" MACRO for the MCS-8 assembler that will rotate (shift) the A register one to seven places as defined by the operand field.

11.   Write an MCS-8 assembly program for ROM testing that will compare the contents of a perfect ROM plugged in for memory locations 1024 through 1380 to a ROM under test in the socket for locations 1381 through 1637. If a failure occurs, the program can just halt with the incorrect data in A, the correct data in B, and the faulty address in L.

12.   Write an MCS-8 assembly program that will continuously watch the binary state of 64 signals for any change. The 64 signals are connected eight to each of the

eight input ports. When a change occurs, output a 1 on the corresponding bit of
the corresponding output port. Use memory locations 256 through 264 to hold
previous values.

13. Write an MCS-8 assembly program that will go after the serial receiver program
of Table 8-3 and store the received characters in memory locations 1024 through
1104 (reverse order), for an 80-character transmission. Instead of using the HLT
and interrupt of Table 8-3 to wait for a start bit, write an input loop with a
timeout. If no start bit is detected for 100 msec, reset the character count. (This
will keep the character count "in sync" with 80-character bursts of data.) When
the eightieth character is received, branch to location 2000.

14. (a) Write an assembly language program for the Intel 8080 for adding the
contents of memory locations 100 through 127 consecutively to the contents
of locations 128 through 155, and putting the result consecutively in locations 156 through 182. (Use the exchange and extended instructions to maintain three index counters.)

    (b) Write the same program for an 8900.

    (c) Write the program in PL/M.

15. Rewrite the program of exercise 5 in PL/M.

16. Rewrite the serial receiver program of Table 8-3 for an 8900. Compare the
number of instructions required and the number of bytes of storage (see exercise 1
for delays).

17. (a) Write a memory diagnostic program, in MCS-8 assembly language, that
cycles continually and Halts if an error is found. A single 1 should be written
into and read out of each bit position of each memory location from 4096
through 3120. The bit written should be shifted over for each consecutive
memory location in a cycle of nine to prevent binary addressing errors from
going undetected.

    (b) Write the same program using 8080 instructions where convenient.

    (c) Write the same program in PL/M.

    (d) Write the program for an 8900 (change pattern cycle length to 17).

18. (a) Divide the ADD macroinstruction mentioned in Section 8.7 into eight steps
instead of four.

    (b) What effect would this have on "inner computer" logical complexity require-
ments?

    (c) What effect would it have on control memory requirements?

    (d) What effect would it have on the execution time of the ADD instruction,
assuming the same microinstruction cycle time?

19. (a) Write an interrupt routine for the 8080 that will begin at location 8 and will
push all registers and flags into the stack and *Call* a subroutine at location
E1F. When the subroutine returns, the program should *Pull* all registers and
flags from the stack and *Return* to the main program.

    (b) Write the same program for the 8900.

20. (a) Write an interrupt routine for the 8080 similar to that described in exercise
21, but call one of eight possible subroutines as determined by the highest

order bit (highest priority) that is true on input port O (from eight devices). The starting addresses for these eight subroutines are stored in locations 8 through 15.

(b)   Write the same routine for the 8900.

(c)   Compare the number of instructions required in *a* and *b*.

21.   What will happen if an interrupt occurs while the program of Table 8-3 is running, and the interrupt routine sets the *not zero* flag state:

(a)   If the interrupt occurs while the instruction in location 5 is executing?

(b)   If the interrupt occurs while the instruction in location 15 (Hex) is executing, and the third bit has just been received?

(c)   If the interrupt occurs while the instruction in location 15 (Hex) is executing, and the fifth bit has just been received?

22.   What will happen if the program of Table 8-3 is interrupted at location 15 (Hex) and the *C* register contents is changed (by an incorrectly written interrupt routine) to FF.

BIBLIOGRAPHY

Programming Techniques

Adams and Haden, *Computers*, Wiley, New York, 1973 (Chapters 1 through 6 cover general programming principles.)

Booth, Taylor, *Digital Networks and Computer Systems*, Wiley, New York, 1971, (Chapter XI, "Assembler Languages," p. 391, Programming Languages and Compilers).

Spencer, D. D., *Computers and Programming Guide for Engineers*, Howard W. Sams, Indianapolis, 1973.

Dahl, O. J., et al, *Structured Programming*, Academic Press, New York, 1972 (On programming with Compilers without using GO TO statements).

Dalton, William, "Design Microcomputer Software Like Other Systems—Systematically," *Electronics*, January 19, 1978, pp. 97–101.

Yourdon, Edward, *Techniques of Program Structure and Design*, Prentice-Hall, Englewood Cliffs, N.J., 1975.

Compilers

Burns and Savitt, "Microprogramming Stack Architecture Ease Minicomputer Programmers Burden," *Electronics*, February 15, 1973, pp. 95–101. (Describes Microdata 32/S, which is designed for MPL compiler.)

Heaps, H. S., *An Introduction to Computer Languages*, Prentice-Hall, Englewood Cliffs, N.J., 1973.

McCameron, Fritz, *Fortran Logic and Programming*, Irwin Inc., Homewood, Ill., 1968.

McCracken, Daniel O., *A Guide to PL/M Programming for Microcomputer Applications*, Addison-Wesley, Reading, Mass., 1978.

Microprogramming

Gorman and Kaufman, "Low Cost Minicomputer Opens Up Many New System Opportunities," *Electronics,* June 7, 1973. (Describes the Computer Automation Alpha LSI 16, which uses a PLA control store.)

House, David L., "Micro Level Architecture in Minicomputer Design," *Computer Design,* October 1973, pp. 75–80. (Microprogramming advantages and Microdata 3200 minicomputer design.)

Husson, *Microprogramming Principles and Practices,* Prentice Hall, Englewood Cliffs, N.J., 1970.

IEEE, *Transactions on Computers,* "Special Issue on Microprogramming," C-20, No. 7, July 1971.

National Semiconductor, *GPC/P Microcoding Manual,* Pub. 4200007, National Semiconductor, Santa Clara, Calif. 1974.

Serkin, Baker, W. Davidow, and Allison, "Microprogramming Made Easy With a 4096 Bit Bipolar ROM," *Electronics,* March 15, 1971.

Signetics Memory Systems, "Design of Microprogrammable Systems," Signetics Memory Systems Application Note #SMS0052AN, Sunnyvale, Calif. (An excellent general survey.)

Tucker, S. G., "Microprogram Control for System/360," *IBM Systems Journal,* Vol. 6, No. 4, 1967, pp. 222.

Clymer, Jim, "Use 4-Bit Slices to Design Powerful Microprogrammed Processors. The 2900 Family of Circuits Lets You Build Machines with Cycle Times of 110ns.," *Electronic Design,* 10, May 10, 1977, pp. 62–71.

Lau, Stephen, "Design High-Performance Processors With Bipolar Bit Slices," *Electronic Design,* 7, March 29, 1977, pp. 86–95.

Mick, John R. and Brick, Jim, *Microprogramming Handbook,* Advanced Micro Devices, Sunnyvale, Calif., 1976. (Techniques for 2900 series 4-bit slices.)

Nemec, John et al., "A Primer on Bit-Slice Processors . . . ," *Electronic Design,* 3, February 1, 1977, pp. 52–60.

Salisbury, Alan B., *Microprogrammable Computer Architecture,* Elsevier, New York, 1976.

Interpreters and Operating Systems

Stiefel, Malcolm L., "Venturing Forth," *Mini-Micro Systems,* August 1976, pp. 46–47.

Forth Inc., *Microforth Primer,* Forth Inc., Manhattan Beach, Calif., 1977.

Kahn, Kelvin C., "A Small Scale Operating System Foundation for Microprocessor Applications," *Proceedings of the IEEE,* Vol. 66, No. 2, February 1978. (Describes Intel's RMX-80 operating system.)

Claggett, Edward, "Interpreters vs. Compilers for On-Line Microcomputers," *Control Engineering,* August 1977, pp. 24–27.

Burzio, Gus, "Operating Systems Enhance μCs . . .", *Electronic Design,* 13, June 21, 1978, pp. 94–99.

Intel, *RMX/80 Real-Time Multitasking Executive,* Pub. #9800577A. Intel Corp., Santa Clara, Calif., 1977.

Townzen, David A., "A Task-Scheduling Executive Program for Microcomputer Systems," *Computer Design,* June 1977, pp. 194–202.

224 PROGRAMMED LOGIC II COMPUTER-AIDED PROGRAMMING

Interrupts

Klingman, Edwin E., *Microprocessor Systems Design,* Prentice-Hall, Englewood Cliffs, N.J., 1977, Chapter 12.

Rattner, Justin, "The Open Channel," *Computer,* March 1975, pp. 19–20. (Discussion on elimination of interrupts.)

CHAPTER NINE

PROGRAMMED LOGIC III
Development Systems

9.1 DEVELOPMENT SYSTEMS

The orderly nature of programmed logic makes the whole programming and checkout process an excellent candidate for computerization. The basic tool for automating this entire process is called a development system (see Fig. 9-1). In addition to assembling or compiling programs the development system automates the process of storing and changing both sources and object programs. It also simulates the ROM, which will eventually hold the program, and assists in debugging both the program and hardware. When program checkout is completed, the development system writes and tests a PROM which can be shipped with the final system.

The very cheapest development system consists of nothing more than a microcomputer system on a single PC board with a serial interface for connecting a printing terminal and two audio cassette recorders. Professional quality systems improve assembly speed considerably by using floppy disk drives for storage. In any case, two removable storage devices are needed so that data can be copied from one device to the other. Without this ability to make "backup copies" one mistake or failure could destroy the results of months of work.

Another big time saver in professional development systems is the use of a separate high-speed printer and CRT display terminal. The CRT terminal makes it possible for the programmer to interactively enter and edit programs. The CRT screen can be quickly "proofread" for errors and corrected on the

Fig. 9-1. Microprocessor development system. Intelligent terminal, dual floppy disk drive, and printer provide an ideal environment for development and checkout of programs.

spot. The *Editor* program makes it possible to quickly change or insert characters or even whole lines of program code. To make a change, the programmer simply selects *replace, insert character, insert line, delete character,* or *delete line* mode and positions a *cursor* where the change is desired. The new data or corrections are then simply entered on the keyboard and the resultant change is immediately displayed for rechecking.

When power is first applied to a development system, it begins by executing a program called an *executive, monitor,* or *operating system* (depending on the manufacturer).* This program responds to commands entered on the keyboard by the programmer. Each of these commands causes execution of a different program such as the *text editor* or *assembler.* When the requested program is finished, control is returned to the *executive* program.

The first step in creating a new program is thus to request the text editor program by keying in a command such as EDIT followed by the name of the new program. The source program is then keyed in under control of the text editor program. If the program is more than one page long, the editor splits it up into pages and stores the pages on the diskette. Special control keys allow *paging* forward or backward through the program by automatically writing the displayed page (with corrections) onto the diskette and reading the next page from the disk.

When the program is completed, the programmer pages through it again and corrects any obvious errors, then returns to the executive program. The next step is to request an assembly by typing a command such as ASMB followed by the source program name, a name for the object file, and a code to

---

* The executive program is normally read into RAM from the diskette by a program in ROM called a *bootstrap* program. This is done automatically whenever power is turned on.

indicate whether to print the results. The assembler program then begins its processing. Since the assembler uses the disk to store the symbolic address names, assembly of a large program may take minutes.

Since printing a listing of the program is time consuming, the first assembly is usually simply displayed on the CRT. If there are any assembly errors (and there usually are), the text editor is again called and used to correct the source program. When all errors are corrected the program is again assembled. Several repetitions of this cycle may be required to get a perfect assembly. At this point a printout can be made by keying in PRINT followed by the object program name.

Since the process of assembly and printout of a large program listing can be very time consuming, large programs should be split up into modules of manageable size. This is practical only when the assembler produces *relocatable* object code. A *linker* program can then combine the separate modules into a large system program in contiguous memory locations. This makes it possible to change the size of an individual program module without having to reassemble and print a listing of the entire system program.

9.2 PROGRAM DEBUGGING

Error-free assembly of a program by no means guarantees that it will work properly. The process of finding and removing program "bugs" is called debugging. The *debug* program in the development system essentially automates this task. It allows programs to be executed one step at a time and displays the register contents after each step of the program. It also allows memory locations and registers to be changed by commands from the keyboard.

Figure 9-2 shows an example of a screen during program debugging on an AMI 6800 development system. Program operation can be checked by stepping through the program and looking for correct results. Input data can be inserted from the keyboard by simply writing into registers at the proper time. For example, just after the program has read inputs into the *A* register, any desired input code can be put in the *A* register from the keyboard and the results can be *traced* as each step of the program is executed.

To save time, long sections of code can be executed quickly with an automatic stop at a particular *breakpoint* location. This causes program execution to stop when the breakpoint address appears on the address bus. After entering some input data, the resultant output can be found by setting the breakpoint at the location of the instruction that would output the result. The display would then show the result ready to be output from the *A* register and the 19 program steps that preceeded that step.

```
D.INTRPT   BRK=F2BC->BD  P=F48E->6D  0027 A=80 B=0C X=FDE5 S=FFEA C=C8   N
N 0000   00 E1 C8 D3 D2 60 D7 B7 C7 D5 85 F5 D6 16 D5 63      Da@SR¹W76U0vU1Uc
P F48E   6D 00 27 F9 4F 39 16 C1 41 25 11 C1 5B 25 0A C1      aD'y09iAAZ0A[Z=A
X FDE5   00 52 FD EF FE 41 FD EF FD EF 3F 98 8F FE FB A0      DR)o•A)o)o?8?•(
S FFEA   8E F2 BF EA 15 EA 3E E1 3D 97 81 FD BE F2 F0 EB      @r?.j•j)a=-4[))rpk

P=F191->39   RTS        ---   A=80 B=0C X=FD88 S=FFE9 C=C4    Z
P=F052->20   BRA    F039 ---  A=80 B=0C X=FD88 S=FFEB C=C4    Z
P=F039->30   TSX        ---   A=80 B=0C X=FD88 S=FFEB C=C4    Z
P=F03A->EE   LDX    00,X ---  A=80 B=0C X=FFEC S=FFEB C=C4    Z
P=F03D->31   INS        ---   A=80 B=0C X=FAF0 S=FFEB C=C8    N
P=F03D->31   INS        ---   A=80 B=0C X=FAF0 S=FFEC C=C8    N
P=F03E->33   PUL B      ---   A=80 B=0C X=FAF0 S=FFED C=C8    N
P=F03F->39   RTS        ---   A=80 B=0C X=FAF0 S=FFEE C=C8    N
P=EA58->2B   BMI    EA69 ---  A=80 B=0C X=FAF0 S=FFF0 C=C8    N
P=EA69->7A   DEC    FD68 ---  A=80 B=0C X=FAF0 S=FFF0 C=C8    N
P=EA6C->26   BNE    EA3C ---  A=80 B=0C X=FAF0 S=FFF0 C=C8
P=EA3C->8D   BSR    EA0E ---  A=80 B=0C X=FAF0 S=FFF0 C=C8
P=EA0E->FE   LDX    FC32 ---  A=80 B=0C X=FAF0 S=FFEE C=C8
P=EA11->EE   LDX    04,X ---  A=80 B=0C X=F126 S=FFEE C=C8    N
P=EA13->AD   JSR    02,X ---  A=80 B=0C X=F2AA S=FFEE C=C8    N
P=F2AC->20   BRA    F2B9 ---  A=80 B=0C X=F2AA S=FFEB C=C8    N
P=F2B9->CE   LDX   #FDE5 ---  A=80 B=0C X=F2AA S=FFEC C=C8    N
P=F2BC->BD   JSR    F48E ---  A=80 B=0C X=FDE5 S=FFEC C=C8    N
P=F48E->6D   TST    00,X ---  A=80 B=0C X=FDE5 S=FFEA C=C8    N  ▮
```

Fig. 9-2. CRT display during debugging of a typical program. The top line indicates that the program was stopped by interrupt, breakpoint is set to location F2BC, program counter is at F48E, contents of A, B, X, S, and C registers. The next four lines each show contents (hex) of 16 memory locations starting with the address on the left. At the right of each of these lines the same memory locations are displayed as ASCII characters. The other 19 lines on the screen show the last 19 program execution steps (most recent on top). Each line shows program counter location(s), mnemonic translation of operation code, address field of instruction, contents of A, B, X, S, C registers and condition codes set.

Any memory locations of interest can also be displayed continuously by simply typing in their hex address on the top of the screen. Data can be written into any of the displayed locations by positioning the cursor there and typing in the desired hex value. Since inputs and outputs are treated as memory locations, the status of all input and output registers can likewise be displayed and changed as desired.

Since the debug program automatically translates "machine language" operation codes into assembler mnemonics, the programmer can forget about machine language bits and think in terms of instruction mnemonics. If an error is found in the program a temporary "patch" can be made without reassembly by simply typing in the location and the desired instruction(s) *in mnemonic form.* Address fields of instructions can be indicated as hex digits or by their symbolic name (if a symbol translation table is previously loaded).

During the program debugging process, the RAM in the development system is used to hold the program so that changes can easily be made. Most microcomputer systems ultimately use PROM or ROM for their program memory so development systems include a socket for writing the program into a PROM. When checkout of a program in RAM is complete, a PROM is plugged into the program socket and control is transferred to the PROM programming program. This program writes and verifies a duplicate of the program in RAM into the PROM. This PROM can be shipped with the ultimate system or used as a master for programming other PROMs or ROMs.

Since the PROMs are automatically verified bit by bit to be identical to a working program, the chances for human error and marginal operation are greatly reduced with programmed logic. Since the program documentation is also automatic, the chances of documentation errors, and reasons for out of date documentation, are virtually eliminated.

9-3 FILE MANAGEMENT

Another function which is automated by diskette-based development systems is the record-keeping function. Each diskette has a directory that lists the names, sizes, and locations of all files stored on it (see Fig. 9-3). These files are generally source or object programs, but they can actually be any kind of text such as manuals or specifications. The text editor can be used to make changes or updates to any kind of text including programs. The development system is, in fact, a "word processor" as well.

Normally each engineer, programmer, or secretary has his or her own

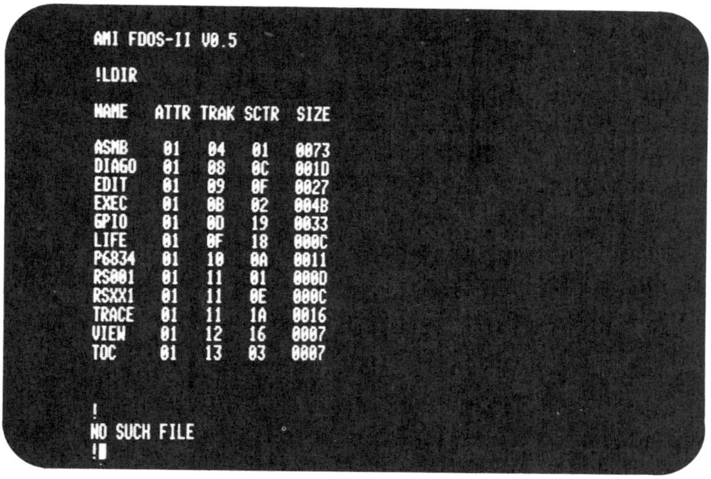

```
AMI FDOS-II V0.5

!LDIR

NAME    ATTR TRAK SCTR  SIZE

ASMB     01   04   01   0073
DIAGO    01   08   0C   001D
EDIT     01   09   0F   0027
EXEC     01   0B   02   004B
GPIO     01   0D   19   0033
LIFE     01   0F   18   000C
P6834    01   10   0A   0011
RS001    01   11   01   000D
RSXX1    01   11   0E   000C
TRACE    01   11   1A   0016
VIEW     01   12   16   0007
TOC      01   13   03   0007


!
NO SUCH FILE
!
```

Fig. 9-3. CRT display of directory of programs stored on diskette. This directory was displayed by the FDOS operating system in response to the typed LDIR command. Any of these names could be edited, assembled, deleted, and so on by typing further commands. The NO SUCH FILE response at the bottom was in response to the accidental typing of a "garbage" character.

diskette which contains the system programs as well as any programs that are currently in development. The operating system has commands for **CREAT**ing new files, **DELET**ing old ones, **RENAM**ing files, **VIEW**ing a file on the CRT, **PRINT**ing a file, **COPY**ing only selected files from one diskette to another, and **MERG**ing several old files under a single new name. A typical list of operating system commands is shown in Table 9-1.

Note that the parameter lists after the commands on Table 9-1 do not necessarily have to be filled in completely. When the command is entered with a carriage return, the program will act on whatever parameters have been entered.

TABLE 9-1   AMI DISK OPERATING SYSTEM COMMANDS

ASMB,sourcefilename,destinationfilename,p,listdevice
    p = 2 object + listing
    p = 3 listing only
    p = 4 object only
CHGAT,filename,newattributes
COPY   (from drive 0 to drive 1)
CREATE,filename,size
DEBUG
DELET:u,filenamel,filename2,....,filenamen
DUMP,filename (onto writer output)
EDIT,inputfilename,outputfilename
EDIT,,outputfilename
HOME:u   (position head to track 0)
INIT:u   (u = 1, 2, 3: initialize user diskette)
INITX   (implies u = 0: initialize system diskette)
LDIR:u,listdevice
LOAD,filename,offset(optional)
MERGE,newfilename,filename1,filename2,....,filenamen
PRINT,filename,listdevice
RENAM,oldfilename,newfilename
RUN,objectfile,inputfilename,outputfilename,parameter
STORE,destinationfilename (from reader device)
VIEW,filename,nlines,startline
XGEN

    :u is unit number. u = 0, 1, 2 or 3
    filename may be filename:u
    all numbers in hex
    listdevice = Disk, Punch, Line printer, Console
    (default is Console)

To use the EDIT command to correct a program we have called STERM2, for example, we would key in the following:

EDIT, STERM2, STERM3 (carriage return)

The text editor program would automatically load the old program (STERM2) and create a new file called STERM3. Any corrections entered would become a part of the new version 3 of the program, while the STERM2 version would be kept on file until canceled by: DELET, STERM2 (carriage return).

Generally it is a good idea to make an occasional backup copy of your diskette so that recovery will be easy in the event of malfunction or accident. The directory shown in Figure 9-3 has only the programs that are delivered with the development system on them. When a program is under development, one would normally have the current and previous versions of both sources and object code on the diskette along with a selection of these system programs.

Program names should be short and descriptive with an "S" or and "O" tacked on to indicate source or object. A revision number should also be put at the end of the name to distinguish successive versions from each other. If the previous version is always retained it will be easier to "retrench" if a new change turns out to be a mistake.

The pages that follow illustrate a typical program development as seen on the screen of an AMI 6800 development system. A program for displaying the alphabet is written, edited, assembled, debugged, corrected, reassembled, and added to the directory file. Although we have not covered enough details for the reader to follow it exactly, it should give a fair idea of the overall process.

9-4 HARDWARE TESTING WITH A DEVELOPMENT SYSTEM

While initial program debugging can be done by manually entering inputs and looking at outputs, the real test of a system requires testing of input and output interface circuits and devices. A powerful option to any development system called *in-circuit emulation* makes it possible to *plug the development system into the microprocessor or microcomputer socket in the product itself.* This makes it possible to run the program at full speed and watch it react normally with the actual input/output devices in the product.

To the product, the 40-pin plug of the in-circuit emulator looks just like a working microcomputer or microprocessor. If the product has separate program memory, the emulator can ignore it and use the RAM in the development system instead. Program changes can thus be made easily and the full powers of the *debug* and *trace* programs are available.

With the program running at full speed, the successive machine states are recorded in a trace memory along with other test points of interest. When a specified address or other trigger signal is detected, the recording process stops with the 1023 machine states prior to that trigger held in memory. It is thus possible to look back through these *prior* machine and test point states *after the fact* and determine the cause of improper operation. The machine states can be displayed in various formats including a "disassembled" format which translates the instruction codes into mnemonics so the program operation can

```
AMI FDOS-II V0.5

!EDIT,,TEST1

FDOS-II/AMI EDITOR VER. 1.1
01 NAM TEST -- PRINT THE ALPHABET
OPT O
ORG $100
DPYOUT EQU $E006 MONITOR DISPLAY OUTPUT ROUTINE
*
START LDA A #13 CARRIAGE RETURN
JSR DPYOUT
LDA A #10 LINEFEED
JSR DPYOUT
LDA A #'A FIRST LETTER TO PRINT
LDA B #26 NUMBER OF LETTERS
LOOP JSR DPYOUT DISPLAY A LETTER
NOP
DEC B COUNT
BNE LOOP CONTINUE UNTIL COUNT IS ZERO
WAI THEN STOP
END
$$
0
```
Enter Editor, type in program

```
!EDIT,,TEST1

FDOS-II/AMI EDITOR VER. 1.1
01 NAM TEST -- PRINT THE ALPHABET
OPT O
ORG $100
DPYOUT EQU $E006 MONITOR DISPLAY OUTPUT ROUTINE
*
START LDA A #13 CARRIAGE RETURN
JSR DPYOUT
LDA A #10 LINEFEED
JSR DPYOUT
LDA A #'A FIRST LETTER TO PRINT
LDA B #26 NUMBER OF LETTERS
LOOP JSR DPYOUT DISPLAY A LETTER
NOP
DEC B COUNT
BNE LOOP CONTINUE UNTIL COUNT IS ZERO
WAI THEN STOP
END
$$
@C PRINT$DISPLAY$$
00T$$
NAM TEST -- DISPLAY THE ALPHABET
0
```
Modify comment line to read "DISPLAY" instead of "PRINT"

```
OPT O
ORG $100
DPYOUT EQU $E006 MONITOR DISPLAY OUTPUT ROUTINE
*
START LDA A #13 CARRIAGE RETURN
JSR DPYOUT
LDA A #10 LINEFEED
JSR DPYOUT
LDA A #'A FIRST LETTER TO PRINT
LDA B #26 NUMBER OF LETTERS
LOOP JSR DPYOUT DISPLAY A LETTER
NOP
DEC B COUNT
BNE LOOP CONTINUE UNTIL COUNT IS ZERO
WAI THEN STOP
END
$$
@C PRINT$DISPLAY$$
00T$$
NAM TEST -- DISPLAY THE ALPHABET
EE$$

!ASMB,TEST1,TESTL,3,D
13
0
```
Exit Editor, assemble for listing

```
!ASMB,TEST1,TESTL,3,D
13
!VIEW,TESTL

$---

PAGE 001 TEST --

00001                 NAM    TEST   -- DISPLAY THE ALPHABET
00002                 OPT    O
00003 0100            ORG    $100
00004      E006 DPYOUT EQU   $E006  MONITOR DISPLAY OUTPUT ROUTI
00005
00006 0100 86 0D START LDA A #13    CARRIAGE RETURN
00007 0102 BD E006      JSR   DPYOUT
00008 0105 86 0A       LDA A  #10   LINEFEED
00009 0107 BD E006      JSR   DPYOUT
00010 010A 86 41       LDA A  #'A   FIRST LETTER TO PRINT
00011 010C C6 1A       LDA B  #26   NUMBER OF LETTERS
00012 010E BD E006 LOOP JSR   DPYOUT DISPLAY A LETTER
00013 0111 01         NOP
00014 0112 5A         DEC B  COUNT
```
VIEW listing, a screenfull at a time

```
00001                 NAM    TEST   -- DISPLAY THE ALPHABET
00002                 OPT    O
00003 0100            ORG    $100
00004      E006 DPYOUT EQU   $E006  MONITOR DISPLAY OUTPUT ROUTI
00005
00006 0100 86 0D START LDA A #13    CARRIAGE RETURN
00007 0102 BD E006      JSR   DPYOUT
00008 0105 86 0A       LDA A  #10   LINEFEED
00009 0107 BD E006      JSR   DPYOUT
00010 010A 86 41       LDA A  #'A   FIRST LETTER TO PRINT
00011 010C C6 1A       LDA B  #26   NUMBER OF LETTERS
00012 010E BD E006 LOOP JSR   DPYOUT DISPLAY A LETTER
00013 0111 01         NOP
00014 0112 5A         DEC B  COUNT
00015 0113 26 F9       BNE    LOOP  CONTINUE UNTIL COUNT IS ZERO
00016 0115 3E         WAI          THEN STOP
00017                 END

TOTAL ERRORS 00000

!ASMB,TEST1,TESTX,4
14
0
```
Assemble for object file

```
00001                 NAM    TEST   -- DISPLAY THE ALPHABET
00002                 OPT    O
00003 0100            ORG    $100
00004      E006 DPYOUT EQU   $E006  MONITOR DISPLAY OUTPUT ROUTI
00005
00006 0100 86 0D START LDA A #13    CARRIAGE RETURN
00007 0102 BD E006      JSR   DPYOUT
00008 0105 86 0A       LDA A  #10   LINEFEED
00009 0107 BD E006      JSR   DPYOUT
00010 010A 86 41       LDA A  #'A   FIRST LETTER TO PRINT
00011 010C C6 1A       LDA B  #26   NUMBER OF LETTERS
00012 010E BD E006 LOOP JSR   DPYOUT DISPLAY A LETTER
00013 0111 01         NOP
00014 0112 5A         DEC B  COUNT
00015 0113 26 F9       BNE    LOOP  CONTINUE UNTIL COUNT IS ZERO
00016 0115 3E         WAI          THEN STOP
00017                 END

TOTAL ERRORS 00000

!ASMB,TEST1,TESTX,4
14
!RUN,TESTX
,AAAAAAAAAAAAAAAAAAAAAAAAA
```
Run program—it doesn't quite work

Fig. 9-4 (a–f). A typical program development sequence is illustrated here and on the next two pages. A program is created, assembled, tested, debugged and verified to run correctly. (AMI 6800 development system.)

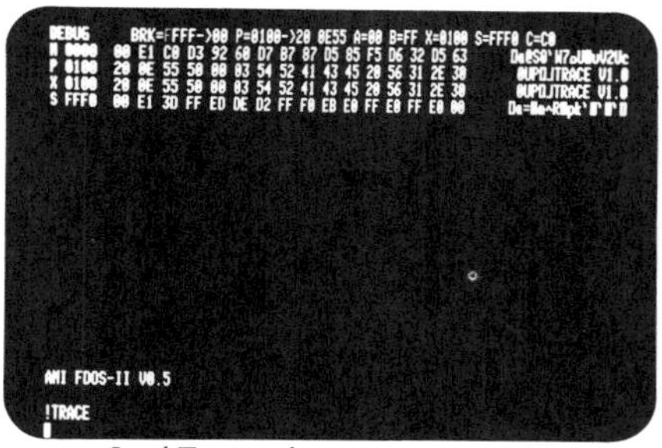

Load Trace package

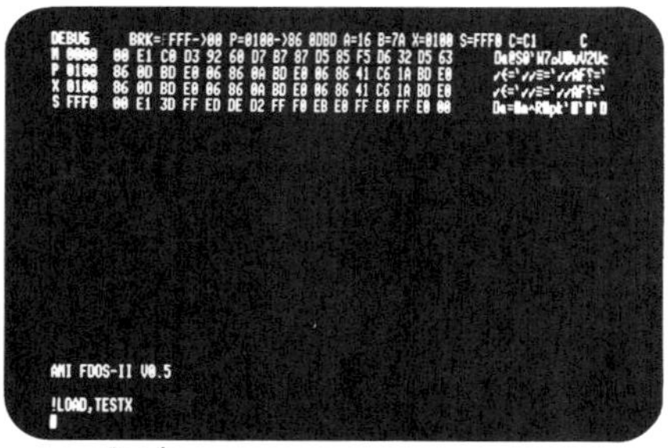

Load program

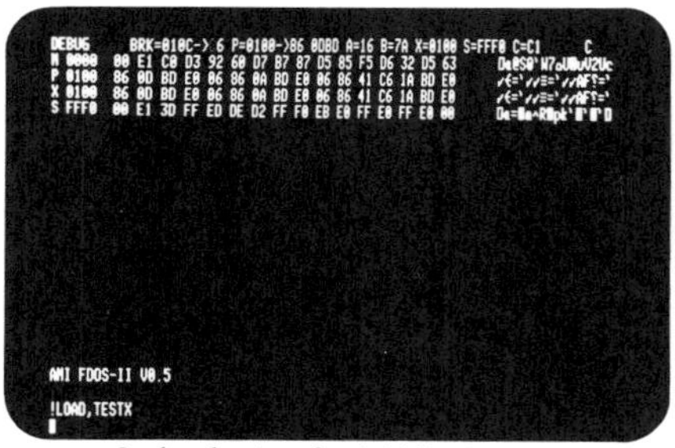

Set breakpoint after initialization

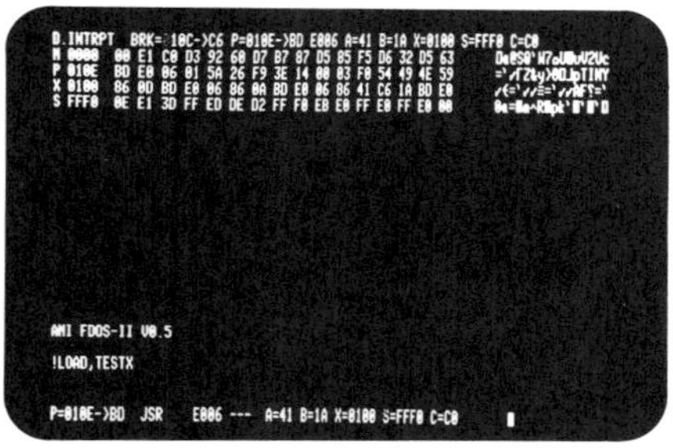

Go. Start Trace after breakpoint

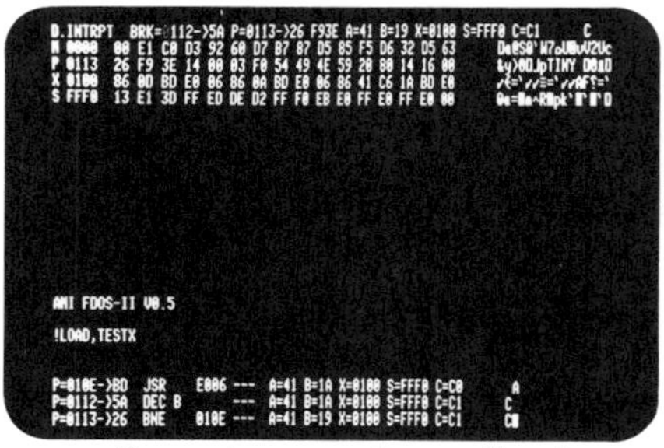

Trace main program. Suppress Trace in subroutines

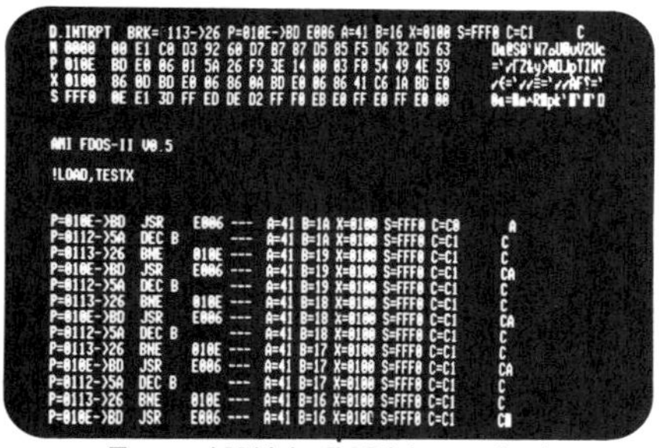

Trace—AHA! An INCA instruction is missing.

Fig. 9-4 (g–l).   Debugging a program.

be traced much as one reads assembly language. Groups of test probe inputs can be decoded and displayed as decimal, ASCII, hex, and so on, as desired.

In the single step mode it is possible to actually watch the input data being read by the program. Output instructions can be executed and their results will be seen on the output devices on the product. In-circuit emulation thus makes it possible to easily connect all of the input and output devices to the development system just as they are connected to the microcomputer in the product.

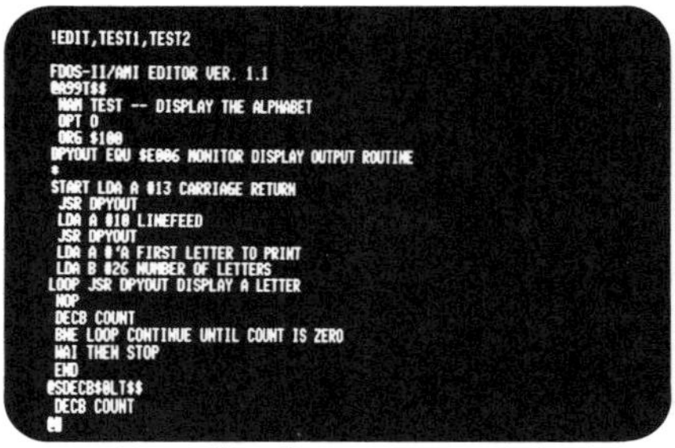

```
!EDIT,TEST1,TEST2
FDOS-11/AMI EDITOR VER. 1.1
0099T$$
NAM TEST -- DISPLAY THE ALPHABET
OPT O
ORG $100
DPYOUT EQU $E806 MONITOR DISPLAY OUTPUT ROUTINE
*
START LDA A #13 CARRIAGE RETURN
JSR DPYOUT
LDA A #10 LINEFEED
JSR DPYOUT
LDA A #'A FIRST LETTER TO PRINT
LDA B #26 NUMBER OF LETTERS
LOOP JSR DPYOUT DISPLAY A LETTER
NOP
DECB COUNT
BNE LOOP CONTINUE UNTIL COUNT IS ZERO
WAI THEN STOP
END
@SDECB$0LT$$
DECB COUNT
@@
```
Enter Editor, list program, search for place to fix

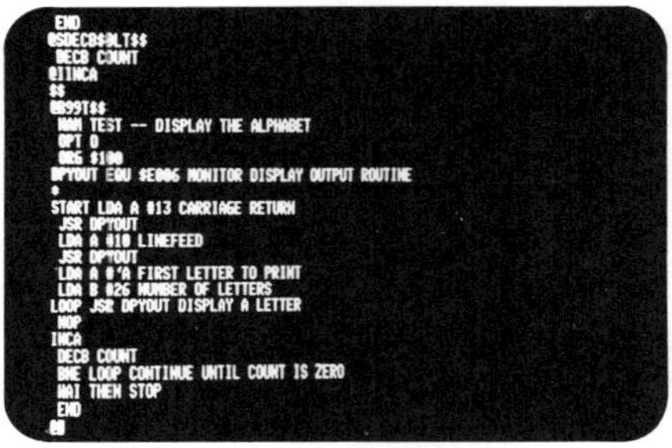

```
END
@SDECB$0LT$$
DECB COUNT
@11NCA
$$
0099T$$
NAM TEST -- DISPLAY THE ALPHABET
OPT O
ORG $100
DPYOUT EQU $E806 MONITOR DISPLAY OUTPUT ROUTINE
*
START LDA A #13 CARRIAGE RETURN
JSR DPYOUT
LDA A #10 LINEFEED
JSR DPYOUT
LDA A #'A FIRST LETTER TO PRINT
LDA B #26 NUMBER OF LETTERS
LOOP JSR DPYOUT DISPLAY A LETTER
NOP
INCA
DECB COUNT
BNE LOOP CONTINUE UNTIL COUNT IS ZERO
WAI THEN STOP
END
@@
```
Insert new line, list program

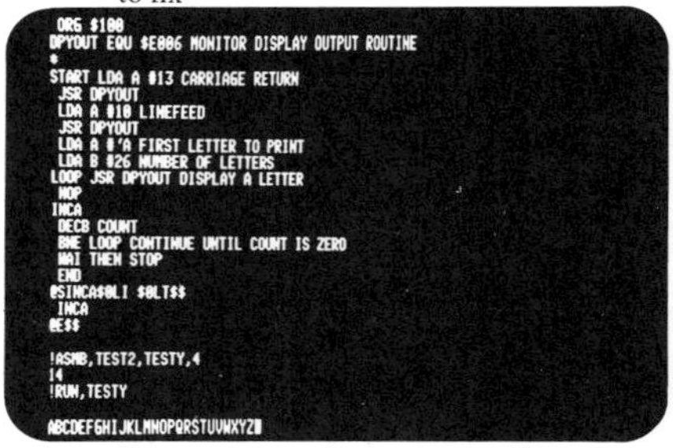

```
ORG $100
DPYOUT EQU $E806 MONITOR DISPLAY OUTPUT ROUTINE
*
START LDA A #13 CARRIAGE RETURN
JSR DPYOUT
LDA A #10 LINEFEED
JSR DPYOUT
LDA A #'A FIRST LETTER TO PRINT
LDA B #26 NUMBER OF LETTERS
LOOP JSR DPYOUT DISPLAY A LETTER
NOP
INCA
DECB COUNT
BNE LOOP CONTINUE UNTIL COUNT IS ZERO
WAI THEN STOP
END
@SINCA$0LI $0LT$$
INCA
@E$$

!ASMB,TEST2,TESTY,4
14
!RUN,TESTY

ABCDEFGHIJKLMNOPQRSTUVWXYZ@
```
Fix error on new line, exit, assemble, run—this time it works

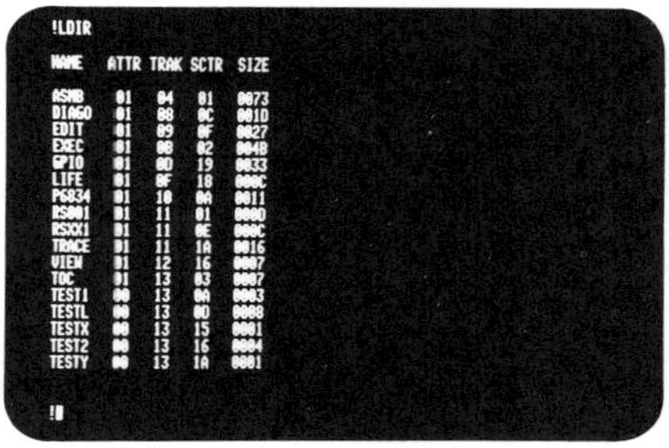

```
!LDIR

NAME  ATTR TRAK SCTR SIZE

ASMB   01   04   01   0073
DIAGO  01   86   0C   001D
EDIT   01   09   0F   0027
EXEC   01   80   02   004B
GPIO   01   00   19   0033
LIFE   01   0F   18   000C
P6834  01   10   00   0011
RS001  01   11   01   000D
RSOCI  01   11   0E   000C
TRACE  01   11   1A   0016
VIEW   01   12   16   0007
TOC    01   13   03   0007
TEST1  00   13   00   0008
TESTL  00   13   15   0001
TESTX  00   13   16   0004
TEST2  00   13   10   0001
TESTY  00   13   1A   0001

!@
```
List directory

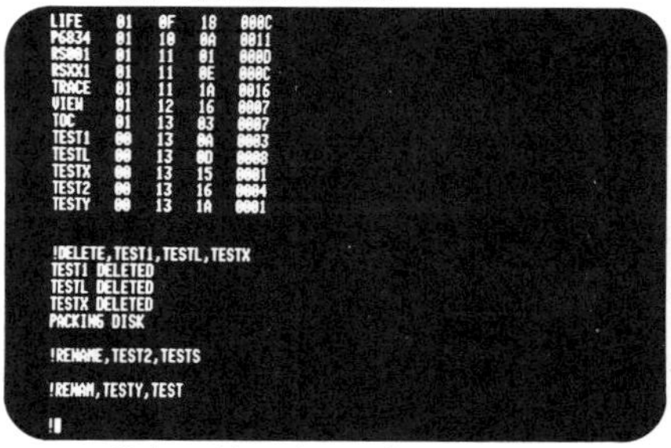

```
LIFE   01   0F   18   000C
P6834  01   10   00   0011
RS001  01   11   01   000D
RSOCI  01   11   0E   000C
TRACE  01   11   1A   0016
VIEW   01   12   16   0007
TOC    01   13   03   0007
TEST1  00   13   00   0008
TESTL  00   13   00   0008
TESTX  00   13   15
TEST2  00   13   16   0004
TESTY  00   13   1A   0001

!DELETE,TEST1,TESTL,TESTX
TEST1 DELETED
TESTL DELETED
TESTX DELETED
PACKING DISK

!RENAME,TEST2,TESTS

!RENAM,TESTY,TEST

!@
```
Delete old versions and rename new ones

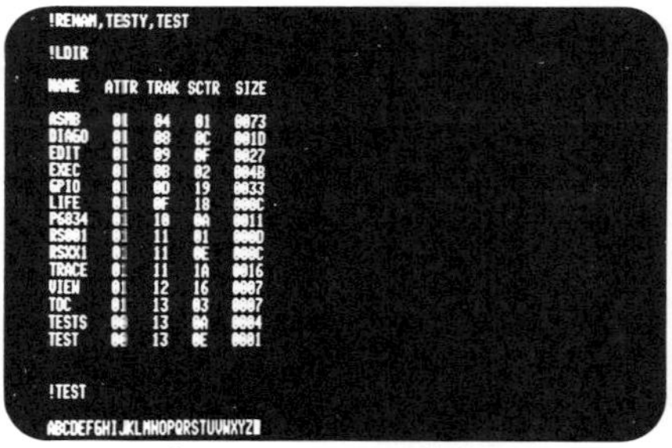

```
!RENAM,TESTY,TEST

!LDIR

NAME  ATTR TRAK SCTR SIZE

ASMB   01   04   01   0073
DIAGO  01   86   0C   001D
EDIT   01   09   0F   0027
EXEC   01   80   02   004B
GPIO   01   00   19   0033
LIFE   01   0F   18   000C
P6834  01   10   00   0011
RS001  01   11   01   000D
RSOCI  01   11   0E   000D
TRACE  01   11   1A   0016
VIEW   01   12   16   0007
TOC    01   13   03   0007
TESTS  00   13   00   0008
TEST   00   13   0E   0001

!TEST

ABCDEFGHIJKLMNOPQRSTUVWXYZ@
```
List directory, run renamed object file

Fig. 9-4 (m–r). Editing and manipulating files.

Subtle problems involving input/output interaction and/or timing can thus be worked on in the automated environment of the development system.

In addition to input and output devices, the in-circuit emulator can be used to test ROM and RAM chips in the product. Individual locations can be written into and read from with the debug program or special diagnostic programs can be run. The RAM diagnostic (part of the system software) writes and reads back worst-case patterns to the RAM in the product. The PROM verification program can be used to compare every bit in the PROM or ROM in the product to a known good PROM. A changable memory map in the development system determines whether devices in the product or their equivalents in the development system will respond to a particular address range. The development system is thus a powerful tool for hardware testing as well as program testing and development.

EXERCISES

1. What is the contents (in hex) of memory location 0003 in Figure 9-2?
2. What is the hex code for the next instruction to be executed in Figure 9-2?
3. Which file in Figure 9-3 is the largest? Which is the smallest?
4. Which memory addresses do the alphabet printing sample program of Figure 9-4 occupy (before the correction)?
5. For the development system of Table 9-1 what must be keyed in to do the following:
    (a) Delete a file called ALPHA3?
    (b) Edit a file called ALPHA3 and produce a new revision #4?
    (c) Assemble the edited file and display the listing on the console?
    (d) Run the object program?

CHAPTER TEN

PROGRAMMED LOGIC IV
Microcomputer Hardware Design

The process of system design is a gradual progression from defining general requirements to defining specific details. In microprocessor-based systems most of the details of operation are defined by programs. Before the programming task can begin, however, a hardware configuration must be defined that is (at least tentatively) capable of handling the system requirements.

Interface circuits must be designed to convert the digital microcomputer signals into the signals required by electromechanical input and output devices. In simple systems this interfacing job may be all there is to the hardware design task. Single-chip microcomputers (see Fig. 8-1) include a microprocessor, memory, timer, and parallel input and latched output lines all on a single chip. If the capabilities of the microcomputer chip are adequate for the job, the hardware design task consists of simply designing a PC board to interconnect the interface circuits and packaging it up with a power supply.

Often simple external logic is needed to handle tasks that would put a heavy load on processor execution time. For example, a serial asynchronous stream of input data can be processed directly by the program if the program examines the input at least every ⅛ *bit time*. With external synchronization logic this can be reduced to one interrupt per *bit* time. With an external shift register the program need only be interrupted once for each *byte* time.

236

To make an intelligent decision as to where to draw the line between hardware and software implementation it is often necessary to write rough versions of the programs required. For example, if the serial data rate is 1200 baud, the ⅛ bit time interrupts would be only 104 μsec apart. Since the interrupt program would take at least that long to execute, we can exclude that possibility. The external synchronization approach reduces the interrupt interval to 833 μsec. If the interrupt routine takes 300 μsec to execute, the input task still takes $300/833 = 36\%$ of the processor's time. If the inputs are done on the byte level, the input routine execution time is reduced to about 150 μsec every 6664 μsec or only 2.25% of the processor's time.

If the application requires a significant amount of other processing, it would be clearly desirable to operate on the byte level. Special LSI serial receiver chips are available for this task. If the application requires very little other processing during data reception, the bit level input may be satisfactory. If an interval timer can be used exclusively by the input program, this can be accomplished without additional hardware. The point is that in these early stages of design preliminary estimates of both hardware cost and program *loading* must often be made to make intelligent hardware–software tradeoff decisions.

Peripheral LSI chips, such as the serial data receiver, or additional RAM and ROM chips can be added to any microcomputer or microprocessor system by simply connecting them in parallel to a common *bus*.

10.1 THE BUS CONCEPT

A bus is really nothing more than a group of wires (or more usually printed circuit traces). Numerous devices are connected in parallel along a bus, but only one device can send on the bus at a time. *The system bus used with microprocessors is conceptually subdivided into three buses:*

1. An *address* bus (8, 12, 16, or 20 bits).
2. A *data* bus (8 or 16 bits).
3. A *control* bus (4 or more signals: READ', WRITE' RESET, etc.).

Generally all devices in the system are connected in parallel to the data bus and at least part of the address bus and control bus. Figure 10-1 shows a microprocessor system bus. The microprocessor is connected to all bus signals. The microprocessor can read or write from any of the devices on the bus by *putting the device's address on the address bus* and sending the proper control signals on the control bus. When the processor writes data to a device, it sends the data over the data bus along with a WRITE strobe on the control bus. When the processor reads data from a device it disables its own tri-state drivers on the data bus, sends a READ strobe on the control bus, and senses

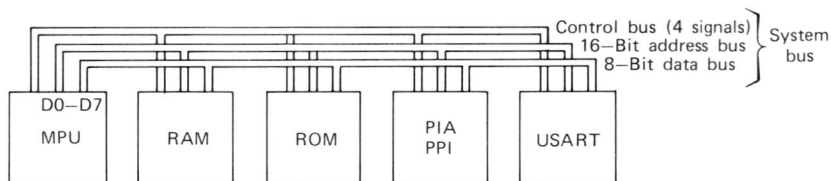

Fig. 10-1   A microprocessor system bus.

the data put on the bus by the tri-state drivers in the device. The other devices on the bus can be ROMs, RAMs, other microprocessors, or any of a large selection of programmable peripheral chips.

10.2   PERIPHERAL INTERFACE CIRCUITS

In the pages to follow we examine a wide range of peripheral LSI circuits that interface the microprocessor bus to the outside world. Some of these devices are special purpose in that they interface to a particular device such as a CRT display, floppy disk, or serial communications link. A few of these devices are so general that they are often packaged on the same chip with a microprocessor, ROM, and RAM to make a *microcomputer* chip.

Even though a microcomputer uses an internal bus to interconnect circuits of this type that are on the chip, it generally also provides an external bus for expansion. Whether the devices are on the chip or on an external bus, their basic operation is the same.

Generally peripheral interface circuits have *data registers* and *control/status registers* which the processor reads or writes into as though they were memory locations (or I/O ports). The control registers make it possible for a single LSI circuit to be used for a wide variety of jobs.

When power is first applied to any microcomputer the program automatically starts running at a special *restart location*. The first order of business of any program is to *initialize* the state of all control and data registers in the system. The bit patterns written into the control registers of LSI peripheral circuits *essentially customize those circuits for the job at hand.*

For example, the control register bits on a CRT display controller (e.g., 6845) specify every detail of screen formatting including number of lines, number of characters per line, character size, cursor type, interlace, and sync pulse characteristics. Customers with widely differing display requirements can thus use the same controller chip. Since the display characteristics are determined by bits in the main program ROM, changes to the display format can be made as easily as any other program change.

This idea of customizing LSI functions with control register bits is another example of how LSI circuits can be made to fill a wide range of jobs through programming. (This is a very different kind of programming, however, because the control register state is like a single, very special, instruction which is continually executed.)

Control registers are also written into *after* initialization when the operating mode of the circuit is changed during system operation. For example, it may be desirable to use a different CRT screen format during different modes of system operation.

10.2.1 General Purpose Parallel Interface Circuits

Certainly the most general type of input/output interface circuit is the *Programmable Peripheral Interface* (also called a Peripheral Interface Adaptor). These circuits create 24 general purpose input/output pins which are interfaced to the data bus in such a way that input and output signals can be treated by the program as bits in memory (or I/O ports). Control registers on the chip can be initialized by the restart program to specify which of the 24 I/O lines will be inputs and which will be outputs.

The two least significant address bus lines are decoded by the chip so that the 24 I/O lines look like three consecutive bytes of memory and the control register looks like the fourth. (The external *chip select* decode logic determines the actual address of these four memory locations.)

To look at an input line, the program simply reads the corresponding memory location then tests the bit that corresponds to the input of interest. When the program writes to one of the three memory locations, all eight bits of the corresponding output data register are updated. If the state of only one output signal is to be changed, the program can simply read the present state of that output register into an accumulator, alter the desired bit, then write it back. Some general purpose interface circuits make it possible to set or reset a single output bit without having to worry about the others.

Some microcomputers (e.g., 8048) have a simplified form of peripheral interface adaptor: by using a *quasi-bidirectional* output circuit, they eliminate the need for a control register to specify pins as inputs or outputs. All lines are *both* inputs *and* outputs. If an output is set high it can still be easily pulled low by an input signal because the output drive circuit can only source a fraction of a milliamp. (It sources more when it first goes high.) True outputs thus look just like an input with a large pullup resistor.

10.2.2 Other Programmable Interface Devices

If speed were not a problem, all input/output interfacing could simply be handled by programmed manipulation of general purpose input/output lines.

Often, however, there are things that simply happen too fast to be handled by the program. When this is the case, external logic must be added to preprocess inputs and simplify output requirements enough that the microcomputer can handle the job. Fortunately, there are many fast peripheral functions so commonly used that special LSI chips have been developed to handle them.

On high-volume applications where upward growth is not expected, peripheral chips should be used only as needed to stay within processor speed limits. On systems where low production volume or upward growth is expected, however, it is often wise to use peripheral chips wherever they are available. This reduces development cost and frees up processor power for future expansion. The pages that follow list some of the available peripheral circuits:

Serial Receiver/Transmitters. Include UARTs (Universal Asynchronous Receiver/Transmitters), USARTs (same thing but for Synchronous and Asynchronous, see Fig. 10.2), ACIAs (Asynchronous Communications Interface

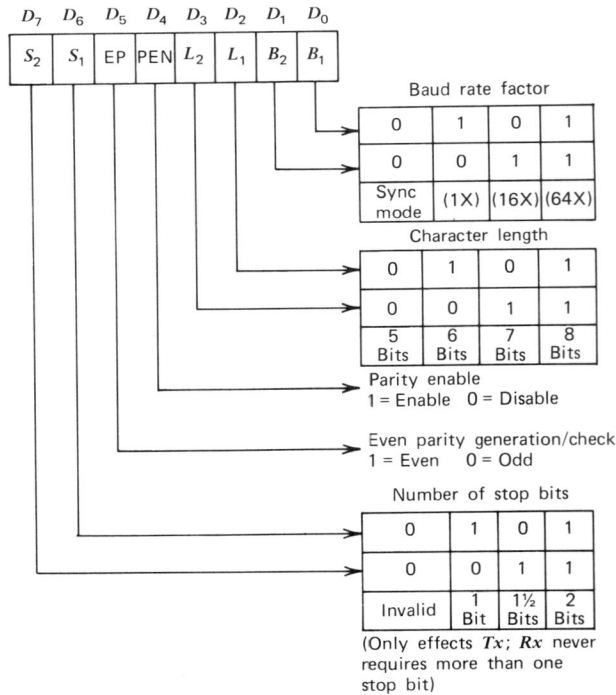

D_7 D_6 D_5 D_4 D_3 D_2 D_1 D_0

| S_2 | S_1 | EP | PEN | L_2 | L_1 | B_2 | B_1 |

Baud rate factor

0	1	0	1
0	0	1	1
Sync mode	(1X)	(16X)	(64X)

Character length

0	1	0	1
0	0	1	1
5 Bits	6 Bits	7 Bits	8 Bits

Parity enable
1 = Enable   0 = Disable

Even parity generation/check
1 = Even   0 = Odd

Number of stop bits

0	1	0	1
0	0	1	1
Invalid	1 Bit	1½ Bits	2 Bits

(Only effects *Tx*; *Rx* never requires more than one stop bit)

Fig. 10-2.   Control register mode instruction format for USART (universal synchronous/ asynchronous receiver/transmitter) in asynchronous mode.

Adapters), PCI (Programmable Communications Interface), and many other names. They send and receive serial data bits and buffer them so the microcomputer can simply read or write complete bytes to and from the data register. The control register programs word length, parity type, number of stop bits, sync codes, and so on. On some chips a control register also programs bit rate. At slow data rates the serial receiver/transmitter function could be done by program (as we did in Fig. 8-1). Some microcomputers (e.g., 6801) include a serial receiver/transmitter on the chip.

Serial Communications Controllers. Handle all of the above jobs plus the special tasks required for SDLC (synchronous data link control) and HDLC (high-level data link control) "packet" communications protocols. The specific protocol is specified by control register bits but such jobs as flag detection, synchronization, 16-bit CRC polynomial error checking, and zero insertion/deletion are also handled. With help from a USART these jobs *could* be done in program (but they are nasty). In fact much of the slower logic on these chips is programmed.

Interrupt Controllers (e.g., 8259). These cascadable chips each handle eight interrupt inputs on a priority basis. Each interrupt can be separately disabled by its own control register bit. The interrupts are automatically *vectored* in that they each cause the program to jump to an address (vector) which is stored in a specific location in ROM. Most microcomputers and some general purpose parallel interface circuits include two or three levels of vectored interrupt.

Interval Timer/Event Counters. Count system clocks to provide timed interrupts. They can also count events. The processor presets the desired count into the counter and receives an interrupt when it has counted down to zero. The counter state can be read at any time. Since this circuit is simple and requires only one input pin, it is included in all microcomputer chips.

Arithmetic Processors (e.g., AMD 9511). Do 16- and 32-bit fixed and floating point arithmetic at high speed. Operations include multiplication, division, logs, powers, trigonometry, and so on. The processor simply writes operands into a push-down stack on this chip, writes a command, then reads out the result later. Can be used with DMA to speed these transfers.

Floppy Disk Controllers. Handle all details of formatting, error control, addressing, and so on for diskette drives. Control register initialization customizes such parameters as track step rate, head settling time, head load time, and so on. Executes (with its own internal programs) macro commands such as single sector and multisector read and write in response to control register commands. DMA transfers data if desired.

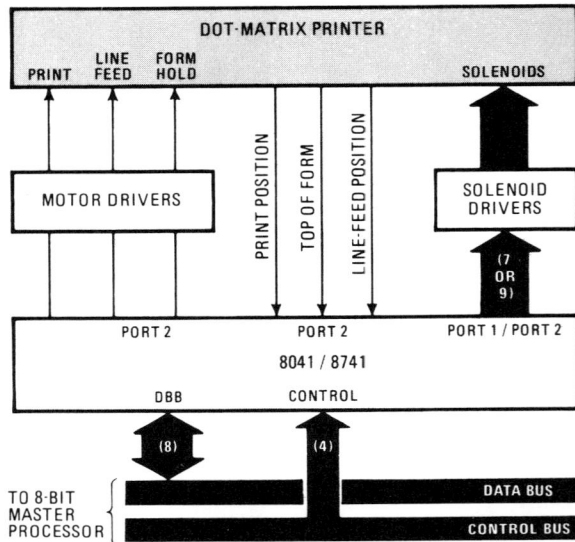

Fig. 10-3. A Universal peripheral interface programmed to act as a buffered printer controller. The main microprocessor can simply transfer an entire message to this controller, then continue with other tasks while the UPI handles the details of dot-matrix printing. Since the UPI is actually a complete microcomputer, the resulting system is actually an easily understood multicomputer system.

Analog Interface Circuits.   Up to six analog inputs can be converted to digital samples by a single analog-to-digital input chip (e.g., μA9708). Analog outputs can be produced by digital-to-analog converter chips (e.g., 5018). Some complete microcomputer chips (e.g., Intel's 8022) have analog inputs in addition to the usual general purpose digital input and output pins. Analog inputs can thus be filtered, compared, remembered, and otherwise processed by digital programming techniques.

Universal Peripheral Interfaces (Fig. 10-3).   Actually just a microcomputer with a special interface register that can be used as a control or data register by the main processor. The user simply programs this microcomputer to do any special interfacing job and defines control and data formats. The result is actually a multicomputer system. But by thinking of the peripheral microcomputers as special purpose interface controllers the system is easier to understand, program, and test. Some products sold as special purpose interface controllers (e.g., the 8294 Data Encriptor) are really just UPIs with a special masked ROM program.

10.2.3 DMA Controllers (Direct Memory Access Controllers)

Normally the preceding peripheral circuits are simply connected in parallel to the main memory bus. Each time they have or need a new word of data they must be "serviced" by the program. This is a time consuming process because the program must answer an interrupt, input or output a word, increment a count, and check to see if there is room to store more characters.

With a DMA controller chip the program is freed of this time consuming task. The program simply sets up a buffer area for each device and the DMA controller does all the work. Each time a peripheral device has or needs a new word, the DMA controller seizes the bus by *hold*ing the processor, puts the correct memory address on the address bus, cycles the memory, then releases the processor again. A *current address counter* is automatically maintained for each peripheral device and incremented and compared to the final address with each transfer. Each peripheral controller uses *its normal connection to the data bus* to transfer data whenever its *DMA request* to the DMA controller is answered by a *DMA acknowledge*.

One DMA controller chip generally handles four prioritized DMA channels, each with its own counter and DMA request/acknowledge logic. Multiple DMA chips can be cascaded to form a priority chain with the processor always taking last priority. A control register option can select a "rotating priority" mode in which simultaneous DMA requests are handled on a rotating basis.

The increase in system capability when a DMA controller is added can be significant if many fast I/O operations are required. Often only the fast, block oriented devices are handled by DMA while slower devices are done by conventional I/O programs. With really fast devices, it is often necessary to either use DMA or provide an external buffer memory (or both).

10.3 ADDRESS DECODER DESIGN

One of the basic tasks of microprocessor hardware design is to decode the address bus to produce a unique *chip select* for each chip on the system bus. When the microprocessor program wants to read or write from a particular memory location or peripheral controller register, it will send a READ or WRITE strobe on the control bus and a specific address on the address bus. The chip select decoding must select only one of the chips on the bus. The hardware designer can minimize the chip select logic by making address assignments which are compatible with simple decoding.

The starting point in making address assignments must be the RAM and ROM memory areas. When a microprocessor restarts after power failure it initializes its program counter to a restart address which it finds by reading from a specific memory address. Since the restart address must be stored in ROM, this essentially defines the address area that must be assigned to ROMs. If more than one ROM chip is needed to store the program the chip selects for the ROMs should be chosen such that a contiguous memory area is created.

For example, a 2 k × 8 bit ROM will internally decode the least significant 11 bits of the address bus (A0–A10) to determine which word to put on the data bus when it receives a chip select and read strobe. The chip select for the second ROM chip must thus differ from the first one only by A11 (the next bit on the address bus). If more than two ROM chips are required, a decoder (e.g., 8205, 74LS138) can be used to generate a total of eight chip selects. If the address inputs to the decoder are connected to A11, A12, and A13 up to eight ROMs can be connected to the bus for a contiguous 16 k byte ROM address space. Note that this kind of memory decoding is similar in principle to the decoding tree of Figure 4-11, but with the final 11 bits of decoding done inside the ROM chip. By connecting the remaining address bits (A14 and A15) to enable inputs of the decoder, complete decoding of the ROM addresses is obtained.

Chip enables for RAMs are handled in the same way. RAMs generally have fewer than eight data bus pins so one chip select can drive more than one RAM. For example a 1 k × 4 bit RAM is connected to 10 bits of the address bus (A0–A9) and only 4 bits of the data bus. They must therefore be used in pairs for an 8-bit data bus. Each pair of chips can have a common chip select since each chip serves only half of the data bus.

An easy way to keep track of address assignments and make sure that all chip enables are mutually exclusive is to tabulate them as shown in Table 10.1. As long as each chip enable differs in at least one bit from all other chip enables there will be no conflict. For example, the ROM and RAM decode ena-

TABLE 10-1 AN ADDRESS DECODING TABLE FOR A SYSTEM WITH 16 k BYTES OF ROM, 8 k BYTES OF RAM, AND ONE PERIPHERAL CHIP

Bus Signals	A15 A14 A13 A12	A11 A10 A9 A8	A7 A6 A5 A4	A3 A2 A1 A0
ROMS		(A10 ——————	—————	————— A0)
ROM Decoder	E1' E2' A2 A1	A0		
RAMS		(A9 —————	—————	————— A0)
RAM Decoder	E3 E2' E1' A2	A1 A0		
Peripheral Chip	CE			(RS1 RS0)

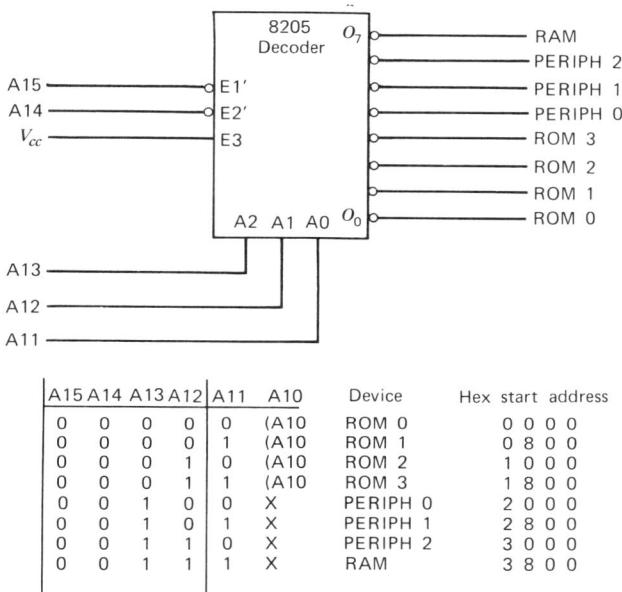

Fig. 10.4 Address decoding for smaller system.

A15	A14	A13	A12	A11	A10	Device	Hex start address
0	0	0	0	0	(A10	ROM 0	0 0 0 0
0	0	0	0	1	(A10	ROM 1	0 8 0 0
0	0	0	1	0	(A10	ROM 2	1 0 0 0
0	0	0	1	1	(A10	ROM 3	1 8 0 0
0	0	1	0	0	X	PERIPH 0	2 0 0 0
0	0	1	0	1	X	PERIPH 1	2 8 0 0
0	0	1	1	0	X	PERIPH 2	3 0 0 0
0	0	1	1	1	X	RAM	3 8 0 0

bles require different A15 states. The peripheral chip enable requires A14 to be true, while both the RAM and ROM decodes require A14 to be false.

By dividing the table in groups of four bits (as in Table 10-1), the hexidecimal addresses of the devices can be read off directly from address states required by the enables. For example, the ROMs cover all states with A14 and A15 false or addresses 0000 through 3FFF. Likewise the RAMs cover all states with $A15 \cdot A14' \cdot A13'$ or a hex range of 8000 through 9FFF. The peripheral chip is incompletely decoded so it would respond to any address with A14 true (32,768 addresses altogether). We can arbitrarily pick hex locations 4000 through 4003 as program addresses for the peripheral chip. Notice that this chip internally decodes A0 and A1 to address its four different control and data registers. The program can thus write or read into each of these four registers as though they were unique memory locations.

10.3.1 Address Decoding in Smaller Systems

As system size gets smaller, address decoding can be increasingly simplified by incompletely decoding the addresses. Again the basic rules are:

TABLE 10-2   ADDRESS DECODING USING CHIP SELECTS ONLY (Linear selection)

Device	A15	A14	A13	A12	A11	A10	A9	A8	A7	A6	A5	A4	A3	A2	A1	A0
ROM					CS'	(A10———										—A0)
RAM				CS		(A9———										—A0)
Periph 0			CS												(A1	A0)
Periph 1	CS'	CS												(A2———		—A0)
Periph 2	CS	CS'													(A1	A0)

1.   All chip selects must be mutually exclusive.

2.   The ROM area should be contiguous and must include the restart address.

3.   The RAM area should be contiguous.

Figure 10-4 shows the decoding for a system that has only one RAM chip. The RAM and the peripheral control chips can thus use other outputs of the same decoder used for the ROMs.

For even smaller systems decoding can be done with no decoder at all. Chip selects are simply assigned to separate address bits as shown in Table 10-2. Peripheral chips with more than one chip select input (like periph 1 and 2 in Table 10-2) can do a small amount of decoding without any separate gates.

Some microprocessors (e.g., Intel) have separate input/output instructions which produce strobes on the control bus that are separate from the memory read and write strobes. This allows addresses in the bottom 256 bytes of memory space to be used for input/output devices as well as memory.

10.4   BUS LOADING AND BUFFERING

Since MOS devices present primarily a capacitive load on the bus, the number of devices that can be connected in parallel to the bus is limited by the capacitive drive capability of the tri-state drivers. Timing specs are usually defined for a load of about 150 pf, but loads of up to 300 pf can be driven if the extra 45 nsec (for an Intel 8085) of delay is taken into consideration. Since worst-case MOS input capacity is often only 5 pf, up to 30 5 pf devices can be connected to a bus at standard speed, or 60 devices at reduced speed!

Other factors probably make it unwise to connect so many devices in parallel to one bus. One problem is that 60 parallel devices mean 60 possible locations for a short circuit on the bus. Troubleshooting on such a large bus can be difficult. On large systems that fit on a single PC board it is often enough to simply buffer the more heavily loaded, least significant address bits. Data bus loading can often be relieved by buffering only the ROM portion of the data bus. Since the ROMs do not receive from the data bus, simple unidirectional buffers can be used.

On really large systems a mother-board interconnects an entire subrack full of circuit boards. In this case inverting buffers are usually used on each board to buffer all inputs and outputs. Though the signals on the mother-board are inverted, they are inverted again by the inverters on each PC board before they are actually used.

The bidirectional nature of the data bus makes it necessary to use special bidirectional drivers in most cases. As Figure 10-5 shows, these devices have

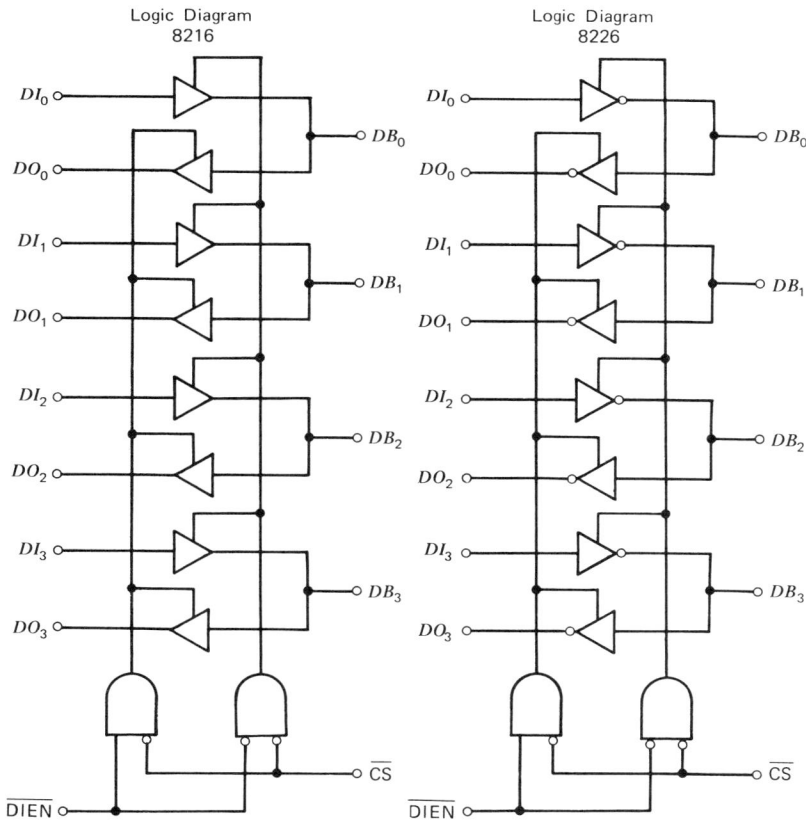

Fig. 10-5. Bi-directional bus drivers.

two control signals that must be dealt with. The direction control signal (DIEN') is generally connected to a control bus signal from the microprocessor (BUSEN' for example). Since the chip select (CS') input actually enables the tri-state drivers, the address decoding logic must also provide this signal. If several devices share a buffer, *the buffer chip select must go true whenever any of the individual device chip selects goes true.*

10.5   BUS COMPATIBILITY AND TIMING

The manufacturers of microprocessors have made it very easy to do designs which use exclusively their products by offering families of bus-compatible devices. While the safest approach is to stay within these matched component families, it often turns out that another manufacturer has a peripheral chip that does just what you need. Usually the hardware required to make one manufacturer's bus look like another is trivial, but timing specifications must be checked carefully.

Data and address bus signals are standard except for the fact that some manufacturers (e.g., Intel 8085) send address bits over the data bus during the first part of the memory cycle. A separate address bus can be created in this case by adding an address latch which holds the bits when they appear on the data bus.

Most bus compatibility problems result from different timing and definition of control bus signals. Although they differ in detailed implementation all microprocessor control busses include the following signals:

1.   A *write strobe* that tells the addressed device to store the data on the data bus.

2.   A *read strobe* that tells the addressed device to gate its data onto the data bus.

3.   A *clock* signal that is synchronized to the processor signals.

4.   A *reset* signal that is used to initialize the devices on the bus when power is first turned on.

5.   A *ready* input to allow slow devices to hold up the processor cycle long enough to satisfy their timing requirements.

6.   A *hold, halt,* or *DMA request* input to allow other devices to take control of the bus away from the processor.

7.   (Usually) a hold *acknowledge* output to allow the processor to indicate that it has relinquished the bus.

8.   An *interrupt request* input to allow external devices to interrupt the program flow for service. Sometimes there are several vectored interrupt inputs.

9. A *trap* or *nonmaskable interrupt* input that is used for servicing or emergency conditions.

10. An *interrupt acknowledge* output over which the processor indicates that it has acknowledged an interrupt.

11. (Sometimes) an *address latch enable* to indicate that address information is on the data bus.

While the manufacturers provide the basic functions listed in their control bus, there is some variation in the actual implementation of the signals. For example, the Motorola 6800 and MOS Technology 6500 series have a

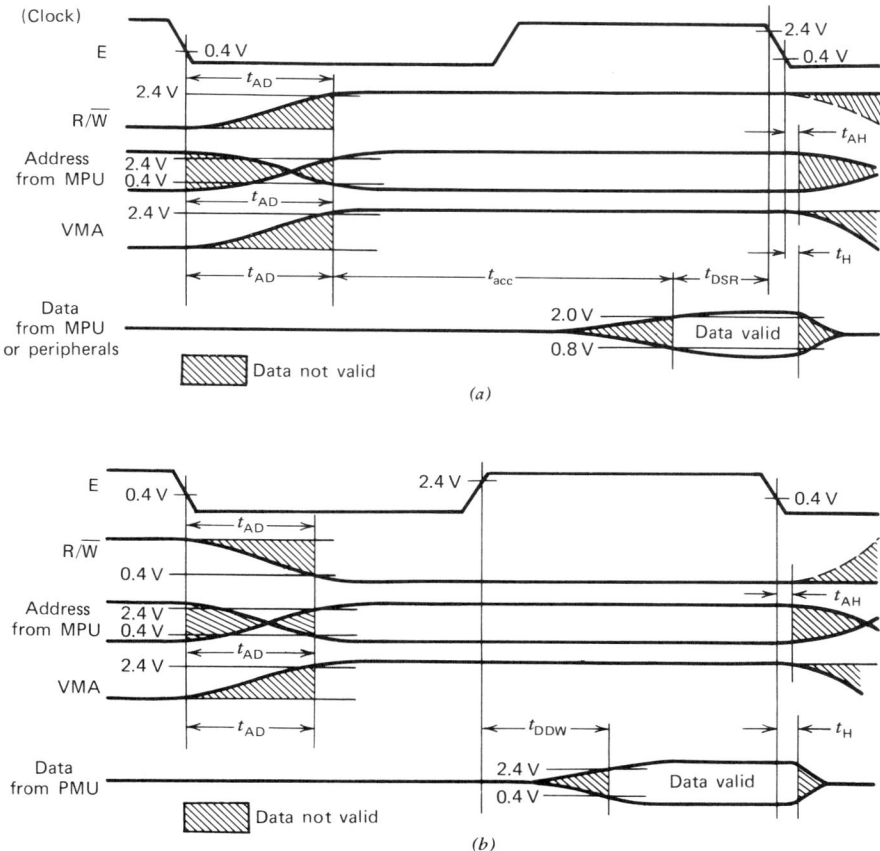

Fig. 10-6. Read and write bus timing for Motorola 6802 microprocessor. $t_{AD} < 270$ nsec, $t_{ACE} < 530$ nsec, $t_{DSR} > 100$ nsec, $t_{H} > 20$ nsec, $t_{AH} > 20$ nsec, $t_{DDW} < 225$ nsec. (a) Read data from memory or peripherals. (b) Write data in memory or peripherals.

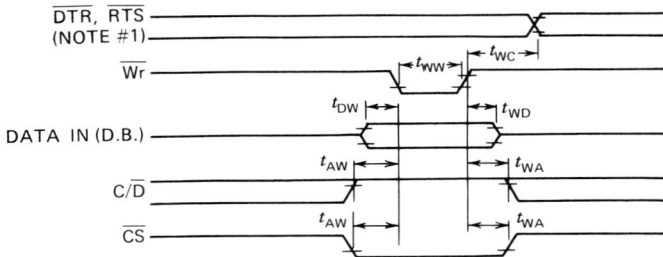

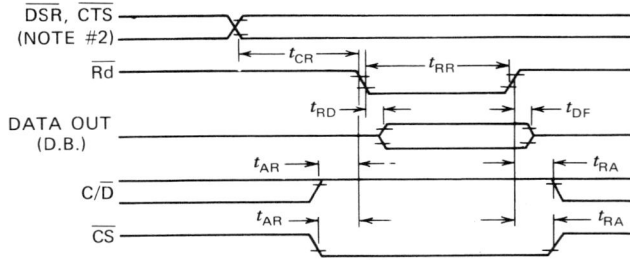

NOTE #1: T_{WC} Includes the response timing of a control byte.
NOTE #2: T_{CR} Includes the effect of CTS on the TxENBL circuitry.

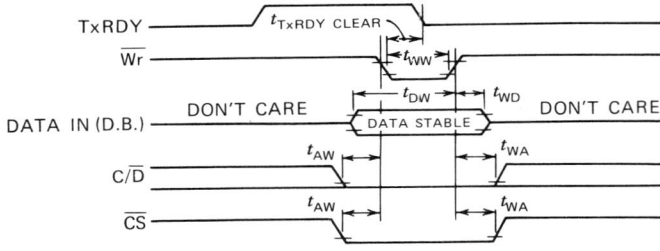

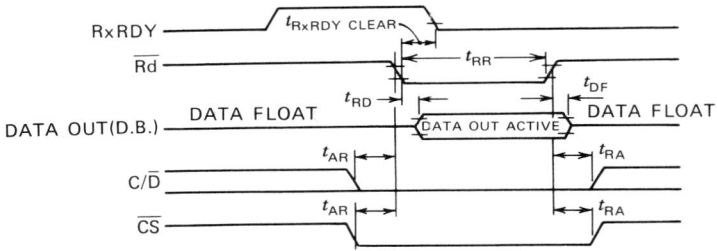

Fig. 10-7. Timing specifications for Intel 8251A Communications Interface (USART). t_{WW} > 250 nsec, $t_{T \times RDY\ CLEAR}$ < 150 nsec, t_{DW} > 150 nsec, t_{WD} > 0 nsec, t_{AW} > 0 nsec, t_{WA} > 0 nsec, $t_{R \times RDY\ CLEAR}$ < 150 nsec, t_{RR} > 250 nsec, t_{RD} < 200 nsec, t_{DF} = 10–100 nsec, t_{AR} > 0 nsec, t_{RA} > 0 nsec, t_{WW} > 250 nsec, t_{WC} < 8 nsec, t_{CR} < 20 nsec. T × RDY, R × RDY, DTR', RTS', DSR', and CTS' are all external communications signals not associated with the bus. (a) Write control or output port cycle (CPU → USART). (b) Read control or input port (CPU ← USART). (c) Write data cycle (CPU → USART). (d) Read data cycle (CPU ← USART).

250

read/write' (R/W') signal and a *valid memory address* (VMA) signal. *Read strobe* and *write strobe* signals similar to Intel's can be generated with an inverter and two 3-input NAND gates as follows:

$$read\ strobe = R/W' \cdot VMA \cdot clock$$
$$write\ strobe = (R/W')' \cdot VMA \cdot clock$$

Whenever devices that are not family matched are used together, the timing specifications of the peripheral device must be carefully compared against those of the processor bus. As Figures 10-6 and 10-7 show, these specifications can be quite complicated. One must compare timing requirements for one transition at a time in the write data cycle, then do the same for the read data cycle, and each of the control cycles (if they are separately specified, as they are in Fig. 10-7). The $C/\bar{D}$ on Figure 10-7 is true for control cycles and false for data cycles. Normally it is simply connected to one of the address bus lines. Since the 8251A has only one data register address and one control register address, the $C/\bar{D}$ input would normally be connected to the least significant address bit so that the control register and data register would look like adjacent memory locations to the processor.

Family microprocessor product lines are designed to be fairly foolproof when used together but caution is still advisable. If, for example, the manufacturer allowed for only one level of gating in the chip select decoding, timing problems could result from the delay of multilevel decoding.

10.6 DESIGNING-IN FLEXIBILITY

One of the biggest advantages of microprocessor based design is its extreme flexibility. Often a product can grow into other products, change to meet competition, and be easily customized if the hardware design provides the needed flexibility. Since memory requirements tend to grow as a product is refined and extended, it is usually a good idea to design it to handle 2–4 times as much memory as the initial concept seems to dictate. Since the memory chips don't have to be installed if they are not needed, the cost of the product is often not increased at all.

Sometimes memory expansion capability can be designed-in by simply providing for the "next generation" memory chip. Every two years or so memory chip capacities have been quadrupling. Provision for the next generation chip can often be made by adding a jumper option to allow moving the chip select decoder inputs over two places. When the new chip actually becomes available, you have quadrupled memory capacity for very little extra cost. Pinouts of new memory chips tend to preserve a continuity with the most popular chip in the previous generation. Information on additional signals required and other changes are usually available more than a year before the part is actually available in quantity.

Another way to improve flexibility is to provide a few jumper or DIP-switch selectable inputs. These can be used to select among several different operating modes for the device. The program simply reads these switch settings during initialization and decides which of several modes of operation are required. Many versions of the same product can thus be identical except for these switch or jumper settings.

10.7   DESIGNING-IN MAINTAINABILITY AND RELIABILITY

Microprocessor-based systems can be either very difficult or very easy to fix depending on how much thought is given to maintenance features. In all but the simplest "throwaway" systems the microprocessor and ROMs should be mounted on plug-in sockets. The microprocessor socket makes it possible to plug in test and development systems, and the ROM sockets make it possible to make major changes or minor corrections by simply plugging in new ROMs. If the system does not have its own input keyboard, at least a couple of switch inputs should be provided for test functions. If the system has no output display, it is generally worthwhile adding at least a few LED indicators for use by test programs.

When power is first applied, the restart program can begin by lighting the first LED. The program can then routinely test the RAM, ROM, and peripheral controllers. If the program displays ascending binary codes on the LEDs as it passes various checkpoints in its self-test, the state of the LEDs will indicate just how much of the system is working. Of course, if the system has its own display, the test program can use that. The main point is that normal system operation proceeds only after all self-tests are passed and that the display indicates the nature of any failure.

ROM testing can be easy if the last word of ROM space (usually unused anyway) contains a *check word*. This check word can be generated by a simple routine in the development system just before writing the PROM. By doing an exclusive-or on every location in ROM and complimenting the result, a check character can be generated which ensures that there is an odd number of ones in each bit position. The ROM testing startup routine in the product then can simply exclusive-or all of the ROM locations together and test for an all 1's result. While this simple vertical odd parity test is quite effective, more sophisticated check polynomials give even better results (see the bibliography).

RAM startup tests can also be effective without being elaborate. For example, here is a simple RAM test sequence:

1. Write into each RAM location its own address.
2. Read each location and compare to its own address.
3. Repeat steps 1 and 2, but use compliment of own address.

Unfortunately, a lot of the possible failure modes make the very first startup test step of simply lighting the first LED impossible. To get that far the power and clock must work, and the data and address bus must be functional. The restart interrupt in the microprocessor simply causes it to read the restart address from a specific location in ROM. For this to work, the ROM chip enable must work and the correct address bits must be sent to the ROM. Since the ROM sends the starting address to the processor over the data bus, any data bus fault will cause the processor to start on the wrong address.

Since there are so many failure modes that can prevent the processor from starting up correctly, it is often helpful to design in the capability of making the processor "free run." When the processor "free runs" it simply counts sequentially through all possible addresses. This binary count sequence on the address bus makes it possible to find faults on the address bus or in the chip enable decoding. Any microprocessor will free run if it is forced to execute the same nonmemory reference, nonbranch instruction over and over again. The trick is thus to force the data bus into a state that looks like the next instruction read from ROM is always the same. For example, if the code for a CLEAR A instruction is forced on the data bus the processor will internally clear the A register and then try to read the next instruction from the next ROM location. This process repeats indefinitely resulting in a continually incrementing address on the address bus.

On a small system without data bus buffers free running can be forced by simply plugging a specially wired plug with 100-Ω resistors to +5 V, and ground in the pattern required to force a CLEAR A instruction into one of the ROM sockets. In systems with a buffered data bus, the buffers should be plugged into sockets so they can be removed and replaced by a similarly wired "free run" plug.

Once the basic functions of the system are working the program can be as sophisticated as desired in automating the troubleshooting function. If the system includes a display, the self-test programs can even directly describe the trouble. If the TRAP or Nonmaskable Interrupt input is made available to the serviceman, a program can be included that displays the program counter and register states whenever the serviceman pushes a button—even if the program is "hung" in a loop. *Debug* programs, which are usually available in ROM form from microprocessor manufacturers, make it possible to use any serial terminal for testing. It thus becomes possible to look at or change the contents of any memory location or register in the system.

10.8   DESIGNING FOR SELF-CLEARING

In systems that must run continuously for months or years on end without attention, the problem of "hang-up states" can be serious. As long as everything works perfectly there is no problem, but a single bad bit due to noise can make the program "blow up." Once this happens the system will not recover without human intervention unless special provisions are made in the hardware.

The safest way to make a system restart itself after an error is to include a "watchdog timer." A watchdog timer is nothing more than a holdover one-shot which is triggered by some basic part of the program. If the program "blows up" and stops triggering the holdover circuit, the holdover output causes a TRAP or Nonmaskable Interrupt input to the microprocessor. The program thus executed should remember the failure then transfer control to the restart program.

A simpler but less effective method is to do a TRAP or NMI whenever any undefined address is accessed or whenever an attempt is made to write to a ROM. This method works on the principle that when a program gets off track, the problem tends to grow into a complete mess as random data are executed as instructions, stack pointers are destroyed, and eventually a state of total chaos is reached. Somewhere in this process an undefined location is accessed or an attempt is made to write into the ROM.

Another approach to automatic restart is to include extensive data checking in the program. Whenever data or address pointers are encountered that are not correct, the program jumps to the TRAP location. None of the three approaches just discussed is foolproof, so they are often used together to improve the chances of catching a runaway processor. Since a hang-up condition can be a sign of program bugs or incipient hardware failure, some means should be provided for logging its occurrence and if possible any details which will make the bug easier to find. If possible a "snapshot" of the state of the program counter, stack pointer, registers, and peripheral devices should be stored somewhere before restarting.

10.9   DOCUMENTATION OF MICROPROCESSOR HARDWARE

Because of the repetitive nature of microprocessor bus connections, schematic diagrams can be greatly simplified by using a few shortcuts. Design of a system can be greatly simplified if a table of bus pin connections similar to Table 10-3

TABLE 10-3. BUS PIN TABLE. SIMPLIFIES THE LOGIC SCHEMATIC BY DEFINING REPETITIVE CONNECTIONS

Bit Number — columns 7–0 form the **Data Bus**; columns 15–0 form the **Address Bus**.

Device	7	6	5	4	3	2	1	0	15	14	13	12	11	10	9	8	7	6	5	4	3	2	1	0	R/W	GND	+5V	φ2	IRQ'
MPU 6800	26	27	28	29	30	31	32	33	25	24	23	22	20	19	18	17	16	15	14	13	12	11	10	9	34	1, 21, 39	8, 2		4
2-PIA 6820	26	27	28	29	30	31	32	33	24														35	36	21	1	20	25	37, 38
ACIA 6850	15	16	17	18	19	20	21	22	10															11	13	1	12	14	7
RAM 4804	5	3	8	6											13	12	16	17	1	2	15	14	10	11	7	9	18		
RAM 4804					5	3	8	6							13	12	16	17	1	2	15	14	10	11	7	9	18		
2-PROM 2716	17	16	15	14	13	11	10	9						19	22	23	1	2	3	4	5	6	7	8		12, 18	24		
Decoder 74LS138									3	2	1															4, 5, 8	16		

255

is made. This table defines all of the connections to *bus* and *power* pins in a neatly organized way. A standard logic diagram can be used to define all of the more random logic and circuit connections. Taken together, the bus pin table and logic diagram define the system.

Since the pin numbers can be copied directly from the IC data sheets onto the table, the effort and chance of error is greatly reduced. Decoding structures are also clearly visible from this format. With the logic diagram freed of the clutter of repetitive connections, the remaining logic connections are more easily understood.

While design is easier and less confusing with the documentation separated into two formats, it may be desirable to have the draftsman draw the pin numbers from the bus pin table in on the logic diagram with circled bus signal names connected to them. This saves the serviceman the trouble of looking up the signals on a separate table. Clutter can still be reduced by drawing in only the circled signal name rather than the actual connections.

10.10   DYNAMIC MEMORIES

In most small microcomputer systems it is practical to use RAM chips which are *static*. Static RAM chips generally use a four-transistor, two-resistor flip-flop circuit internally for each bit cell. While clever layout tricks have greatly reduced the static memory cell size, large scale memories are still more economically done with *dynamic* memory.

Dynamic RAM chips use a single transistor for each bit cell and store data as a charge on about 0.03 pf of circuit capacity.* This makes it possible to squeeze about four times as many bits on a dynamic RAM as on a static RAM of the same chip size. Power dissipation is also greatly reduced with dynamic RAMs because unselected chips draw very little power. The disadvantage with dynamic RAM is that the charge on the tiny capacitors used to store data must be periodically *refreshed* to keep it from gradually fading away.

Whenever a read or write cycle is done the circuitry on the chip automatically refreshes an entire *row* of bits. Since the chip is organized as an *X-Y* array of rows and columns, *external logic must ensure that every row on the chip is addressed at least every 2 msec or so.* This refresh function is done automatically by microcode in some microprocessors (e.g., Zilog). It may also be inherently taken care of by some applications (such as CRT displays) where blocks of data are read out periodically anyway.

---

* While this may not sound like much, the charge representing a single bit still amounts to some 3,000,000 electrons!

The important thing is that every binary combination of the six or seven least significant address bits must be addressed every refresh interval. When refresh is done as a separate function, it can be done in *burst mode* where the normal program stops and 64 (or 128) consecutive locations' are refreshed. Burst refresh can be done by the program in response to a 2 msec interrupt. However, if there are other interrupts which require fast service, this pause may be intolerable.

Distributed refresh does only one row at a time and therefore never ties up more than one memory cycle. It generally requires separate hardware because the cycles occur so often. For example, to do 128 single row cycles every 2 msec there must be one cycle every $2000/128 = 15.6$ μsec. Special RAM support chips (see Fig. 10-8) are available which simplify this task.

Since dynamic RAMs are optimized for building large memory arrays, they generally have multiplexed address inputs to reduce the number of connections and reduce package size. During the first part of a normal read or write cycle the least significant (row address) half of the address is stored in a latch on the RAM chip by the leading edge of the *Row Address Strobe*. The most signifi-

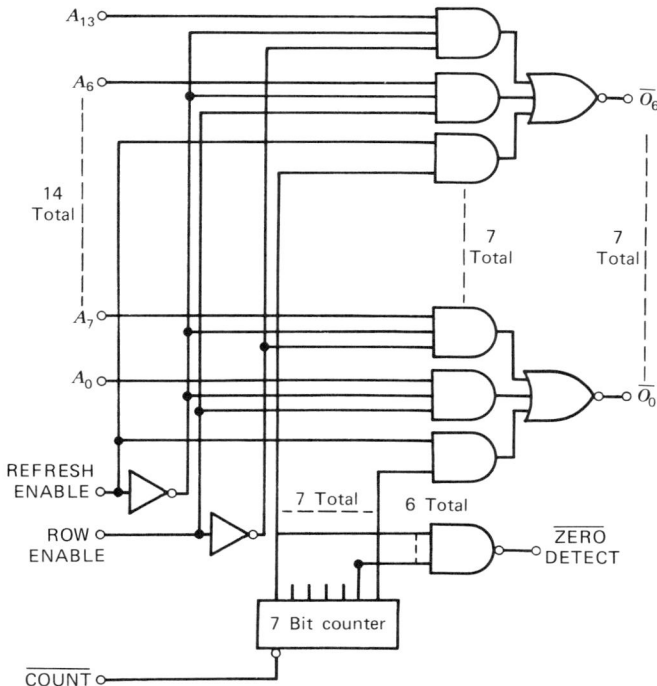

Fig. 10-8. Dynamic RAM support chip (Intel 3242).

cant address bits are then connected to the same pins and strobed by the *Column Address Strobe.* Double use is thus made of the address pins.

The RAM support chip in Figure 10-8 includes a three-input selector which gates the row address, column address, or an internal refresh counter onto the address lines. Its outputs can drive 250 pf or about 50 RAM chips. Note that refresh can be accomplished without going through a complete memory cycle or even chip-selecting the RAM chips.

Other dynamic memory designs provide even more economical data storage by further sacrificing speed and convenience. Charge-coupled devices (CCD) and bubble memories both access data serially—like a long shift register. As we shall see in the next chapter, this simply adds another dimension to the address: the time dimension.

EXERCISES

1. Design (hardware only) a microcomputer-based terminal with 16 key X-Y matrix of key switches (see Fig. 13-23), 16 digit, 7 segment LED display (see Fig. 13-25) and serial asynchronous input/output line.

2. What type of peripheral chip would you add if the microcomputer speed proved inadequate to keep up with the 9600 baud serial input/output?

3. Using Fig. 10-2 what would the hex value of the initialization control word be for:
   (a) Two stop bits, odd parity, five-bit character, $16 \times$ clock rate?
   (b) One stop bit, odd parity, eight-bit character, $64 \times$ clock rate?

4. What are the hexidecimal addresses of the five devices in Table 10-2? (Give the lowest set of address for each device.)

5. By adding another eight-output decoder only, modify the decoding logic of Fig. 10-2 so that it can handle seven additional peripheral devices.

6. (a) Design address decoding for a system using four 2 k $\times$ 8 bit PROMs, three 2 k $\times$ 4 bit RAMs, and eight peripheral controllers. (Package count should be two.) Assume a restart address of $2C_{16}$ (Intel).
   (b) Make a table of hex address ranges for all devices. Give only the lowest address where more than one address is possible.
   (c) Redesign the decoding for the Motorola microprocessors which have a restart address of $FFFF_{16}$. Take advantage of the 6802's faster addressing modes in the base page (256 bytes) of memory space by placing the peripheral controller addresses in the base page but above the 128 bytes of on-chip RAM.

7. Design-in a chip select signal for a buffer which will isolate the four ROMs from the data bus in Fig. 10-2.

8. (a) Design logic to interface eight Intel 8251A communications devices (Fig. 10-7) to the Motorola 6802 bus (Fig. 10-6). Check all timing specifications for compatibility and add logic (if necessary) to make them compatible.

(b) What is the maximum allowable delay in the chip select decoder for this combination?

(c) What is the maximum delay allowable in the gates that generate Wr' and Rd'?

(d) Assuming an increase in signal delay of 0.5 nsec/pf for capacitive loads above the 150 pf specified, how much total capacitive load would cause marginal operation? Assume 32 nsec maximum delay for all gates used.

9. Write a simple RAM test program
   (a) in 8900 assembly language.
   (b) in 8080 assembly language.

10. Write a ROM parity word generator and a startup parity test program.
    (a) in 8900 assembly language.
    (b) in 8080 assembly language.

BIBLIOGRAPHY

Error Checking Characters

Vasa, Suresh, "Calculating an Error Checking Character in Software," *Computer Design,* May 1976.

Bus Specifications

Force, Gordon, "Microprocessor Bus Standard Could Cure Designers Woes," *Electronics,* July 20, 1978.

Universal Peripheral Interface

Phillips, Don and Goodman, "Slave Microcomputer Lightens Main Microprocessor Load," *Electronics,* July 7, 1977, pp. 109–112.

Smith, Lionel, *Printer Control with the UPI-41,* Intel, Santa Clara, Calif., 1977, Appendix Note #AP-27.

Peripheral ICs

Metzger, Jerry, "Peripheral IC's for Microprocessors," *Electronics Products,* April 1978, pp. 69–72.

Peripheral Control Chips

Weissberger, Alan, "Data-link Control Chips: Bringing Order to Data Protocols," *Electronics,* June 8, 1978, pp. 104–111.

The following Intel application notes give extensive application details on specific chips. They can be purchased for $1.00 each from Intel Corporation, 3065 Bowers Ave., Santa Clara, Calif., 95051.

AP-15 8255A *Programmable Peripheral Interface Applications*.
AP-16 *Using the 8251 USART*.
AP-31 *Using the 8259 Programmable Interrupt Controller*.
AP-29 *Using the 8085 Serial I/O Lines*.
AP-25 *Simplify Your Dynamic RAM/Microprocessor Interface*.
UPI-41 *Users Manual* Pub. #9800504A ($5).
Memory Design Handbook, 1977 ($5).

Memory Chips

Hnatek, Eugene, "Current Semiconductor Memories," *Computer Design,* April 1978, pp. 115–126.

Analog Input Microcomputers

Ittner, William and Miller, Jeffrey A., "Microcomputer's On-Chip Functions Ease Users' Programming Chores," *Electronics,* July 20, 1978, pp. 129–133. Also see *Electronics,* May 25, 1978, pp. 122.

Peripheral Chips, Hardware, and Software Design

Motorola Semiconductor, *Microprocessor Applications Manual,* McGraw-Hill, New York, 1975.

Dynamic RAMs

Coker, Derrell, "16-k RAM Eases Memory Design for Mainframes Microcomputers," *Electronics,* April 28, 1977, pp. 115–120.

Brown, J. Reese Jr., "Timing Peculiarities of Multiplexed RAMs," *Computer Design,* July 1977, pp. 85–92.

CHAPTER ELEVEN

THE TIME
Dimension

11.1   REPEATING CIRCUITS IN TIME

One reason we can do so much with so little logic using the programmed
approach is that *we use the same hardware at different times for different things.*
In this chapter we examine other methods of making multiple use of hardware
by utilizing *the time dimension.*

When we send eight signals over eight separate wires, we have information
distributed in space. If we define eight consecutive *time slots,* we can send the
same eight signals, one at a time, over a *single wire.* This is information *distributed in time.* On a clocked system each clock would start a new time slot,
and every eighth clock is equivalent to one clock with the signals distributed in
space. We can send the eight signals in many different ways, depending on
how we choose the time and space dimensions. For example, we could use two
wires with four time slots or four wires with two time slots. The time and
space dimensions can thus be traded one for the other.

We can also use the time dimension with logic gates. *A single AND gate,
for example, can be equivalent to eight separate AND gates* if the inputs are
defined as eight different signals during eight time slots. The output of the gate
during each time slot will be the AND of the inputs for that time slot. As a
matter of fact, any combinational logic circuit can do the job of n repetitions of
the same circuit, if time is divided into n time slots. Since we frequently need

repetitive logic structures, we can often significantly reduce the amount of logic required to do a job by *repeating the structure in time instead of space.*

A single flip-flop, of course, *cannot* be made to look like eight flip-flops by simply assigning eight time slots. An eight-bit shift register, however, *is* equivalent to a single D flip-flop with eight time slots. With each clock the input to the shift register is stored in a flip-flop that will appear on the output again, eight clock transitions later (the next clock for that time slot). A *shift register can thus be thought of as a time-shared D flip-flop.* Just as with a D flip-flop, the output during clock $n + 1$ simply equals the input during clock n. The only difference is that, between clock n and clock $n + 1$ for a particular time slot, seven other clocks occur, one for each of the seven other time slots.

11.2 SERIAL OPERATION

Figure 11-1 shows three possible ways to AND together two 8-bit numbers stored in registers and put the result in one of the registers. In Figure 11-1a the gates are repeated in space in that there are actually eight separate gates. In Figure 11-1b the time dimension is utilized to make use of the same gate during eight time slots. Figure 11-1c shows a combined approach using both the time and space dimensions. Approach (b) is commonly called *serial operation* because the bits are handled one at a time. It is certainly the most economical approach, since only one gate is required and the two 8-bit shift registers are available in a single IC package (e.g., 9328). The total package count is thus 1.25 versus 6 for approach a and 4.5 for approach c. Though approach b is by far the most economical, there is a catch. Since eight clocks are required to complete the operation, the maximum possible speed is one-eighth that of the *parallel* approach a. In many cases we do not need to do an operation at maximum speed so we can make a significant savings of hardware by using the time dimension (serial operation).

As Figure 11-1c shows, we can use any combination of the time and space dimensions. This type of operation is called *serial-parallel operation.* If using eight time slots makes the operation take too long, this approach will do it twice as fast (four time slots) and still save hardware when compared to the parallel approach (a). We can thus organize the time and space dimensions of repetitive logic to make full use of the speed of the logic being used.

Whenever logic circuits are operated at less than their maximum possible speed, capability is wasted. In a clocked logic system decisions are made with each clock that result in a new set of states for all the memory elements in the system. When the system clock is slow, the logic decisions are made and then the entire system just sits idle until the next clock. *Whenever logic is clocked at less than the maximum rate, the designer should be alert for logic that can be*

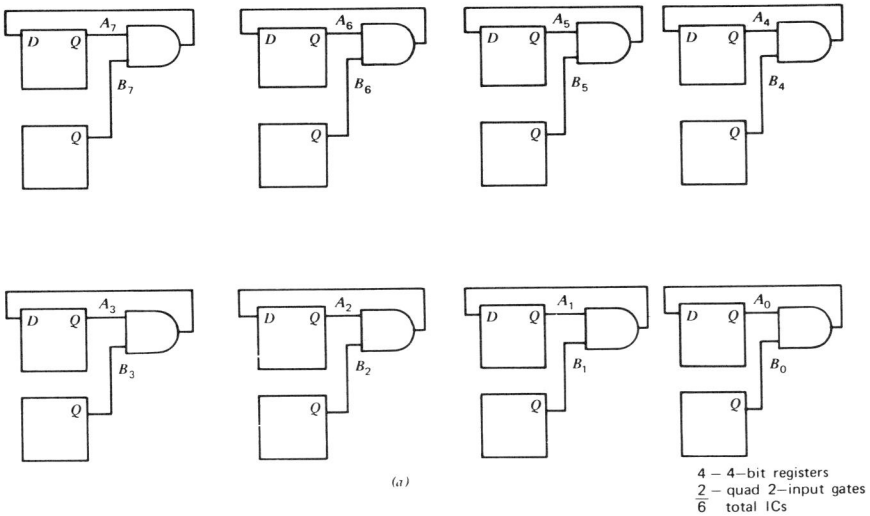

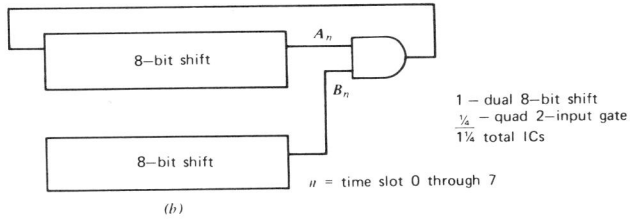

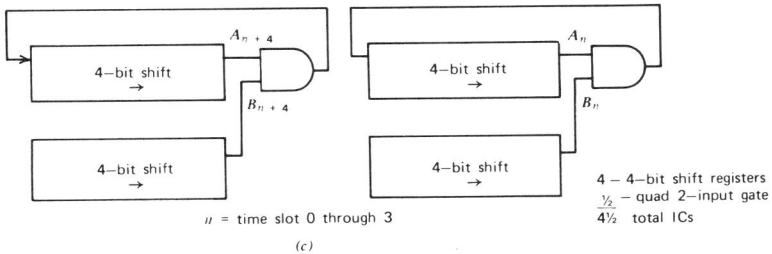

Fig. 11-1. Three approaches to ANDing two 8-bit registers: (a) eight bits of AND in space; (b) eight bits of AND in time (eight time slots); (c) eight bits of AND 4 × 2 in time and space.

263

repeated in time instead of space. Often, where data are being handled as words or characters, the entire system can be organized to handle the data one bit at a time (serial). If speed requirements do not allow one time slot for each bit, serial–parallel operation can be used with as many time slots as time allows. Actually, because of the MSI shift register packages available, the number of time slots is usually four, eight, 16, or some other standard shift register length. The savings in package count can be great if hex shift register packages are used (e.g., the 2518 contains six 32-bit shift registers).

With a serial organization, the design is essentially identical to a system with one-bit words, except that D flip-flops are replaced by n-bit shift registers and the clock runs n times as fast for a word length of n bits. One exception is operations such as addition, subtraction, and shifting which require a carry from the *adjacent bit.* Since the adjacent bit was in the previous time slot, it is not available unless it is retained by a *carry flip-flop.* Any "interconnection" between consecutive time slots can thus be made with a single additional D flip-flop.

Serial–parallel organizations can be used for higher speed. For example a four-bit adder, or arithmetic logic unit, can do 16-bit operations in four clock times. A 16-bit system can thus be built with one-fourth the combinational logic, if one-fourth the speed is acceptable. The system is identical to a four-bit system, except that four-bit shift registers are used in place of D flip-flops, and each system clock actually consists of four clocks. Again, a carry flip-flop is added (see Fig. 11-2) to pass the carry from one time slot to the next.

Sometimes other types of interconnections are needed between circuits repeated in time besides simply passing a signal from one time slot to the next. For example, we may want to know if the result of all eight AND operations in Figure 11-1 is zero. In the parallel approach of Figure 11-1a this would require an eight-input AND gate. With the serial approach of Figure 11-1b we can do the same job with a flip-flop and a two-input gate, as shown in Figure

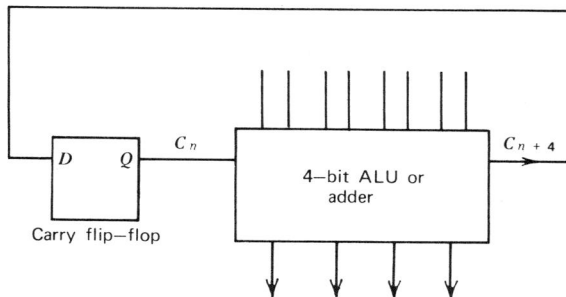

Fig. 11-2. Passing CARRY from one time slot to the next.

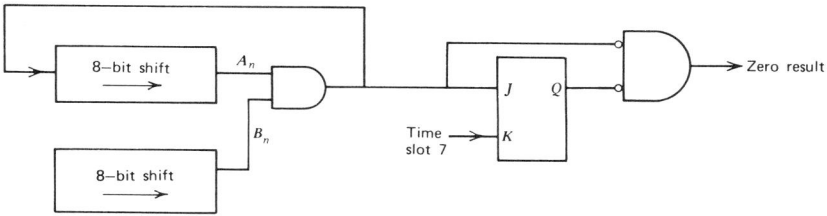

Fig. 11-3. Detecting all-zero result with a flip-flop.

11-3. If any of the first seven results are 1, the flip-flop will be set before time slot seven. The output of the gate thus indicates, during time slot seven, whether all eight results were zero. Normally a system will have a *time slot counter* that keeps track of the time slots. The final time slot (time slot seven in Fig. 11-3) is equivalent to the system clock in a parallel system.

More complex logic functions between time slots can be done by writing an equation as a function of the time slot counter signals and the serial data signal, and then using that function to set a flip-flop. For example, the circuit in Figure 11-4 will serially detect a result of 11110000. Of course, either of these functions could be detected in parallel during time slot seven by a normal eight-input gate if shift register *A* was a serial-in parallel-out type. It is thus possible to combine serial and parallel operation within the same system.

11.3 TIME-SHARED CIRCUITS

There are many practical problems that require *many repetitions of the same complete logic circuit.* For example, a computer system with many keyboards, communications lines, or other I/O devices normally needs interface logic for

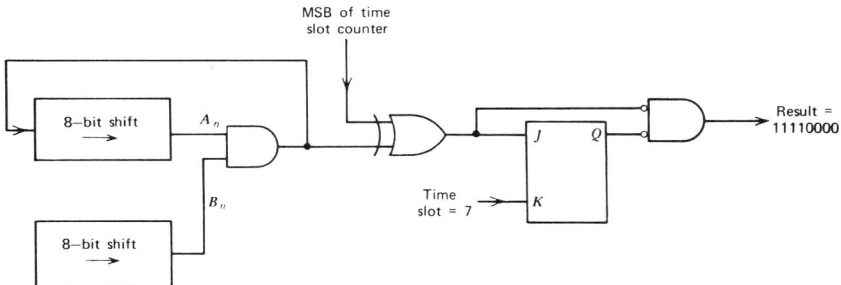

Fig. 11-4. Detecting "11110000" result with a flip-flop.

each circuit. The Telex serial data format previously used as a programming example (Fig. 8-1*a*) is a perfect example. Some computer systems require interfacing to hundreds or thousands of such lines. Although it is possible to write a program to receive several such lines bit by bit, the full processor capability is used up in receiving only a small number of such signals. To handle many such inputs, and leave processor time for other jobs, separate logic must be used to receive the serial input signals and present *complete characters* to the processor input.

Figure 11-5*a* shows a system for doing this with a separate serial data receiver circuit for each line. Since all these circuits are identical, *we can just as well repeat the circuits in time instead of space* (see Fig. 11-5*b*). With the separate receiver approach of Figure 11-5*a*, the outputs of the many receivers are essentially multiplexed together at the processor input bus (since each one inputs at a different time under program control). With the time-shared approach the multiplexing is essentially moved out further from the processor. The many lines are multiplexed to one, using a multiplexer tree as shown in Figure 11-6. This tree is controlled by a *time slot counter* so that each line is connected to the input to the time-shared receiver during a different time slot.

As an example of repeating complete logic subsystems in time, we will design a serial data receiver for Telex signals, then time share it among 2048

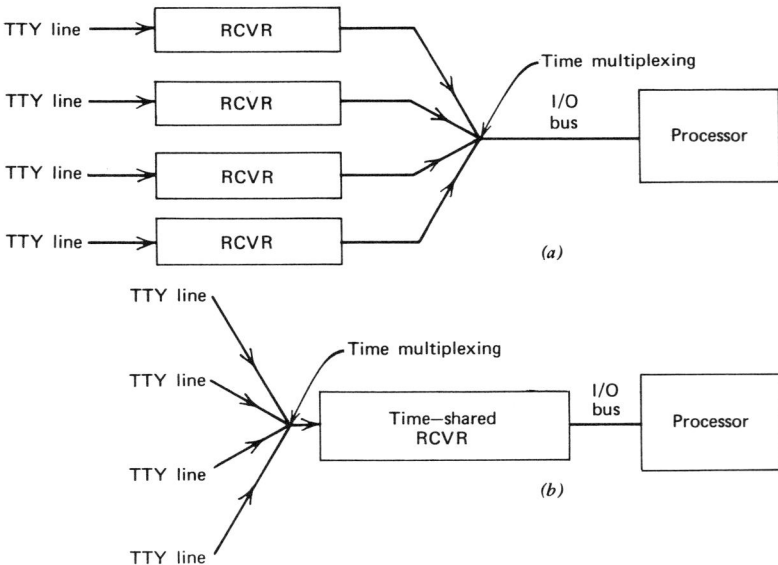

Fig. 11-5.   (*a*) Separate approach. (*b*) Time-shared approach.

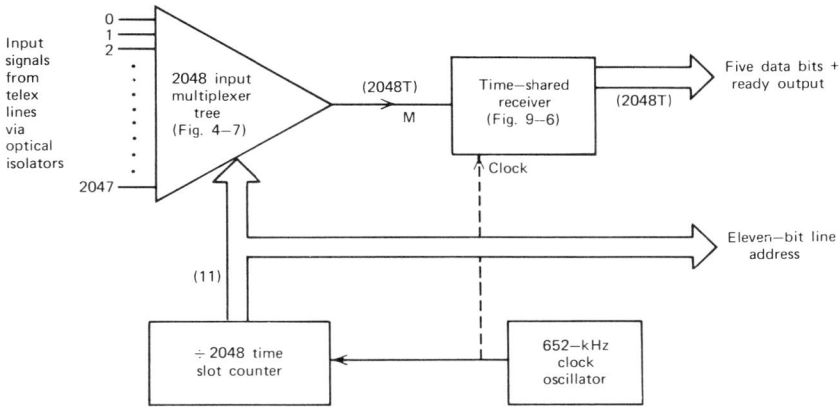

Fig. 11-6. Using the time-shared circuit.

separate lines. We will design a receiver for a single channel, then repeat it in time 2048 times by simply using 2048-bit shift registers in place of the flip-flops in the conventional design. The clock rate will be 2048 times as fast as for a single channel receiver; and the 2048 inputs will be connected, in turn, to the input of the circuit by a 2048-input multiplexer tree addressed by a 11-bit binary time slot counter clocked by the same clock.

The design of the conventional serial data receiver is altered somewhat by the fact that *we must design it using only simple D flip-flops,* with no PRESET or CLEAR inputs. Of course, we can use gating to make a D flip-flop into any other flip-flop type (see Fig. 5-2) if that makes the design easier.

11.4 EXAMPLE: A TIME-SHARED SERIAL DATA RECEIVER

Figure 11-7 shows a block diagram of our serial data receiver. Note that the technique we will use is quite general (except for the use of D flip-flops for counting) and can also be used for non time-shared receivers for serial, start–stop data. In fact, the same principle can be used for writing a program to receive several such lines with programmed logic. A longer data word can, of course, be received by simply lengthening the shift register (e.g., a nine-bit register for eight data bits). Since combinational logic becomes insignificant when time shared, the entire design emphasis is on minimizing the number of flip-flops in the circuit. Instead of counting bits as they shift in, we will use the arrival of the *start* bit in the extra shift register stage to indicate that the entire character has been received.

Since the incoming serial signal is completely asynchronous and we want to "strobe in" the data bits in their center, we clock the circuit at seven times the data rate (7 × 45 bits/sec = 318 clocks/sec) and use a ÷7 *clock phase counter* to strobe in the data bits on every seventh clock, with phasing determined by the start transition. Figure 11-8a is a state diagram for the clock phase counter. When there is a start bit in the final shift register stage (ST' = 1) and the line is idle (M = 1), the counter just stays in the WAIT state. A *start* transition (M = 0) advances the counter to the START PRESET state. This causes the entire shift register to be reset. For the rest of the character, the counter counts in a cycle of seven counts (skipping around START PRESET), generating a shift clock each time it goes through the CLOCK ($P1$) state. As the timing diagram (Fig. 11-8b) shows, this will be at the middle of the start bit and each data bit time.

When all the bits have shifted in, the zero start bit arrives at the last shift register stage (ST) and prevents the P counter from leaving the WAIT state until another start bit is received. To prevent premature detection of a start bit, the P counter is set back to state 0 when the last bit is strobed in. This makes it insensitive to start bits until the middle of the stop bit.

Since the clock rate is exactly determined by a crystal oscillator, the strobe point is exact within ±0.5 clock time. Since the clock time is one-seventh of a bit time, the strobe point for receiving data bits is accurate within ±½ × ⅐ = ±$\frac{1}{14}$ or ±7% for all bits.

Noise pulses on an idle line are rejected by a special NOISE gate that restores the logic to the normal idle state if the start pulse is not at least one-half bit in duration. This operation is shown in Figure 11-8c.

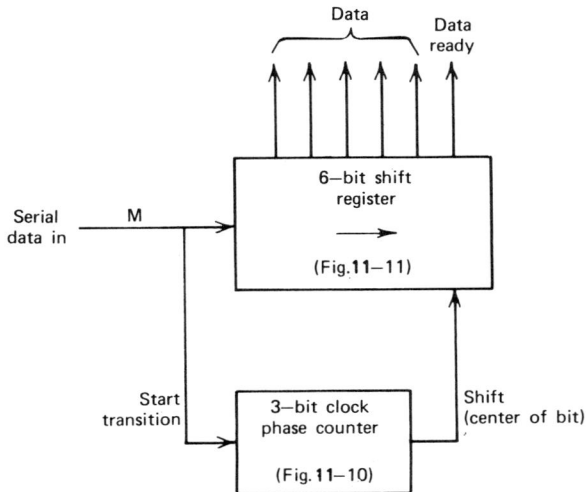

Fig. 11-7.   Serial data receiver block diagram.

PO
000

Preset
P7
111

ST' · M'
(Start)

BI' · ST
(Last bit)

Clock
P1
001

ST
Normal
cycle

Wait
P6
110

ST' · M

P2
010

P5
101

P3
011

P4
100

(a)

CAN TAKE NEW START

M

| Start | B1 | B2 | B3 | B4 | B5 | Stop |

Clock

P ... 6 6 7 0 1 2 3 4 5 6 0 1 2 3 4 5 6 0 1 2 3 4 5 6 0 1 2 3 4 5 6 0 1 2 3 4 5 6 0 1 0 1 2 3 4 5 6 ...

DATA READY PULSE

FF	Contents →		B1	B2	B3	B4	B5
S	B5	X X X 1 1 0	0 B1	B1 B2	B2 B3	B3 B4	B4 B5
h	B4	X X X 1 1 1	1 0	0 B1	B1 B2	B2 B3	B3 B4
i	B3	X X X 1 1 1	1 1	1 1	0 B1	B1 B2	B2 B3
f	B2	X X X 1 1 1	No change	1 1	1 1	1 1	0 B1	B1 B2
	B1	X X X 1 1 1	1 1	1 1	1 1	1 1	0 B1
t	ST	0 0 0 1 1 1	1 1	1 1	1 1	1 1	1 0

(b)

M

P ...6 6 6 6 6 7 0 1 2 3 4 5 6 6 6 ...

ST ...0 0 0 0 0 0 1 1 0 0 0 0 0 0 0 ...

(c)

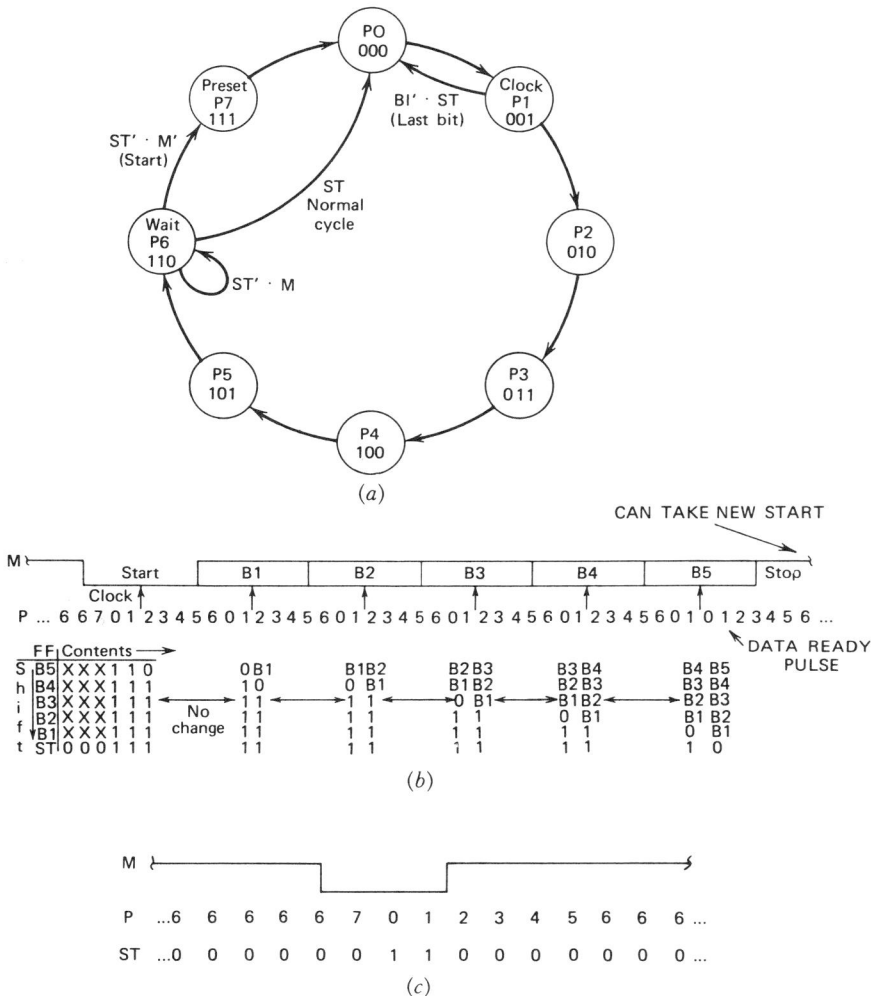

Fig. 11-8. Serial data receiver. (a) State diagram for clock phase counter (seven different phases, hence three bits). (b) Normal character—timing diagram. (c) Noise on idle line—timing diagram.

The preceding discussion is for a conventional logic circuit, but it is also valid, during each time slot, for the time-shared circuit. The only difference in the hardware, in fact, is that shift registers are used instead of D flip-flops. Since the clock rate for each circuit is only 318 clocks per second, we could easily time share the circuit among *tens of thousands of lines*. For example, if we used 10,192-bit shift registers for the flip-flops, the required clock rate would still be only $10{,}192 \times 318 = 3.24$ mHz. A single circuit could thus handle 10,192 input lines! For our example, we will time share the circuit among 2048 input lines. We will therefore need a 2048-input multiplexer tree as shown in Figure 11-6. The input address for this multiplexer comes from an 11-bit binary counter that is clocked by the same $2048 \times 318 = 652$ kHz clock that clocks the 2048-bit shift registers. Each bit of the shift registers thus corresponds to a flip-flop state for one of the 2048 input lines.

Figure 11-9 shows the Veitch diagrams used to design the clock phase counter. In the position corresponding to each state, we must write an expression for the *next state* (D input) of each of the three flip-flops. For example, from state 0 we always go to state 1, so we write 0, 0, and 1, respectively, in the lower right corner of the three diagrams. From state 1 we will go to state 0 or state 2 depending on whether it is the last bit (B1$'$·ST). Since PF4 and PF1 will be 0 in either case, we can write 0's in the state 1 position of these two diagrams. Since PF2 will set only if B1$'$·ST is *false,* we write B1 + ST$'$ in the state 1 position of that diagram (B1 + ST$'$ = (B1$'$·ST)$'$ by De Morgan's theorem). We continue mechanizing the arrows *from* each state on the state diagram on Figure 11-8*a* until we have the completed Veitch diagrams.

Examining the resulting diagrams, we see that they are the kind of messy functions that require many gates. This, of course, is why we usually use *J–K*

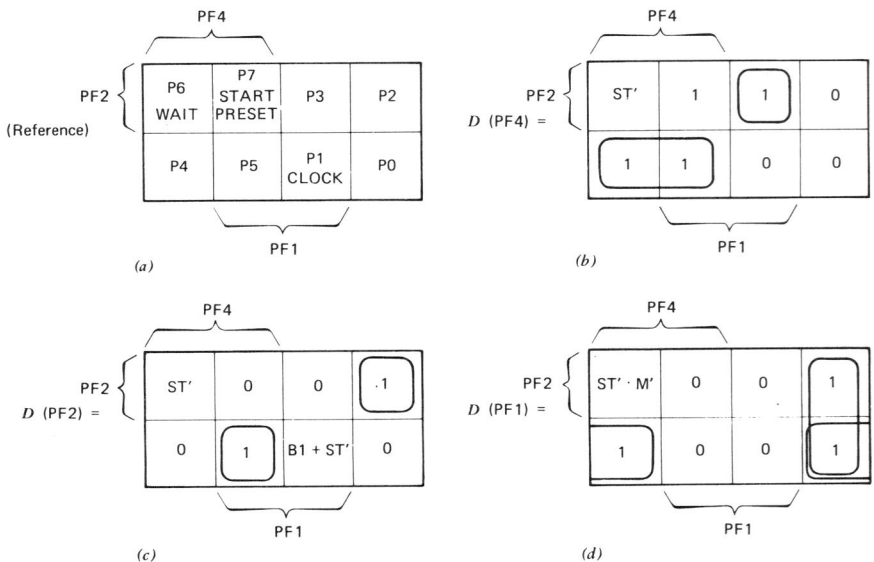

Fig. 11-9. Clock phase counter design.

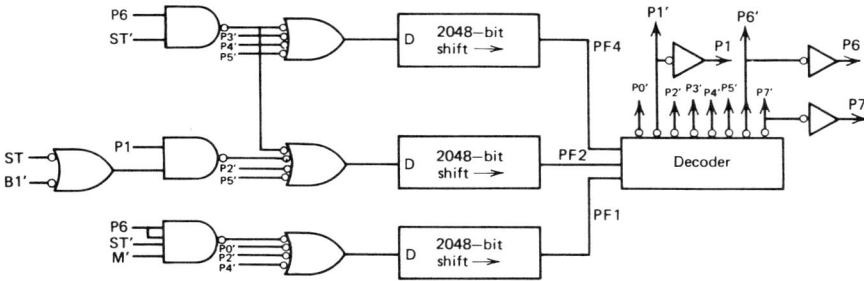

Fig. 11-10. Clock phase counter (time-shared 2048X).

flip-flops for this type of counter. We are using D flip-flops because large shift registers virtually all have D inputs (though some small shift registers, such as the 74195, have J and K' inputs). To simplify our logic, we will use an MSI decoder to decode the states of the counter. Our logic equations, in terms of decoder outputs P0–P7, will then be

$$D_{(PF_4)} = P6 \cdot ST' + P3 + P4 + P5 \qquad (11\text{-}1)$$

$$D_{(PF_2)} = P6 \cdot ST' + P1(B1 + ST') + P2 + P5 \qquad (11\text{-}2)$$

$$D_{(PF_1)} = P6 \cdot ST' \cdot M' + P0 + P2 + P4 \qquad (11\text{-}3)$$

These equations are mechanized in Figure 11-10. Note that instead of D flip-flops, we have shown 2048-bit shift registers. This gives the equivalent of 2048 independent clock phase counters—each active during a different time slot for a different input line.

The six-bit shift register is, likewise, designed as if it were to be made of D flip-flops. Shifting is enabled only during state P1 when ST is set. Since we must shift only when P1·ST is true, we need logic to keep the previous state of each shift register stage when P1·ST is false. Figure 11-11 shows how this can be done with quad two-input multiplexers. Each shift register state input comes from *itself* if P1·ST is false, or from the next stage on the left if P1·ST is true. It will thus shift only if P1·ST is true, and remain unchanged otherwise. The STROBE input to the multiplexers makes their outputs all go false during the P7 (START PRESET) state. Since it is so easy to force all zeros into the register, the data are shifted in inverted. This makes the (zero) start bit shift in as a 1. When this 1 appears in the last stage (ST'), it is gated with P1 and used as a DATA READY indication. This gives us a DATA READY signal that goes true only during one clock time (the second P1 state during bit five).

If a brief noise pulse made an idle line go to state 0 for less than one-half bit

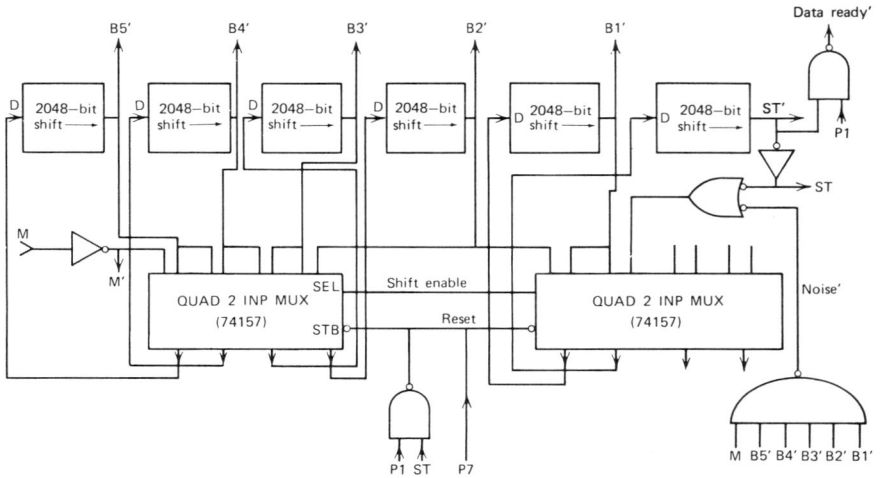

Fig. 11-11.   Shift register for serial data receiver.

time, the shift register could preset, then just keep shifting in the stop code. Since there would be no *start* bit, ST' would just stay reset so the phase counter would keep on counting. When a real character did come, it would be clocked in at whatever phase the phase counter happened to be counting in, possibly producing errors. To prevent this, an extra (noise) gate is added to set ST' if a 1 start bit is detected. This immediately stops the sequence without generation of a DATA READY, as shown in Figure 11-7*b*.

When the DATA READY signal goes true, the five data bits and the 11-bit line address (from the time slot counter) are loaded into a **buffer** for input to the processor. Since DATA READY signals occasionally come on consecutive clock pulses (if two lines assigned to adjacent time slots receive characters simultaneously), a data buffer or DMA connection to the processor memory is required. Another approach is to let the clock pause whenever a character is ready until the character can be taken. To maintain timing accuracy, the clock can be run in bursts at a higher speed, with a fixed amount of time between the start of the bursts.

The M input to the circuit is produced by a 2048-input multiplexer tree (see Fig. 11-6). Part of this tree is on the same module with the receiver logic, and part of it is on the line interface modules with the optical isolators. If we put a 64-input multiplexer (as shown in Fig. 4-7) on the time-shared receiver module, then each of the 64 inputs can come from a 32-input multiplexer on a different line module. The system is therefore expandable in increments of 32 by plugging only as many line modules as required.

To indicate the number of time slots on a wire or circuit shown on a diagram, the number of time slots, followed by T, should be written in parentheses next to the wire. This is similar to the notation sometimes used in block diagrams, where, for example, 32 wires are shown as one wire with (32) written next to it. A wire with 32 time slots is thus indicated by writing $(32T)$ next to it. Each of the 64 inputs to the multiplexer on the line receiver module could thus be labeled $(32T)$. The M signal out of that multiplexer would be then labeled $(2048T)$, as shown in Figure 11-8.

If we count the IC packages for our time-shared line receiver example, we see that *only eight* logic packages are shown on Figures 11-10 and 11-11. Since the circuit handles 2048 lines, the logic cost per line is only a fraction of a cent. Nine shift registers are required with one bit per line to be handled. Since large shift registers cost only a fraction of a cent per bit, *the total cost per line is only a few cents!*

This approach is thus even more economical than programmed logic, where the same job must be done many times. The reason is that with programmed logic we time share logic in a general way, which is somewhat inefficient for any particular job. (Compare, for example, the efficiency of this logic with the program on Table 8-3, Section 8.1.1.) By designing special, time-shared logic for a particular job that must be done many times, we can achieve much greater efficiency.

11.5   OTHER POSSIBLE TIME-SHARING ORGANIZATIONS

There are many other possible ways to time share circuits. We need not have one shift register per bit, for example. We could instead use a single 1024-bit shift register as *all* of the memory elements for 115 Telex line receivers; Figure 11-12 shows such an arrangement. The nine bits (six shift registers plus three phase counter bits) required for each channel would simply be stored in consecutive locations of the 1024-bit shift register. With nine shift clocks, a new counter and shift register state could then be set up by shifting the bits into a single receive logic circuit mechanized with MSI circuits. On the tenth clock the receive logic would function, incrementing the clock phase count and/or shifting in a new bit, just as a single line receiver would do. On the next nine clocks the previous states for the next channel shift in, while the new state of the other channel shifts out into the 1024-bit shift register. Since $3 + 8 + 1024 = 1035$ bits are in the circular shifting path, $1035/9 = 115$ Telex lines can be handled by the circuit.

Of course, various combinations of serial and parallel bit storage could be used. Choice of organization depends on speed requirements, number of circuits required, and shift register types available. The important thing is to

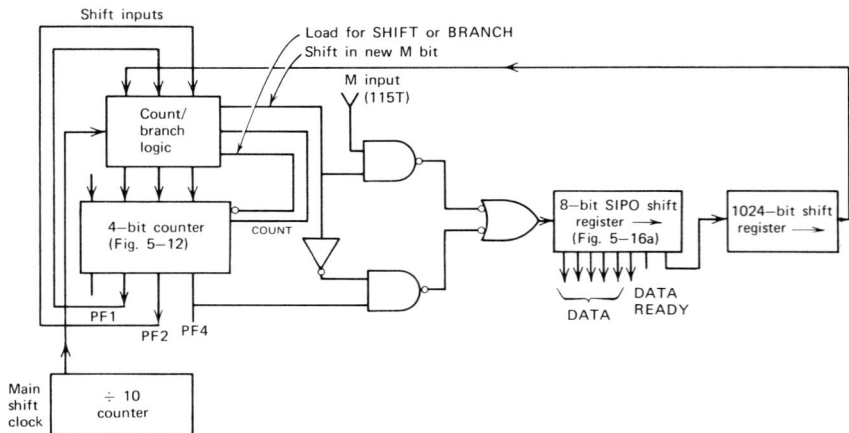

Fig. 11-12.   Time-shared line receiver with one shift register.

learn to use the time dimension as fluently as we use the space dimension. The
organization in Figure 11-12, for example, is essentially an extension of the
previous approach. In addition to time sharing the channels, we time share the
nine bits. We thus use one MOS shift register and nine time slots, instead of
nine separate shift registers "repeated in space."

Another thing to remember is that a time-shared circuit can be designed to
do *several jobs*. For example, if we design a serial data receiver that could be
"programmed" with two control bits to receive one of four formats (e.g., different data rates, or number of bits), we could time share it and effectively
have any desired mixture of the four receiver types. We can thus have the efficiency of a specially designed circuit with some of the flexibility of the
programmed approach. Another possibility would be to design a combination
send/receive circuit with a send/receive control bit. We could thus use the
same shift register bits for sending or receiving on a particular line. Since the
combinational logic cost becomes negligible in a time-shared circuit, the added
logic complexity is of no consequence.

Recirculating shift registers (e.g., the Intel 2401 dual 1024-bit) have a built-
in multiplexer to recirculate internally the previous contents except when a
WRITE input goes true. This can sometimes be used to simplify control logic,
as the effect of the WRITE input in a time-shared circuit is as if it was a flip-
flop *clock enable*. Except for the need to reset on P7, the time-shared shift
register in Figure 11-11 could be done without the external multiplexers if
recirculating registers were used. A time-shared counter designed with recirculating shift registers will count only when the WRITE input is true.

11.6 TIME-SHARED OUTPUT CIRCUITS

Outputs from time-shared circuits can be converted from a sequence of brief signals separated in time to signals separated in space by a *demultiplexer* tree similar to the one shown in Figure 4-11. The addresses for the tree would, of course, come from the time slot counter. The tree in Figure 4-11, however, will simply produce a pulse, or no pulse (for 0 or 1), on each input during the corresponding time slot. What we normally want is a continuous logic level output whose state is updated each clock time. To obtain continuous outputs, we must have a flip-flop for each output. An *addressable latch* (e.g., a 9334) is like an eight-output demultiplexer with storage on each output. We can therefore get continuous outputs by simply using addressable latches for the *final stage* of a demultiplexer tree.

We could, therefore, build a time-shared transmitter for sending serial data signals like those shown on Figure 11-7a. A demultiplexer tree, partly on the time-shared logic board and partly scattered on the boards containing the send relays, would then make the time-space transformation. The final level of the tree would consist of addressable latches.

Output circuits also differ from input circuits in that the processor (or control logic) *initiates* the output. Whenever the processor wants to initiate an output character, it must put the character into the desired time slot of the output shift register. The output circuit then takes care of sending the start bit and shifting out the character at the proper rate. Since the processor may output to any line at any time, it would have to wait for the proper time slot to come up if we simply used a shift register for the time-shared flip-flop as we did on the input circuit. This problem is easily solved by just using *a dynamic RAM, addressed by the time slot counter, as the time-shared flip-flops*. Since each bit will be addressed in turn, the RAM will look just like a shift register to the time-shared circuitry.

As Figure 11-13 shows, two input multiplexers can be used to switch the data and address inputs such that the processor can output to any address at any time, yet the time-shared circuit can continue to use the RAM as a time-shared flip-flop. The address multiplexer can, of course, serve all time-shared flip-flops, but one-fourth of a quad two-input multiplexer is needed for each data input. A similar circuit can be used for time-shared *input* circuits, where it is desirable to read the states of the inputs randomly, under program control, instead of immediately when each input becomes ready.

Because of the complexity of these circuits, they can be represented by a simple *D* flip-flop diagram on the logic diagram for the time-shared logic device. A note on the schematic can show the actual circuit (Fig. 11-13) and indicate that the *D* flip-flop signals actually indicate that circuit.

11.7   TROUBLESHOOTING TECHNIQUES

Once a few simple principles are learned, troubleshooting time-shared and serial circuits is no more difficult than fixing the same circuit in conventional form. In fact, the principle of treating the circuit as a conventional circuit can be used in troubleshooting as well as in design.

Most gross faults, of course, can be found by just looking for an activity at some point in the circuit. This is actually *aided* by the time sharing, since the activity on thousands of circuits (time slots) can be seen at one time so there is always some activity.

As a servicing aid, an extra *D* flip-flop can be put on one of the modules in the system. The desired time slot decode is connected to the clock input, and a test lead is connected to the *D* input. An oscilloscope connected to the flip-flop output will show waveforms *exactly like those that would be seen on a conventional circuit for a single channel*. The flip-flop input test lead is touched to any point in the circuit just as is normally done with a scope probe. The flip-flop holds the signal value at that point during the desired time slot and allows the circuit to be repaired as an ordinary circuit (see Fig. 11-14).

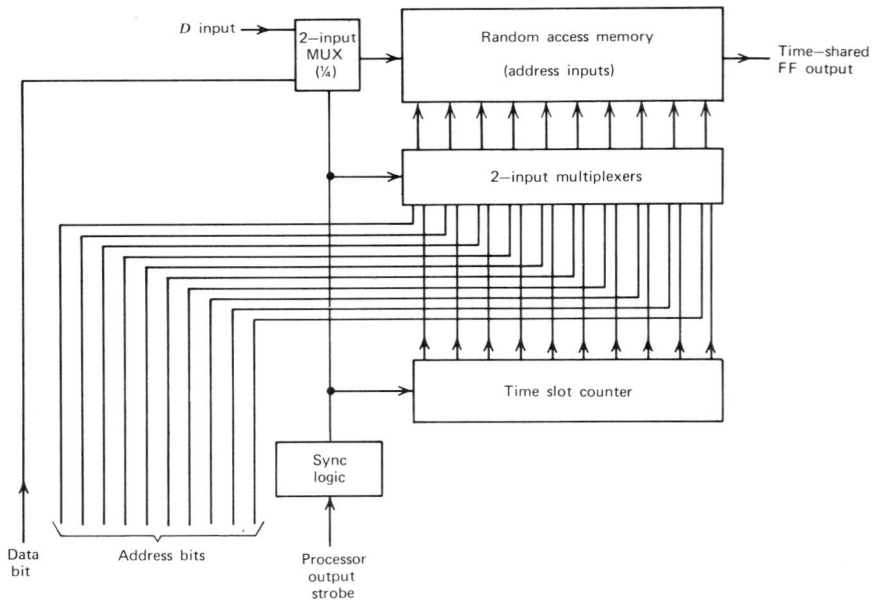

Fig. 11-13.   Random access, time-shared *D* flip-flop.

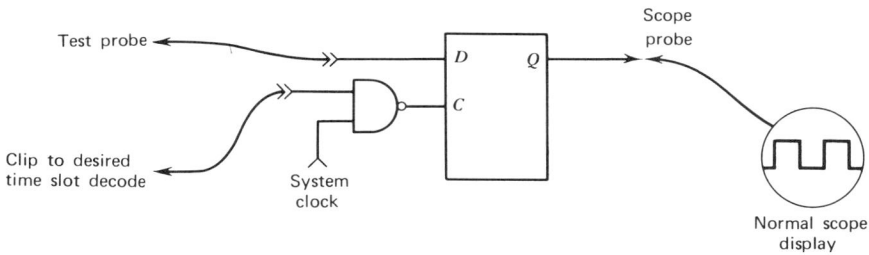

Fig. 11-14. Test flip-flop.

Of course, the biggest service aid of all is that much less servicing is required because only one board must be kept working instead of the large number replaced.

11.8 SENDING DATA SERIALLY

Serial operation is often thought of as a way to save *logic gates* by effectively time sharing them among many bits. Almost more important, serial operation *time shares interconnecting wires*. With the availability of MSI and LSI, the cost and failure rate of a system have increasingly become related to the number of *interconnections* required. By sending one bit at a time, many bits can be sent over a single interconnection.

Saving interconnections becomes even more important when signals are sent over a long distance because the longer the distance, the more expensive the wire. Not only the wire, but the connector pins and send and receive circuits, can be time shared by serial operation. When signals are sent over really long distances, such as over telephone lines, the line and send and receive circuitry becomes so expensive that serial operation is virtually always used.

As the logic required to convert a signal from parallel to serial becomes cheaper, it becomes more and more attractive to send signals serially—even over relatively short system cables. Figure 11-15a shows a typical system for sending parallel signals over a cable. Separate send and receive circuits, wires, and connector pins are needed for each data bit and for the clock.

Figure 11-15b shows a serial system for sending the same data using a serial start/stop format (like the one shown in Fig. 11-7a). We thus have a tradeoff between the cost and reliability of the added logic, and the senders, receivers, connector pins, and wires eliminated. Often, there is actually a *net savings* in logic required with the serial approach since the serial operation can be carried as far as desired into the subsystem at each end, causing a savings in logic

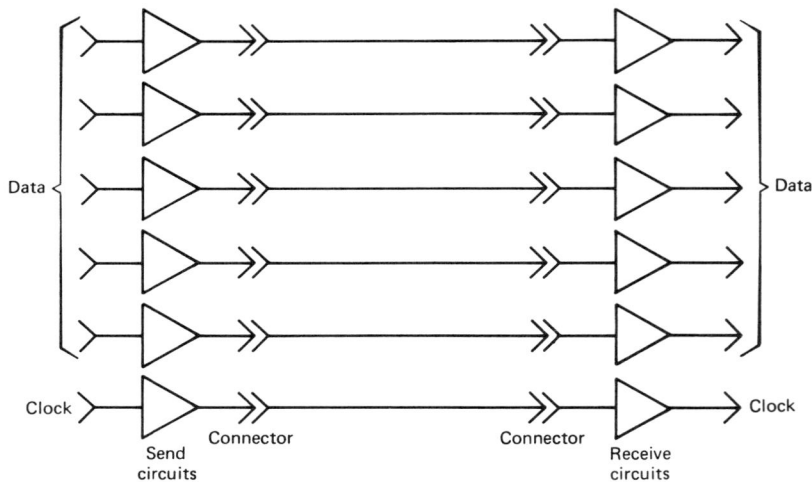

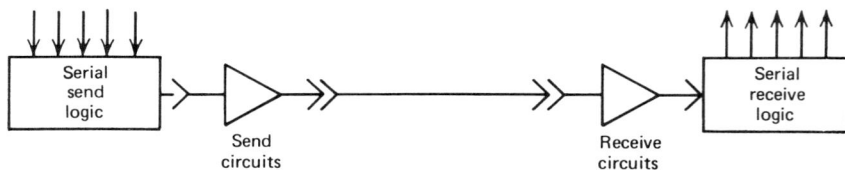

Fig. 11-15.   (a) Parallel data transmission. (b) Serial data transmission.

that offsets the extra logic required to add start and stop bits and generate receive clock. *Parity,* for example, can be generated, or checked, with a single *J-K* flip-flop that toggles for each 1 received when the data are serial. With parallel data a complex logic circuit is required to do the same job. Since data are normally sent from, and received into, flip-flop registers, serial operation reduces the number of packages required for these registers. Instead of using, parallel-in parallel-out, *four-bit* registers (Fig. 5-16c) at both ends, eight-bit registers can be used at each end (a parallel-in serial out at the send end, and a' serial-in parallel-out register at the receive end).

By using a crystal-controlled clock at each end, the start/stop format (Fig. 11-7a) can be extended to essentially any number of bits by simply lengthening the shift registers at each end. A crystal-controlled oscillator gives ±0.001% accuracy for practically no extra cost. The cumulative error over 30 bits with 0.001% accuracy is therefore only 30 × 0.001 = 0.03%. Figure 11-16 shows an example of a computer output format that includes 16 data bits, a six-bit de-

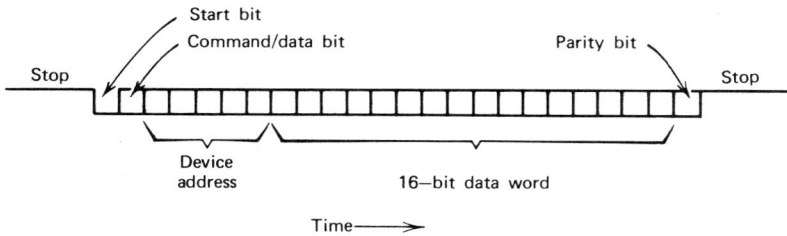

Fig. 11-16.   Long start/stop serial output word.

vice address, a control bit, and a parity check bit. *A single driver and receiver thus replace 25 separate ones*; cost and reliability are therefore improved by a factor of 25. The savings in parity check logic, and use of eight-bit registers instead of four-bit, more than offsets the additional logic required.

It might seem at first glance that speed of data transfer is 25 times slower with the serial interface. In reality, however, it can actually be *faster*. The reason is that parallel transfers are usually limited in speed, *not* by speed of the sender-receiver link, but by *skew* problems. Variations in driver and receiver delay, and variations in the length of the wires in the cable (since they are twisted about a center), make it necessary to provide a *timing guard band* on either side of the clock transition. With serial operation, the data produce their own clock, so skew is a meaningless concept. Simple send and receive circuits can easily send data at 20 mHz (see Section 12.9). All 25 bits can thus be sent in $25/20 = 1.2$ μsec. Skew requirements make this a minimum transfer time for a parallel system with a cable of any length.

If even greater speed is needed, a *serial/parallel* system can be used. Since each path is self-clocking, there is still no skew problem. When all start bits arrive at the end of the shift registers, the data are ready.

Another benefit of serial transmission is elimination of *cross talk*. Since cross talk is additive, large noise signals can be generated on a parallel interface with only a small amount of coupling from *each* line. For example, if 24 of the 25 lines of a parallel system change simultaneously, the twenty-fifth line will have 24 additive cross talk signals coupled to it. If each line produces only 0.04 V of noise, the total effect is $24 \times 0.04 = 1$ *V of noise*. Certain patterns can thus cause incorrect operation on a *parallel* transmission system with very little cross talk.

11.9   SERIAL STORAGE DEVICES AND ERROR CHECKING

Just as we can save hardware by sending data over wires serially, storage devices also often write and read data serially. A single head on a disc memory can read or write tens of thousands of bits on one magnetic track. Each of these bits can be thought of as having *an address in the time dimension* as well as the space dimension.

To make it possible to read and write a reasonable amount of data at one time, the serial data are usually divided into *sectors* numbered from an *index* mark on the disk. Each sector includes hundreds of data bytes plus extra bytes for error checking and sector identification. Since all bits are recorded serially, a single read and write amplifier can be used for all bits. All error checking, track selection, buffering, and even transmission to the computer interface logic can also be done serially. By running a counter in synchronism with the rotation of the disk, as shown in Figure 11-17, any desired sector can be accessed.

Magnetic tapes and data communications systems use an almost identical system. The blocks of data, in this case, are called *records*. Records are usually given numbers and identified by special characters at the beginning of the record. The *program* then usually checks or searches for the desired record. If an error is detected on a particular record after writing or transmission, it is *rewritten* or sent again.

Parity checking and generation with serial data are very easy. Figure 11-18*a* shows a *D* flip-flop connected so it will toggle every time the DATA input is a 1. If this flip-flop is set at the beginning of each byte, it will contain the *odd parity* bit at the end of the byte (since it will be 1 if it has toggled an even

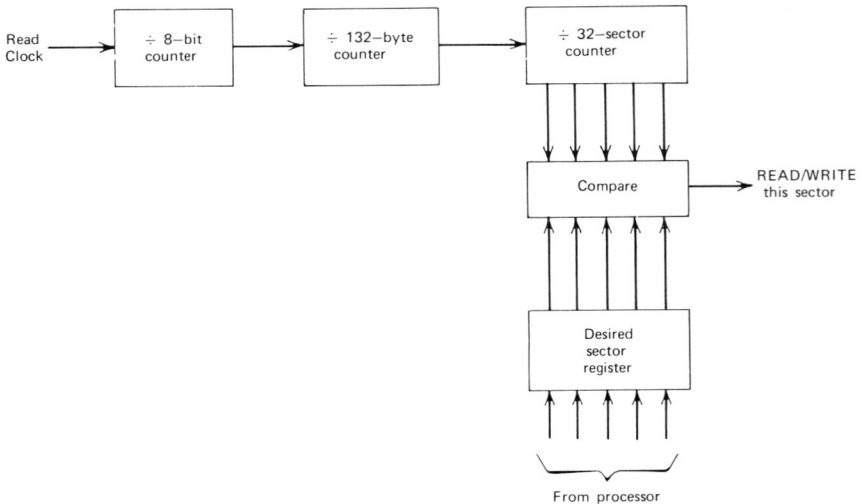

Fig. 11-17. Finding the desired sector.

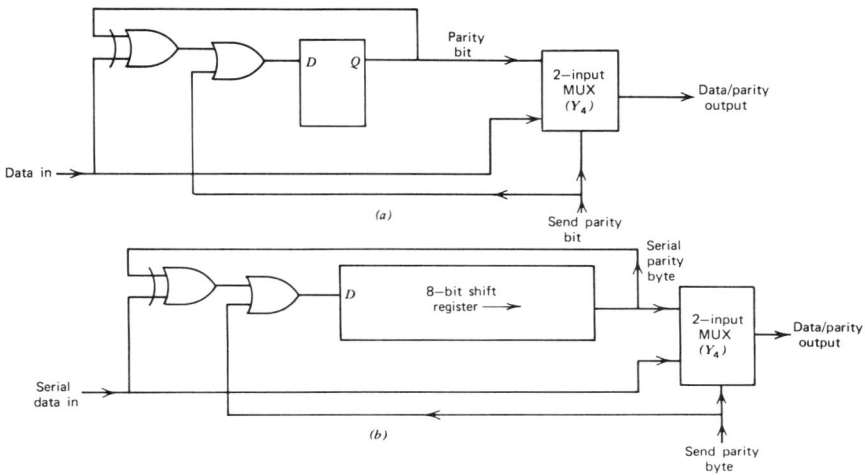

Fig. 11-18.   Serial parity generation: (a) parity bit logic; (b) parity byte logic.

number of times). A two-input multiplexer on the output causes the parity bit to be output at the end of each serial byte. At the same time a 1 input is forced by the input OR gate.

For a long message more than one check bit is usually used. One common technique is called a *block check code* or *longitudinal parity byte*. This consists of a byte at the end of the message, each bit of which indicates parity for the corresponding bit of all bytes in the message. With parallel logic this would require a circuit as shown in Figure 11-18a *for each bit*. Since the bits are serial, we can *repeat the circuit in time instead of space* by simply using an eight-bit shift register in place of the flip-flop, as shown in Figure 11-18b. During each bit time, the bit shifted into the register will be inverted if the corresponding data bit is 1. At the end of the message the parity byte is simply shifted out of the register via the two-input multiplexer. At the same time eight 1's are shifted in to initialize the register for the next record.

EXERCISES

1. Design a circuit that serially inverts all 16 bits in a shift register.

2. Design an eight-input AND similar to Figure 11-1a but with only two time slots.

3. (a) Design a 32-bit serial-parallel adder using a single four-bit adder circuit. The circuit should add register A to register B and put the result in register C. Include clock source, time slot counter, and carry flip-flop control.

(b) How long does the addition take if #9328 dual eight-bit shift registers (with a maximum propagation delay of 39 nsec and a maximum setup time of 16 nsec) are used? The four-bit adder has a maximum delay time of 39 nsec.

(c) What is the package count?

(d) Design a faster 32-bit adder using two 4-bit adders.

(e) What is the total add time of this version?

(f) What is the package count?

4. Design an adder similar to the one in exercise 3a and b but with any of four registers usable as the addend, augend, or destination, as indicated by three 2-bit address fields.

5. Design a circuit for comparing the contents of two 8-bit registers using a single Exclusive OR gate and producing an output during the eighth time slot.

6. Design a circuit for transferring the contents of one of eight 8-bit shift registers into eight consecutive bit locations in a 1024 × 1 bit RAM. A seven-bit address indicates the desired RAM byte location, and a three-bit address indicates the desired source register.

7. Design a circuit, similar to Figure 11-4, for detecting a 00110000 result.

8. Design a circuit to serially ADD eight consecutive bit locations in a 1024 × 1 RAM to the contents of an eight-bit accumulator register. The result should go to the accumulator. A seven-bit address indicates the desired RAM byte.

9. (a) Design a circuit for checking the state of an input signal every 29 msec and giving a true output if that state has changed since the previous clock.

(b) Make a time-shared equivalent of 1024 such circuits that will indicate whenever any of the 1024 inputs change and indicate the binary address of that input.

10. Design a system for keeping a Gray code count (Fig. 5-6a) of the number of times each of 1024 inputs is true. Each input should be sampled once every second. Use three 1024-bit *recirculating* shift registers and the same logic shown in Figure 5-6f.

11. Redesign the serial data receiver example (Figs. 11-10 and 11-11) to receive eight-bit characters, with 10 msec per bit, from 1024 lines. Use *recirculating* shift registers with two input AND gates instead of multiplexers between shift register stages. Simplify the D inputs to the counter by using the WRITE input to enable state changes. The D input equations then will be simplified because of "don't cares" in the states where WRITE is false.

12. Design the "count/branch logic" required for the 115-line serial receiver in Figure 11-12.

13. Design a 2048-line, time-shared, serial data transmitter that will produce the start/stop format of Figure 11-8b when a character is loaded in the shift register. While a RAM arrangement as shown in Figure 11-13 will be used for each time-shared flip-flop, use a simple D flip-flop symbol with (2048T) written inside it on the logic diagram.

14. Design a time-shared circuit for scanning telephone lines for changes in state (picking up and hanging up receiver) and counting dial pulses. Use seven time-shared flip-flops as follows: one to remember the previous state of the line, two to

count time since the last change, and four to count dial pulses. A *data ready* signal should indicate that a four-bit code from the dial pulse counter is available for input. Digits should be indicated by their normal BCD code. When a telephone is picked up, a 1111 code should be input. Lines should be scanned every 20 msec. A dialed digit is indicated by the number of times the line current is broken for 60 msec, with a 40-msec spacing between breaks. If the line breaks for more than 300 msec the party is assumed to have hung up.

15. Design a 1024-output demultiplexer for driving send relays. Instead of using addressable latches, use eight-bit serial-in parallel-out shift registers. Take advantage of the slow response of relays by periodically fast-shifting the new states in. Draw the time slot counter, a tree for generating 64 different shift clocks, and one of the eight-bit output stages.

16. Design a seven-wire transmission system that will light 64-remote LEDs to indicate the position of 64 toggle switches. The switch end should send the states of the 64 switches in order by sensing the switches using a matrix like the one in Figure 4-12b, with a switch at each intersection. This matrix should be driven by a counter counting at 10 kHz. The six-counter bits are sent to the LED end and used to drive a matrix like the one in Figure 4-12a, with an LED at each intersection.

17. Design a system similar to exercise 16 but with a single connection. Each end should have its own counter. The counter at the LED end should be kept in synchronization by a start bit from the switch end. The line should be idle, with a stop bit on it, 50% of the time.

18. Design a serial transmission system using the format shown in Figure 11-16. The send end should have 16 parallel *data* lines, five *address* lines, a *command strobe,* and a *data strobe* as inputs. The receive end should provide similar outputs but with an additional *parity error* output.

19. Design a first-in/first-out buffer to hold characters received by the time-shared serial data receiver (Figs. 11-10 and 11-11). Whenever DATA READY goes true, the five data bits and 11 address bits should be loaded into two 8-bit parallel-in serial-out shift registers. The 16 bits should then be immediately shifted out at 6 mHz and written, one bit at a time, into a 256×1 bit bipolar RAM. A four-bit write address counter and a four-bit read address counter should keep track of the 16 possible write and read "word" locations. Data can be independently (on alternate clocks) read out, under computer control, and presented as a 16-bit parallel word to the computer input. If the buffer is full, old data should simply be written over.

20. Design an error checking system for serial data transmission of 16-bit words. The word is first sent inverted, then non-inverted. At the receive end a bit-by-bit comparison is made of both transmissions of the word.

CHAPTER TWELVE

NASTY REALITIES II
Noise and Reflections

12.1 INTRODUCTION

When we design logic, we work in the pure world of mathematics, where equations perfectly express the functions we want. When we try to mechanize logic equations in the real world, we find that the gates we use have certain limitations, and interconnecting wires are far from perfect. Instead of simply transmitting a perfect replica of the signal, an interconnecting wire adds signal reflections and noise that may cause the system to malfunction. The malfunctions are of the marginal type that show up only with certain ICs and certain patterns of signals. We must therefore be very careful in interconnecting logic for *we have no guarantee that we will find noise and reflection problems in checkout.* In most cases we can avoid problems by simply following certain *wiring rules.* The remainder of this chapter discusses the physical effects that make these rules necessary.

Low-speed logic circuits, such as the CMOS family, approach the mathematical ideal. *Because the circuits are slow, they are essentially unaffected by noise and reflections.* Even their DC characteristics are virtually ideal. Since gate inputs draw no current, **fan-out** (the number of inputs that can be driven by one output) is essentially unlimited. Although most of the problems discussed in this chapter are significant only where long cables are used with CMOS circuits, they are significant for normal logic interconnections for TTL circuits. With faster logic families (e.g., 74S or ECL) the problems can become significant with interconnections only *a few inches* long.

A sequential logic system (Fig. 3-1*a*) consists of flip-flops and gates interconnected so that the next state of the flip-flops is a function of the present state and the system inputs. As long as these state changes go according to the design, the system is working properly. Noise and reflections are thus a problem only if they cause a flip-flop to go to the wrong state. Generally, if a flop-flop input is changed by noise, for less than the *setup time* of the flip-flop, it still goes to the correct state. The slower a flip-flop is then, the longer the duration of noise it can tolerate.

The maximum duration of noise or reflections that can be produced on a piece of wire is related to its length. Depending on the type of insulation, *this duration varies between 3 and 4 nsec/ft of length*. For any given logic type we can thus build systems up to a certain physical size and essentially forget about noise and reflection problems. For example, a TTL flip-flop (e.g., 7474) has a *typical* setup time of 15 nsec. If we assume a minimum setup time of 10 nsec, we can use it in a system with interconnections of up to $10/4 = 2.5$ ft long without much worry. If, however, we use the faster, Schottky TTL flip-flops (e.g., 74S74) the typical setup time is only 3 nsec, so we may have problems with wires only $2/4 = \frac{1}{2}$ ft long! Even faster ECL can have problems with wires only 3 in. long.

12.2 WIRING AS AN ANTENNA OR TRANSMISSION LINE

The relationship between maximum noise pulse duration and wiring length is related to the fact that a piece of wire makes a poor *antenna* unless it is at least one-fourth wave long. System wiring can be thought of as a receiving antenna for external noise. *Cross talk* (coupling of signal components from one wire to another nearby) can be thought of as radio transmission, where one wire acts as a transmitting antenna and the other as a receiver. Since the wires make poor antennas when less than one-fourth wave long, only the higher frequency components of the signal are coupled.

An even more useful way to look at high-speed system wiring is *as a transmission line*. Simple transmission line concepts allow us to predict reflection waveforms with amazing accuracy. The transmission line concept becomes not only useful but necessary if we are to operate logic with connections longer than the maximum lengths just calculated for the various logic types (e.g., 3 in. for ECL).

Most logic wiring does not resemble our idea of a "transmission line" at all, but even random wire-wrap wiring has a fairly uniform *impedance* and *propa-*

gation velocity. The propagation velocity is largely determined by the type of insulation. Almost all standard wire insulations give a delay of 1.54 nsec/ft; glass-epoxy PC boards give a delay of about 1.8 nsec/ft. Capacitive gate loads along a wire can increase the delay a little more, so that 2 nsec/ft is a pretty good worst-case number to use. In the real world it is very difficult to make a transmission line impedance below 30 Ω or above 600 Ω. For example, to make a 600 Ω impedance, we must suspend an insulated No. 18 wire *20 ft* above the ground. If we tape a No. 30 insulated wire down flat to a grounded copper sheet, its impedance is still about 70 Ω. The same wire about 0.1 in. above ground gives an impedance of about 200 Ω. We can thus assume that random wire-wrap wires will have an impedance that varies from about 100 to 200 Ω, or 150 Ω ± 33%. This is sufficiently accurate for the kind of analysis we need if we are not pushing things. By using twisted-pair wire we can control impedance within about ± 10%. Depending on the particular twisted-pair wire, impedance will be about 100 Ω. Printed circuitry conductors over a copper "ground plane" can be designed for a fairly consistent 50 Ω if very high performance is desired.

12.3   REFLECTIONS AND RINGING

The characteristic impedance (Z_0) of a wire is the ratio of voltage/current through the wire for high-frequency signals. If a wire is terminated by a resistor that is equal to the impedance of the wire (see Fig. 12-1*a*), there will be no problem when a signal propagates to the end of the wire. If, however, some other impedance is at the end of the wire, something has to give. The result is that a *reflection* is produced such that all boundary conditions can be satisfied.

Figure 12-1*b* shows an example of a *series-terminated* line. In this case only half the voltage step appears on the line initially, because the voltage step divides equally across the series resistor and the line impedance. An $E/2$ voltage step and an $E/2Z_0$ current step thus travel down the line. The open circuit on the far end means that the current at the end of the wire *must be zero* at all times. The current will go to zero when the step reaches the end of the wire if a reflection, consisting of a current step of $-E/2Z_0$, is produced the instant the step arrives at the far end. This reflection signal produces a voltage of $E/2Z_0 \times Z_0 = E/2$, which adds to the original $E/2$ step to produce a total output voltage of $E/2 + E/2 = E$. When the reflection reaches the sending end, the voltage there also goes to E. This kind of a linear problem can be solved easily by solving for the *reflection coefficient* as follows:

$$\text{reflection coefficient} = \Gamma_R = \frac{Z_r - Z_0}{Z_r + Z_0} \qquad (12\text{-}1)$$

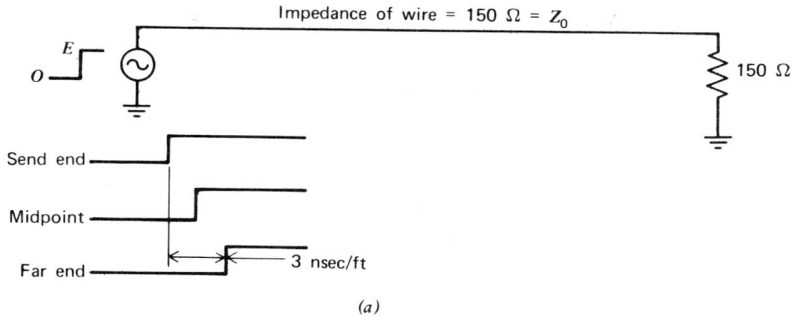

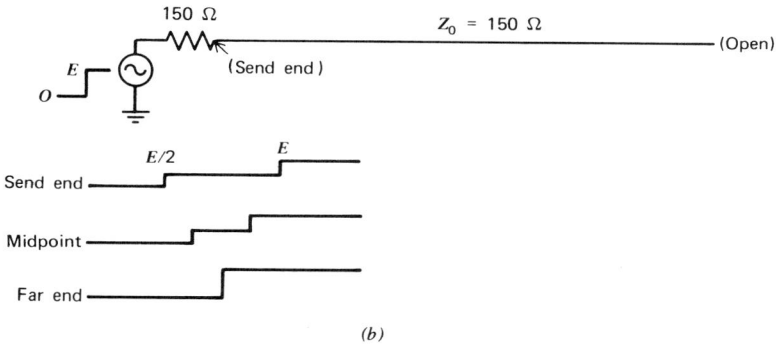

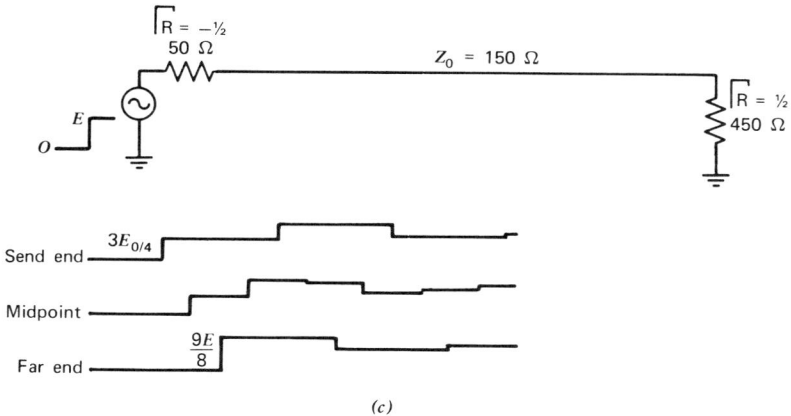

Fig. 12-1. (a) Parallel-terminated line ($R = Z_0$). (b) Series-terminated line ($R = Z_0$). (c) Improperly terminated line ($R_1 = Z_0/3$, $R_2 = 3Z_0$).

The reflected signal is thus the arriving signal multiplied by the reflection coefficient. The reflection coefficient at the open end in Figure 12-1b is thus

$$\Gamma_R = \frac{\infty - Z_0}{\infty + Z_0} = 1 \qquad (12\text{-}2)$$

On the sending end $Z_r = Z_0$, so

$$\Gamma_R = \frac{Z_0 - Z_0}{Z_0 + Z_0} = 0 \qquad (12\text{-}3)$$

Thus there is no reflection from the sending end. If there are reflections from both ends, the original signal will continue to bounce back and forth between the two ends with a period equal to the round-trip time down the wire (3 to 4 nsec/ft) as shown in Figure 12-1c. Each time the step "bounces off" the discontinuity at the end of the line, a new amplitude is calculated by multiplying the arriving step by $1 + \Gamma^R$ (arriving step plus reflection produced).

In the next section we develop a more generally useful method of predicting reflections. However, it is interesting to consider the reflection coefficient produced by various terminations. A short circuit produces a reflection coefficient of -1 (for zero net voltage). All terminations having less than the characteristic impedance produce negative reflection coefficients (reducing the output voltage), and all terminations above the characteristic impedance produce positive reflections (increasing the output voltage), asymptotically approaching a reflection coefficient of $+1$ (doubled voltage) with an infinite impedance. A termination three times the characteristic impedance produces a 50% reflection (50% overshoot).

Although equation 12-1 can also be used with complex terminations (including capacitance or inductance), a more intuitive approach yields good results too. For example, a capacitance across the terminator draws a current

$$I = C \frac{dv}{dt} \qquad (12\text{-}4)$$

during the rising slope of arriving signal. This current can be thought of as a reflection current waveform that will produce a voltage reflection pulse of $-IZ_0$, superimposed on any reflection produced by the resistive termination. A capacitive termination can thus be thought of as producing a reflective pulse during the rise or fall of the arriving signal, which opposes that signal and therefore delays its rise. Capacitive loads along a line (such as gate inputs) produce a similar reflection pulse propagating in *both directions* down the line. If such loads are spaced evenly along the line such that the round-trip propagation time between them is *greater than* the rise time of the signal, the reflections will not be additive.

Unterminated "stubs," as shown in Figure 12-2, create a discontinuity with *a duration equal to the round-trip delay of the stub.* During this time the line impedance is $Z_0/2$ since the stub and the line are driven in parallel. A reflection of

$$\Gamma_R = \frac{Z_0/2 - Z_0}{Z_0/2 + Z_0} = \frac{-Z_0/2}{3Z_0/2} = -\frac{1}{3} \qquad (12\text{-}5)$$

is thus produced *until the reflection from the end of the stub returns the signal voltage to normal.* (The $(2/3)$ E signal reflects off the open stub end as $(4/3)$ E, then is reduced to E by the $-1/3$ reflection from driving the main line in both directions.) Again, the effects of these reflections will not be additive if the stubs are sufficiently widely spaced along the line that the reflections will not be superimposed. It is possible, however, that reflections will overlap if stubs are spaced evenly (e.g., the *reflection off the send end* of the first stub reflection could add to the second stub's reflection). If the *duration* of the reflections is less than the minimum setup time of the logic used, their effect can usually be ignored. Connector impedance discontinuities likewise *can usually be ignored* because the duration of the reflections they produce (see Fig. 12-3) is equal to the round-trip time it takes the signal to travel through them. Actually, if the duration of the reflection is less than the rise time of the signal, the effect will not be visible as a notch. Instead, the signal *slope* will just be reflected, giving a much less noticeable effect. The important thing is that the flip-flop input effectively *integrates out the effect of the reflections* if their period is short enough.

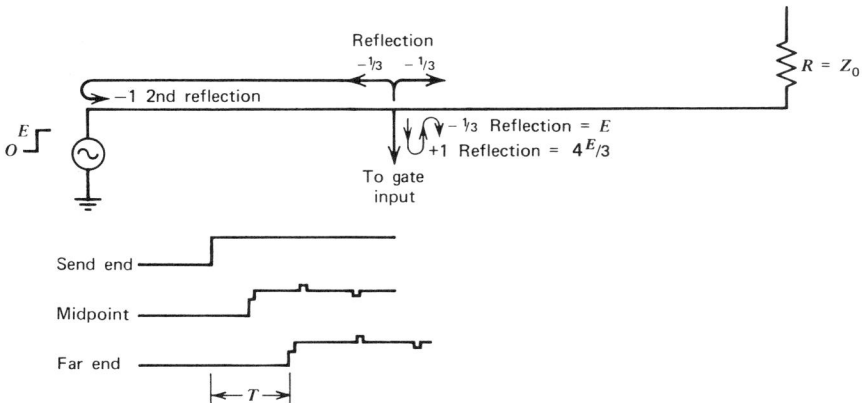

Fig. 12-2. Reflections from stub connections.

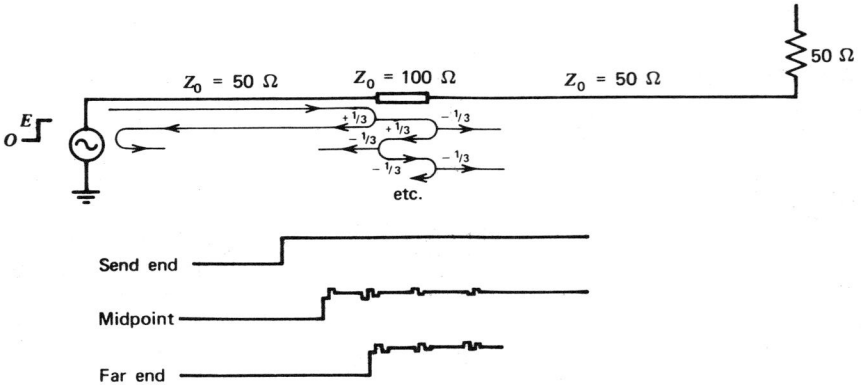

Fig. 12-3.   Reflections from a mismatched connector.

12.4   GRAPHICAL SOLUTION OF REFLECTION PROBLEMS

Though the concept of reflection coefficient is usable with linear terminations, most logic circuits have *nonlinear* voltage/current characteristics. We can still analyze reflection problems very effectively by using a simple graphical technique. With this technique we can find the voltages that will satisfy the $Z_0 = E/I$ relationship required by the transmission line and, at the same time, satisfy the E/I relationship required by the terminations. We will thus be able to read the voltages at each transition time directly off the graph.

Figure 12-4 shows a graphical solution of the resistive termination problem of Figure 12-1c with $E = 1$ V. We simply draw an *impedance plane* with voltage on the vertical axis and current on the horizontal axis. For convenience, we will *always make the relative scale of the voltage and current axes such that any line drawn at 45° represents voltage and current values with a ratio equal to the characteristic impedance ($Z_0 = E/I$) of the wiring.* Since the impedance is 150 Ω in Figure 12-1c, 1 V, on the Y axis, represents the same displacement as $1/150 = 6.7$ mA on the X axis.

We now plot "load lines" for the circuits at each end of the wire, just after the transition being analyzed. Since both ends are linear impedances, we can draw load lines by just joining two voltage/current points with straight lines. The far end would be zero current at zero voltage and $1/450 = 2.2$ mA at 1 V. The send end will have zero current at 1 V and 6.7 mA with $E = IR = 0.067 \times 50 = 0.33$ V drop (a voltage of $+0.67$ V). Joining these points, we now have load lines indicating all possible combinations of voltage and current at the send and the far end.

Since any line drawn at 45° can represent possible voltage and current values on the transmission line, we can now graphically find the actual voltages. The initial voltage at the send end is found by drawing a 45° line from the initial value (zero voltage, zero current) till it intersects the send end load line. V_0 is thus the initial voltage at the send end because it satisfies the E/I requirements of both the transmission line and the 50-Ω resistor to +1 V. We now find the voltage at the far end (when the signal reaches there) by drawing a 45° line from V_0 on the send end line till it intersects the load line for the far end, at V_1. From this point we draw another 45° line till it intersects the send end load line again at V_2, which is thus the voltage at the send end after the reflection returns. Another 45° line, intersecting the far end load line, gives us the voltage at the far end (V_3) at the end of three transition times ($3T$) when the reflection again reaches that end. Each voltage found can be immediately transferred to the voltage waveforms on Figure 12-4b. We could continue finding new values indefinitely, but at a certain point it becomes obvious that they are approaching a final value.

It is interesting to note that, if we averaged out the bumps in the waveform, this approach to the final value would be an exponential $V = \exp(-t/RC)$, where R is the lumped parallel resistance and C the total capacitance of the line. When we work with slow signals (as with CMOS), we can sometimes forget reflections and transmission lines and just think in terms of *charging the wiring capacitance*. It is reassuring to see that the two approaches give similar results.

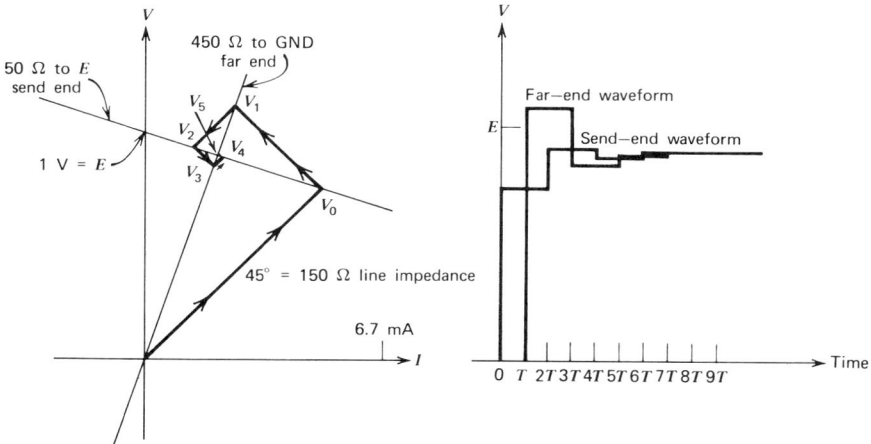

Fig. 12-4. Graphical solution of Figure 12-1c with $E = 1$ V: (a) impedance plane; (b) voltage waveforms.

The assumption of lumped capacitance and resistance just ignores fine details that are unimportant with slow circuits.

The real power of the graphical approach comes when we work with *nonlinear* load lines. Figure 12-5 shows the load lines of various real IC inputs and outputs. Note that since plus current is *into* the line at the send end and *out* of the line at the far end, the same resistor value will look different de-

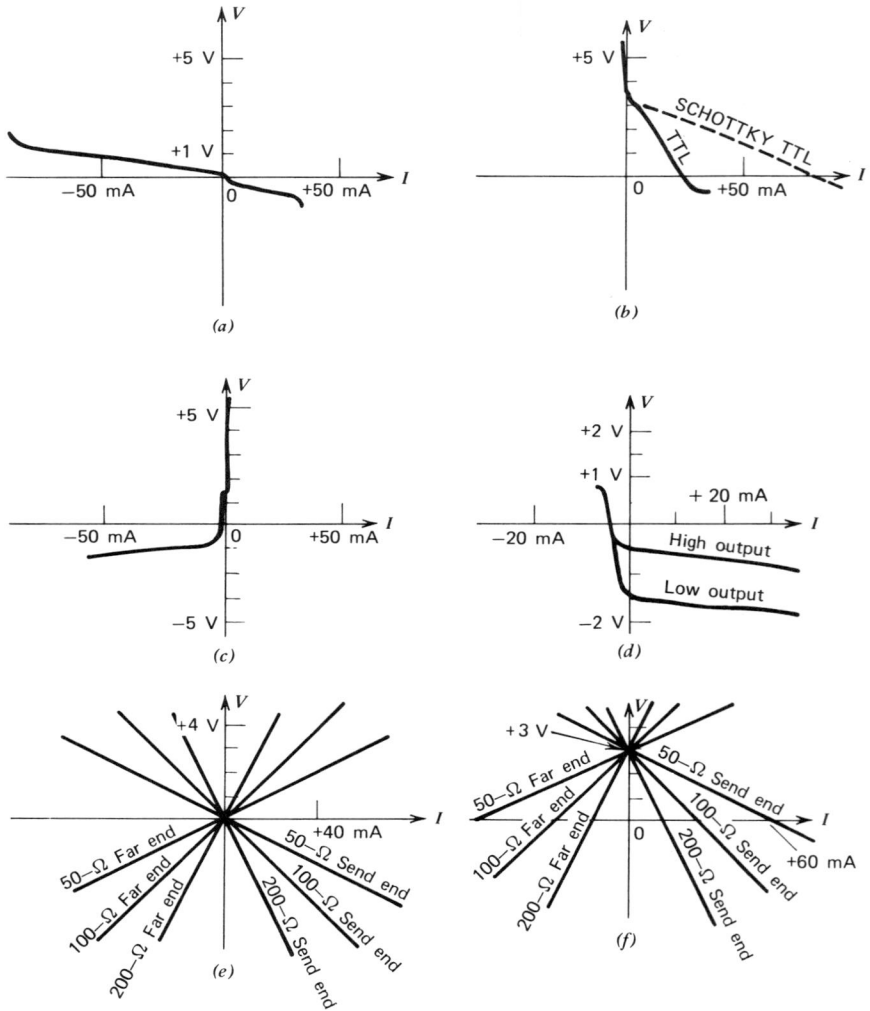

Fig. 12-5. Load line examples ($Z_0 = 100$ is 45°): (a) TTL low output; (b) TTL high output; (c) TTL gate input; (d) ECL high and low outputs; (e) resistors to ground; (f) resistors to +3 V.

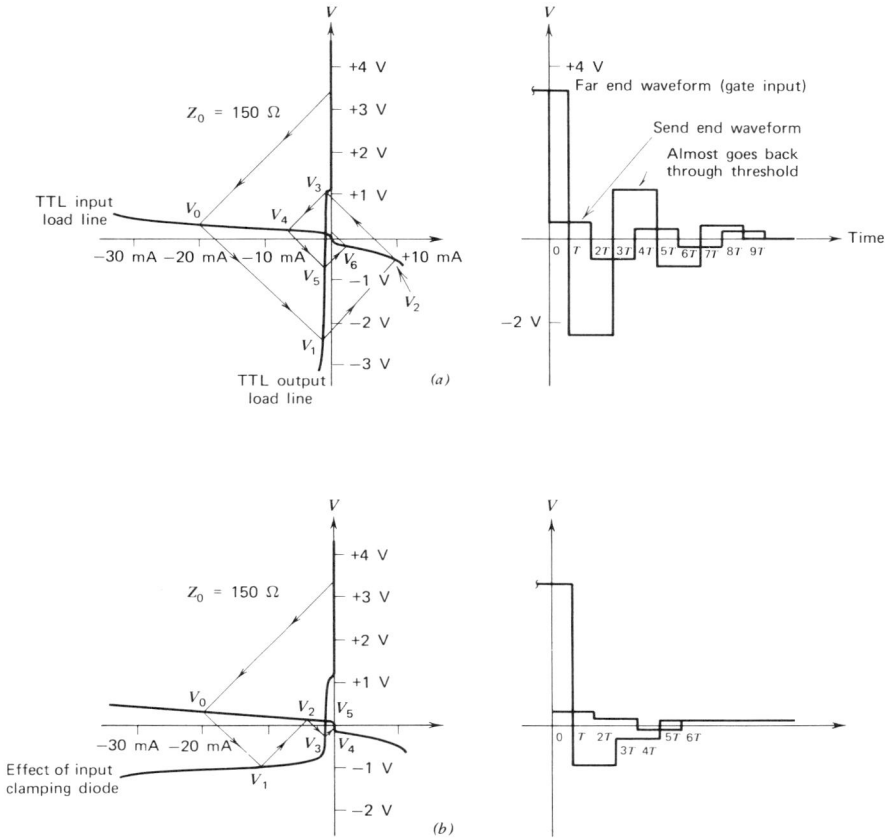

Fig. 12-6.   (a) TTL transition to low without diode clamp. (b) TTL transition to low with diode-clamped input.

pending on which end of the line it is on (Figs. 12-5e and f). The procedure for finding reflection waveforms is exactly the same with the nonlinear termina- tions as with the simple resistors of Figure 12-4.

In the real world, a designer wants to design a signal transmission system that will work. One of the real strengths of the graphical approach is that *it is easy to see what effect changes in the termination, such as added clamping diodes or resistors, will have on the resulting waveform*. The designer can thus "juggle the load lines around" until a satisfactory result is obtained.

Figure 12-6 illustrates a practical example: when TTL was first introduced, the gates had no clamping diodes on their inputs. As Figure 12-6a shows, this means that there will be ringing, after a minus transition, which can go through the logic threshold and cause improper operation. To cure this problem, an input clamping diode to ground was designed into all TTL circuits after about 1970. As Figure 12-6b shows, this diode reduces the overshoot enough to prevent improper logic operation. It is fairly easy to visualize what the effect of adding the diode will be by just looking at Figure 12-6a.

12.5   EFFECT OF REFLECTIONS ON SYSTEM OPERATION

Reflections like those in Figure 12-6a can cause problems *even when a very slow clock is used,* if they appear on the *clock* signal. If ringing on a clock signal has a long enough period, *a single clock edge can look like two or more clocks.* The system will thus *marginally* go through more than one clock step. Since minimum clock width to cause flip-flops to change is about the same as the setup time for a particular logic family, care must be used with clock lines longer than the lengths mentioned in Section 12-1 at *any* clock rate.

Logic signals are much less critical if the clock rate is not pushing the limits for the logic family being used. If the logic levels settle to their final values well before the next clock, reflections and even cross talk will cause no problem. If the clock rate is pushed to the limits, wiring delay must be considered in determining maximum clock rate. In this case ringing due to reflections can cause unreliable operation, because all flip-flop decisions are made just after the signal changes. If, for example, a flip-flop input looks like the "far end waveform" in Figure 12-6a, and the clock transition comes at $5T$, the slightest noise will cause incorrect operation.

Noise margin is normally considered to be the difference between the guaranteed DC output voltage of a logic family and its worst-case input threshold voltage. With TTL the "high" *output* level is guaranteed greater than 2.4 V, and the high *threshold* level is guaranteed less than 2 V. The worst-case DC noise margin is thus $2.4 - 2 = 0.4$ V. As Figure 12-7 shows, a gate on the send end of a 100-Ω TTL transmission line will not necessarily see a high level until $2T$ (2×1.6 nsec/ft = 3.2 nsec/ft of wire). Also, a gate at the far end will see only 3.2 V/2.75 V = 86% of the final voltage until $3T$ (4.8 nsec/ft). The worst case high output voltage before $3T$ would thus be only 2.4 V $\times$ 86% = 2.06 V, giving a **dynamic** *noise margin* of only 0.06 V. For reliable worst-case operation with ordinary TTL logic, we must therefore assume a 4.8 nsec delay per foot of wire ($3T$). We can assume *first pass* switching, and therefore 1.6 nsec/ft wiring delay if we use circuits with a lower output impedance for true output. For example, Schottky TTL (74S series) has a high-level output impedance (see Fig. 12-5b) that is about one-half that of normal TTL. It can thus be depended on to produce a good high level on the first pass.

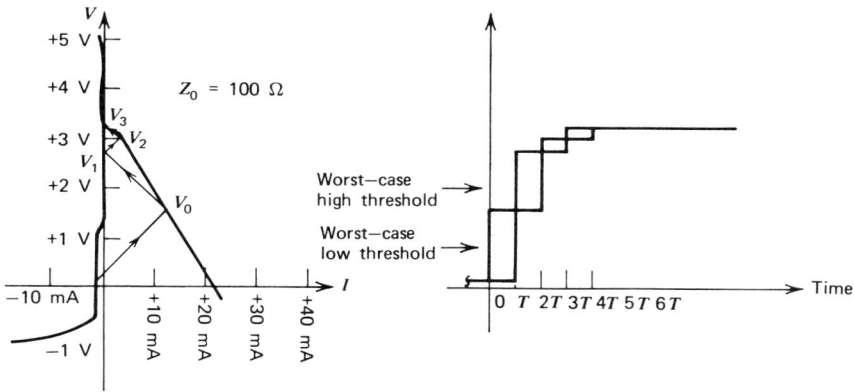

Fig. 12-7.   TTL transition to high.

12.6   GROUND, POWER WIRING, AND CURRENT DUMPING NOISE

One interesting thing we can observe from Figures 12-1, 12-6, and 12-7 is that it matters little what is at the far end of a piece of wire until the round-trip propagation delay $(2T)$ has elapsed. *No matter what is on the far end, the piece of wire simply looks like a resistor equal to the characteristic impedance, until enough time has elapsed for a signal to travel to the far end and back.* If ordinary wiring is used for power and ground connections, this can cause serious problems with high-speed logic circuits—even if the clock rate is slow.

   Figure 12-8 shows how ground noise could cause improper flip-flop operation. For simplicity, it is assumed that the end of the flip-flop output wire is terminated to eliminate reflections and that the flip-flop does not malfunction. When a clock causes the flip-flop to reset, the output signal initially *divides equally* between the ground wire and the Q output wire, because both look like the same impedance $Z_0 = 100 \ \Omega$. This happens because any current into the Q output wire must *come from* the ground wire. After the round-trip delay time has elapsed, the ground voltage again returns to 0 V (due to the -1 reflection). In the meantime, the clock input voltage to the flip-flop has looked like one-half of its true level due to the $+V/2$ voltage on the ground terminal of the flip-flop (see Fig. 12-8b). If the duration of this round-trip delay is long enough, the flip-flop can actually be clocked twice.

   With TTL circuits this effect is even worse because the circuit itself produces a surge of current as it switches because *both* the "pull up" and "pull down" output transistors in the IC are momentarily turned on. This current surge thus produces a voltage of $E = I/Z_0$, which adds to the one shown in Figure 12-8d. A similar minus going pulse is also produced in the (+) voltage supply line. If the duration (round-trip delay) and amplitude of this pulse are excessive, flip-flop states could be lost due to power interruption.

Although CMOS logic has a similar power surge, there is no problem with normal wiring lengths since the circuits are much slower. When the logic circuits are slow, noise pulses such as the one shown (dotted) in Figure 12-8b cause no problems because the flip-flops simply cannot respond that fast. One foot of PC connection, for example, produces a noise pulse of only 2 × 2 nsec = 4 nsec long. Although this has no effect on CMOS, it can cause trouble with faster circuits, such as 74S or ECL.

One very effective way to reduce ground noise is to lower the impedance of the ground wiring by actually making it a *ground plane*. A ground plane is a continuous sheet of copper, usually one side of a PC board. The impedance of the ground connection thus becomes so low that the signal appears almost entirely on the signal wire. Often a *multilayer* PC board, as shown in Figure 12-9, is used. This type of board provides an excellent ground, but is quite expensive because it is actually made by gluing together one board with copper on both sides (a *double-sided* board) and one board with copper on one side. An even better but more expensive method is to laminate an insulating layer between two double-sided boards, giving a total of four layers. One of the in-

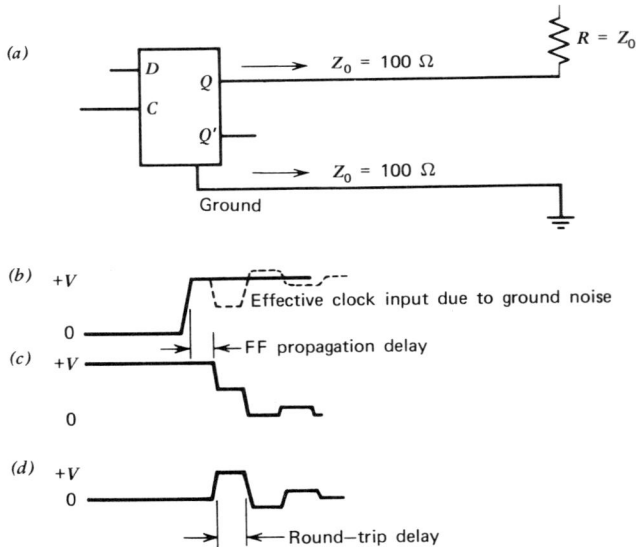

Fig. 12-8. Effect of long ground line: (a) Circuit with grounding problem; (b) Clock; (c) Q output; (d) ground.

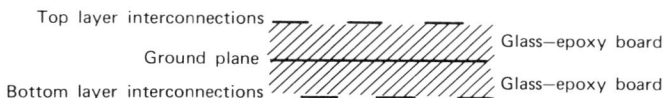

Fig. 12-9.   Cross section of multilayer PC board.

ner layers is used as a *ground plane* and the other as a *power plane*. Noise pulses on both power and ground can thus be greatly reduced.

With careful design it is possible to save the higher (about three to five times) cost of multilayer boards and simply use double-sided PC boards for interconnections. Modern minicomputers use large, 15 × 15-in. double-sided boards to interconnect TTL and Schottky TTL circuits. A fairly effective ground plane is formed by the grid of power and ground connections that go to each circuit. High-frequency bypass capacitors between the power and ground are scattered all over the board—often averaging one 0.01-μF capacitor for every two to five ICs. The result is that instead of a single ground path, as shown in Figure 12-8a, *a large number of parallel paths* run through the network of ground and power connections. The farther apart two circuits are, the more effective parallel ground paths there are. The power and ground connections are made as wide as room allows to further lower the impedance. By using careful layout, circuit board cost can be greatly reduced in this way. Additional checkout and re-layout cost may offset these savings if production volume is small. Since noise problems are marginal, proper operation of the prototype is not enough. All ground and power points must be also checked for noise with a very high-speed oscilloscope during full operation, preferably with diagnostic programs running.

Often ground noise will or will not be a problem depending on a specific layout. For example, Figure 12-10 shows the circuit of Figure 12-8a with a clock driver ground connected directly to the flip-flop ground. There will be no problem here because the noise pulse on ground appears on the clock signal also, so there is no net noise. Flip-flop B, however, will have a noise problem because the noise pulse generated by flip-flop A appears on the clock signal but not on the ground for flip-flop B.

Another type of noise problem, called *current dumping*, is illustrated in Figure 12-11. A TTL gate input takes up to 2 mA (for Schottky) to hold it false. Assume inputs A and B are both low, but input B is initially taking most of the 2 mA. When input B goes high, input A will receive a current step of 2 mA. If the impedance of the wire is 150 Ω, this will produce a signal voltage of $E = IZ = 0.002 \times 150 = 0.3$ V, which will last for the duration of the round-trip delay back to the gate output driving signal A. Since 0.3 V is less than the TTL noise margin specification, it will cause no problem. However, if many

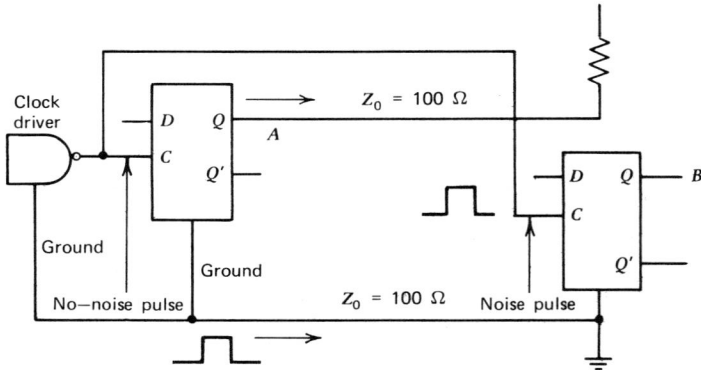

Fig. 12-10.   Effect of layout on ground noise.

similar currents are dumped simultaneously onto the same line, they can cause a problem. For example, if all the *D inputs* to a register made of individual *D* flip-flops or latches change at once, current can be dumped onto the clock input of each flip-flop. The *total* current dumped will be the *sum* of the currents from the individual flip-flops. If the round-trip delay back to the clock driver is sufficient, the flip-flops can be clocked by the noise pulse, even though the clock driver output stays false.

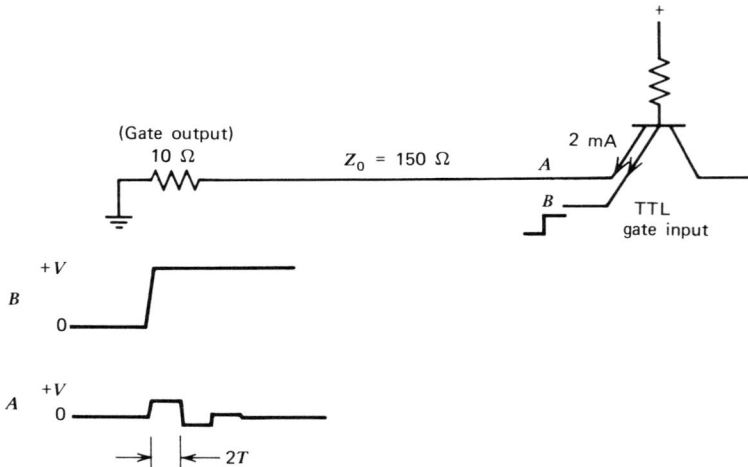

Fig. 12-11.   Noise due to current dumping.

12.7 CROSSTALK IN LOGIC INTERCONNECTIONS

Another problem requiring serious consideration when high-speed logic is used is electromagnetic coupling between signal wires, or *crosstalk*. As with reflections, crosstalk to *clock lines* is a problem at *any* clock speed with high-speed logic circuits. If the clock speed capability of the logic circuits is not pushed *crosstalk can be ignored* on the *logic* signal lines; this is because crosstalk occurs only from signal *transitions*. In a clocked system all signal transitions occur just after the clock. If the logic signals have time to settle after the clock, they will reach the correct final value before the next clock, in spite of temporary incorrect outputs due to crosstalk. One exception is crosstalk *back to previous logic levels* on ECL systems. This type of crosstalk can cause oscillations due to formation of positive feedback loops. With TTL this is less likely, since very little time is spent in the transition region, where forward gain exists.

One easy way to reduce crosstalk is to use a multilayer PC board with a ground plane as shown in Figure 12-9. The ground plane lowers the impedance of the individual conductors and partially intercepts the electrostatic (capacitive) and electromagnetic (inductive) fields linking the two conductors. The crosstalk voltage produced on a passive line is essentially the result of a *mutual impedance* from a noise-producing line, producing a voltage across the characteristic impedance (to ground) of the line. We therefore want as low an impedance to ground as possible and as high a mutual impedance to the noise-producing line as possible. When the distance to the ground plane becomes comparable to the distance between lines, crosstalk rapidly decreases.

Although a multilayer board with ground plane certainly reduces crosstalk problems, the much less expensive, double-sided board is often acceptable with careful layout. The reason is that, in addition to all of the ground and power conductors acting as a ground plane, the *signal conductors also help form a ground plane*. As a conductor crosses many other conductors on the other side of the board, it usually crosses them at right angles, so negligible crosstalk is produced to each one. However, since most of the signals are not changing on any given clock, most of the connections are essentially at AC ground and therefore act as a fairly good ground plane. The only real problems come when long runs are made with two conductors side by side. In this case a ground conductor can often be added between the two wires. Really difficult problems, such as clock distribution, can often be handled by enclosing the clock conductor between the ground conductors, or even soldering twisted-pair wire onto the board.

In low-volume applications the ground plane (and possibly power plane) is probably justified because its cost will be less than the cost of additional layout problems with no ground plane. Either way, checkout should not be

considered complete until clock waveforms have been checked on a high-speed
oscilloscope. If the clock speed capability of the ICs is pushed, the signal waveforms should also be checked for adequate dynamic noise margins.

12.8   CROSSTALK AND NOISE IN CABLES

Although it is usually possible to lay out *random logic* interconnections so that
the same two wires never run next to each other for any distance, this is impractical in long *cable* interconnections. Whenever the length of the cable is
great enough that the signal delay in the cable is comparable to the signal rise
time, the simple concept of the signal dividing across the mutual impedance
and impedance to ground is no longer useful.

As Figure 12-12 shows, mutual inductance (L_{12}) and mutual capacitance
(C_{12}) distributed along the line cause crosstalk signals to be coupled *as the
pulse travels along the line*. The capacitive noise propagates in both directions
from where it is produced in the *same* polarity as the exciting signal, while the
inductive crosstalk propagates backward with the same polarity but *forward
with the opposite polarity*. The net effect is that the capacitive and inductive
effects *add* in producing *near-end crosstalk* but *subtract* in producing *far-end
crosstalk*. If the round-trip delay on the wire is equal to the rise time of the
signal, the near-end crosstalk will be a maximum. Since far-end crosstalk is
the sum of two opposite polarity signals, it will be less, and it may even be
zero or of opposite polarity.

By using coaxial cable or dual dielectric flat cable (see Fig. 12-13) signals
can be sent up to about 30 ft using ordinary gates as senders or receivers. With
ECL circuits a terminating resistor is required, but Schottky TTL circuits can

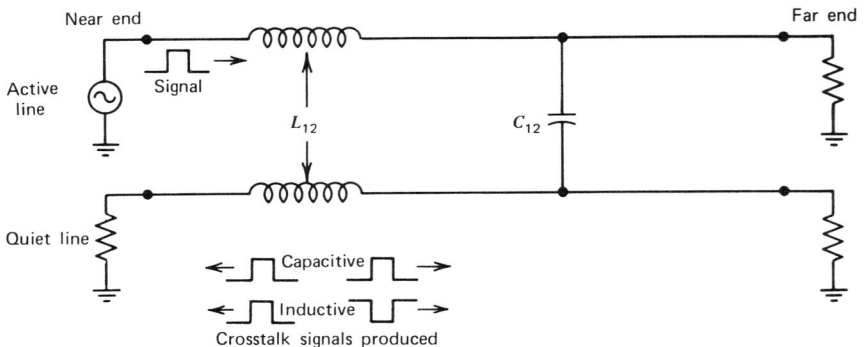

Fig. 12-12.   Crosstalk in a long cable.

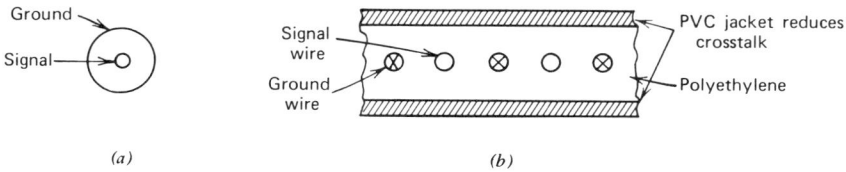

Fig. 12-13. Interconnecting cables: (a) coaxial cable; (b) dual-dielectric flat cable.

be used without terminators. A special high-current driver (e.g., 74S140) should be used, however, when driving 50 Ω coax with TTL.

When cables are run any distance, *external* noise problems can become significant. Obtaining a good common ground becomes more and more difficult as distance is increased. Long ground wires tend to act as antennas and pick up radio stations, power line noise, noise from automobile ignition, electric motor brushes, and so on. If two separated pieces of equipment are plugged into the power line at separate points, the green "safety ground" wire must not be connected to signal ground at both ends. If both ends are grounded to the line as shown in Figure 12-14, noise currents from motors and other devices elsewhere on the power line will also flow through the logic cable ground as a parallel path. This current can generate a significant noise voltage. If *one end or the other* is not connected to the power line ground, the parallel path is eliminated. The metal *cabinet* can still be attached to safety ground as long as signal ground is isolated.

12.9 DIFFERENTIAL SIGNAL TRANSMISSION

When signal ground is allowed to float at one end, noise (with respect to earth ground) induced on the ground connection in a cable is unimportant *so long as*

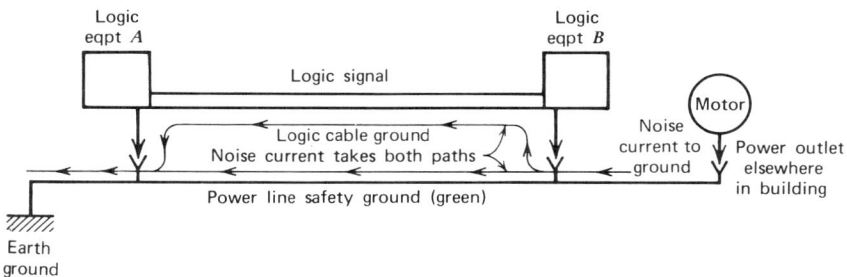

Fig. 12-14. A powerline ground loop.

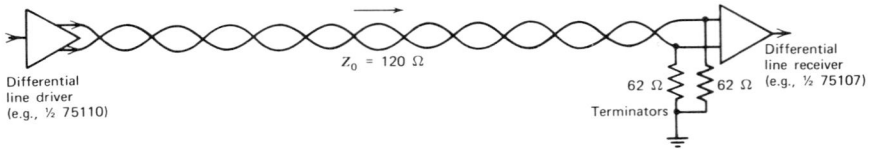

Fig. 12-15.   Differential transmission system.

identical noise is induced in all signal cables. Since the signal ground is also connected to the rest of the logic in the system and each signal line goes only to a gate input, there is an *unbalanced load* that tends to make noise cause faulty signal reception. The higher the frequency, the greater the imbalance, so with high-speed logic systems *differential-mode* signal transmission is often used.

With differential-mode a separate ground is associated with each signal line. By twisting the associated ground with the signal wire, induced noise and crosstalk will be almost identical for both the ground and the signal line. By using a *differential input line receiver*, *common-mode* signals are rejected and only the *difference* between the signal and ground voltage is detected. Since both inputs to the differential receiver have the same impedance, loading on the two wires is balanced. Although grounds from many twisted pairs can be connected at a single point on the driver end with only a slight sacrifice in balance, *if grounds are connected at both ends, the crosstalk reducing benefits of differential-mode operation disappear.* If grounds are common at both ends, ground currents will flow in parallel through all the ground leads. Crosstalk to signal lines will vary, however, depending on how close together the pairs are in the cable. The resulting imbalance essentially generates a differential signal, causing incorrect signal reception.

Figure 12-15 shows a true differential transmission system. The differential transmitter feeds current to one output or the other to send a 1 or 0. The differential receiver responds only to *differences* in voltage between the two inputs. All lines in a cable are separate, balanced pairs. With extremely high-speed logic families, differential operation is required for all but very short interconnections. For example, ECL wiring rules recommend coax or differential operation for lengths of *10 in.* or more.

By putting terminators at both ends, a two-way system can be made with a sender and receiver at each end. In fact, a *party-line system* can be made with senders and receivers scattered along the line so that several separate devices can be served by a single cable. The logic grounds of the various systems should be tied together with a *separate* heavy wire. Noise voltages between the various grounds will cause no trouble *as long as they are less than the rated common-mode input voltage range of the differential line receiver* ($\pm$ 3 V for the 75107).

If the cable is very long or the system is in a very noisy environment (e.g., near a radar antenna), the common-mode noise can easily be too much for the differential receiver input. Though voltage dividers can be used on the differential receiver input to increase common-mode range to a point, *transformer coupling* should be used for really long lines. Small, shielded pulse transformers are inexpensive and *can reject common-mode noise of thousands of volts.* The only problem is that only high-frequency signals can be passed through the transformer, so the transmitted signals must be encoded in such a way that DC response is not needed. Optical isolators provide similar common-mode noise rejection and offer response down to DC. High-frequency response of optical isolators is presently somewhat limited, but they should be very useful for high-speed data transmission in the future.

Transformer-coupled data transmission over long distances has been highly developed by the telephone industry for pulse code modulation (PCM) voice transmission. Ordinary polyvinyl chloride (PVC), or paper-insulated, twisted-pair, telephone cable is routinely used to send digital signals, at a 6.3 MHz rate, over distances up to 4000 ft. With higher quality twisted-pair cables the distance between digital repeaters is increased to 3 mi. With coaxial cables 60 MHz digital signals are sent with a repeater spacing of 5100 ft. The transmission format used in these telephone systems is shown in Figure 12-16a. A pulse is sent to indicate a 1, and no pulse is sent to indicate 0. To prevent a DC component from developing, the pulses are alternately sent with a (+) or a (−) polarity. The system is thus called alternate mark inversion (AMI).

Although the telephone system uses a 50% duty cycle, a simple system can be built using 100% duty cycle signals as shown in Figure 12-16b. Because the average signal level is always zero, the transformer used only needs to have enough low-frequency response to go one bit time without significant "droop." A tiny "H core" or toroidal-core ferrite transformer with only a few turns can thus be used for high-speed transmission. By using a center-tapped winding on

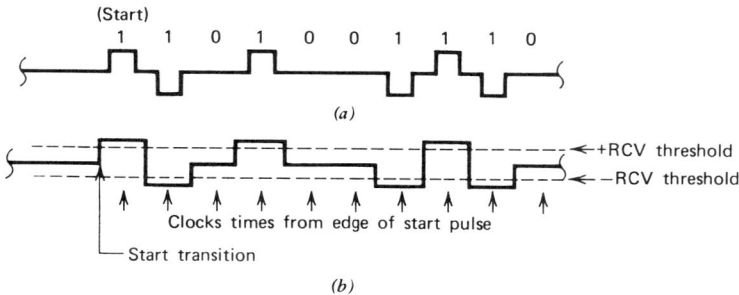

Fig. 12-16. AMI transmission format: (a) 50% duty cycle; (b) 100% duty cycle.

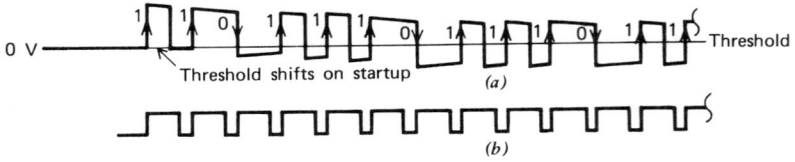

Fig. 12-17. Phase modulation via transformers: (a) signal; (b) clock one shot, period =
3/4 bit.

one side of the transformer, the driver and threshold circuits need not be
bipolar. Driving one side of the center-tapped winding thus gives a (+) pulse
output, while driving the other side gives a (−) pulse output.

By beginning each stream of crystal-clocked data with a "start pulse," a sequence of clocks, centered on each bit, can be digitally generated to strobe the
ORed output of the (+) and (−) threshold circuits into a shift register input. A
simple system using TTL can send serial data at 20 MHz (as described in Ref.
2). Only one pair of transformers and sender and receiver circuits can thus
send many bits of data serially. A bidirectional, party-line system can easily be
built by just putting terminators at each end of the line and making sure the
senders represent a high impedance when they are not sending.

If data are sent continuously in one direction, and even simpler phase
modulation system, as shown in Figure 12-17, can be used. In this system the
send signal is simply the data Exclusive ORed with a 50% duty cycle clock.
This type of signal is self-clocking because there is a transition for each bit. By
firing a "one-shot" timing circuit, with a period of three-fourths bit time, on
each bit, the intermediate transitions can be ignored. Decoding requires simply
looking at the direction of each clock transition (or the polarity of the signal
before each clock transition). Unlike the AMI system, this method cannot be
started and stopped. As Figure 12-17a shows, a DC component is produced
when the signal first starts up, causing unreliable operation of the threshold
circuit until transformer "droop" removes the DC component. One advantage
of this system, when used with continuous signals, is that the threshold is set at
0 V, so the signal can be *any* amplitude. Other systems, such as the AMI
system, require the threshold to be set at the one-half amplitude point for optimum performance with noise.

EXERCISES

1. (a) Calculate reflection coefficients and the first three voltage levels for the circuit
        in Figure 12-1c but with $Z_0 = 100\ \Omega$.
    (b) Graphically calculate the waveforms up to $5T$.

2. Draw a load line for a diode to $+3$ V at the far end.

3. Redraw all the load lines on Figure 12-5 such that a 45° angle is $Z_0 = 50\ \Omega$.

4. Draw the voltage waveform at the midpoint of the line for Figure 12-6b. Show the time scale in nanoseconds (instead of T) for a 3-ft wire with 1.6 nsec/ft delay.

5. Find the send-end and far-end waveforms graphically for a Schottky TTL transition to high with $Z_0 = 100\ \Omega$.

6. Find the send-end and far-end waveforms graphically for an ECL gate driving a wire with an impedance of 100 Ω terminated with a 100-Ω resistor to -2 V.
   (*a*)   For the transition to high.
   (*b*)   For the transition to low.

7. Find the send-end and far-end waveforms graphically for an ECL gate driving a wire with an impedance of 100 Ω with no termination. The ECL gate draws about 160-μ A input current to minus.
   (*a*)   For the transition to high.
   (*b*)   For the transition to low.

8. Find the send-end, middle, and far-end waveforms graphically for an ECL gate driving a wire with an impedance of 100 Ω with no termination of the far end, but with a 50-Ω resistor in *series* with the send end and a 180-Ω resistor to -5 V at the IC output.

9. Find the send-end and far-end waveforms graphically for an open-collector TTL gate driving a 100-Ω line terminated with 150 Ω to $+5$ V:
   (*a*)   Terminator at the far end.
   (*b*)   Terminator at the send end.

10. Draw the *effective* clock waveform for Figure 12-8 if the impedance of the ground line is 50 Ω.

11. If the flip-flop in Figure 12-8 can change with a clock width of 3 nsec, how long could the ground connection be if it is a PC board connection?

12. Redraw Figure 12-10 so that both flip-flops will work properly (keeping relative positions of the gate and flip-flops the same).

13. (*a*)   If a 16-bit register, made of separate D flip-flops, is driven by a long clock line with $Z_0 - 150\ \Omega$, how much current would have to be dumped on the clock line by each flip-flop to cause a problem? Assume noise margin is 1.5 V.
    (*b*)   How long would a wire clock line have to be for trouble, assuming the flip-flops can be clocked by a 5-nsec pulse?
    (*c*)   Suggest a solution to the problem.

14. Redraw Figure 12-15 such that a $+6$-V common-mode noise level can be tolerated.

15. Draw a party-line system similar to Figure 12-15 with a sender and receiver at each end and in the middle.

16. Design a serial data receiver to receive a 20-mHz signal as shown in Figure 12-16b. An 80-mHz crystal clock is available. The message format is as shown in Figure 12-16.

17. Design the logic for generating the signal in Figure 12-16b by driving a 1 : 1 center-tapped transformer from two ECL gates. A 100-Ω line must be driven with 1-V pulses.

18. Draw a schematic of the logic to generate and detect the signals shown in Figure 12-17a.

REFERENCE

1. T. Blakeslee, "Transformer Coupled Transceiver Speeds Two-Way Data Transmission," *Electronics,* March 1, 1973, pp. 94–96.

BIBLIOGRAPHY

Balph, Tom, "Use ECL 10,000 Layout Rules," *Electronic Design,* August 17, 1972, pp. 72–76.

Blakeslee, T., "Transformer Coupled Transceiver Speeds Two-Way Data Transmission," *Electronics,* March 1, 1973, pp. 94–96.

Blood, William, Jr., *MECL System Design Handbook,* Motorola Semiconductor Products, Inc., Mesa, Ariz., 1972.

Davis, J. H., "T2: A 6.3 Mb/s Digital Repeatered Line," *Proceedings of IEEE International Conference on Communications,* 69CP369-COM, 1969, pp. 34–9 to 34–16.

De Falco, John, "Reflections and Crosstalk in Logic Circuit Interconnections," *IEEE Spectrum,* July 1970.

Dreher, Thad, "Cabling Fast Pulses? Don't Trip on the Steps," *The Electric Engineer,* August 1969, pp. 71–75 (coax transmission).

Epand, Donald and Liddane, Ken, "Selecting Capacitors Properly," *Electronic Design* 13, June 21, 1977, pp. 66–71.

Fairchild Staff, *The TTL Applications Handbook,* Fairchild Semiconductor, Mountain View, Calif., 1973, Chapter 14, "Transmission Line Interface Elements".

Gray, Harry J., *Digital Computer Engineering,* Prentice-Hall, Englewood Cliffs, N.J., 1963, Chapter 9, "Pulse and DC Power Transmission Systems.

Marshall, Joseph, "Flat Cable Aids Transfer of Data," *Electronics* July 5, 1973, pp. 89–92.

Sear, Brian E., "Packaging for High Frequency," *The Electronic Engineer,* July 1972, pp. 20–23.

Texas Instruments Staff, *Designing with TTL Integrated Circuits,* McGraw-Hill, New York, 1971, pp. 25–45 and 83–105.

Texas Instruments Staff, *Series 54S/74S Schottky TTL Applications,* Bulletin #CA-176, Texas Instruments, Dallas, Tex., 1973.

Torrero, Edward, "Focus on Fast Logic," *Electronic Design,* June 8, 1972, pp. 50–57 (survey of ECL and Schottky TTL).

Trompeter, Ed, "Cleaning Up Signals with Coax," *Electronic Products,* July 16, 1973, pp. 67–69. (Good discussion of ground noise with coax.)

CHAPTER THIRTEEN

INPUT/OUTPUT
Devices

13.1 INTRODUCTION

Because of the fantastic improvements in cost and reliability of ICs, *system cost and failure rate today are strongly dominated by the mechanical input and output devices in the system.* Since all complete, useful systems start and end with something *mechanical* (in the broad sense, including optical, thermal, etc.), there is no way to avoid having mechanical input and output devices in every system. Integrated circuits have become such a bargain that it has become practical to move the electrical interface "outward" so that *the mechanical devices have been made simpler by increasing the complexity of the electronics.*

In this chapter we do not attempt a complete survey of input/output device types. Instead, we look at some clever, successful examples of using modern electronic techniques to simplify the required mechanical hardware. Although most of the examples are standard computer peripheral devices, the principles they illustrate are equally useful for special-purpose devices. We emphasize three basic "tricks" that have proven to be effective ways to simplify input and output devices: (1) negative feedback, (2) incremental operation, and (3) time sharing.

13.2 POSITION SERVOS

Negative feedback (Fig. 13.1) makes it possible to obtain high accuracy and/or fast response with relatively crude mechanical components. By sensing the

307

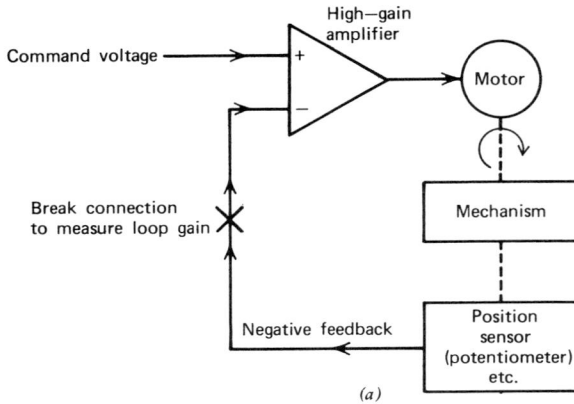

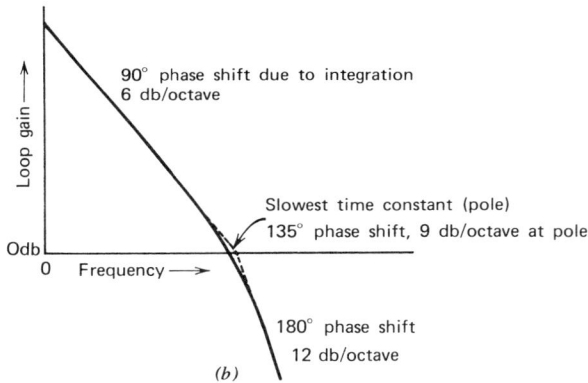

Fig. 13.1. Position servo.

actual position, an *error signal* can be produced, which is used to drive a relatively crude electric motor. The motor is thus driven very hard at first, causing
fast response. As the final position is approached, the error signal gradually
gets smaller until it reaches zero. If the device *overshoots* (coasts past the
desired position), a reverse-polarity error signal drives the device back to the
exactly correct position. Since the sensor can sense the position of the *device to
be positioned,* rather than the motor shaft, *tolerance variations in the linkage
will not cause any error.* Friction, inertia, and other external forces also have
little effect on system performance, since the error signal will just drive the
motor as hard as necessary to get the proper movement. The higher the amplifier gain, the *stiffer* the system performance, since high gain means a
stronger drive signal will be produced by a small error signal.

There is a limitation to the amount of amplifier gain, however, because if it is too high the servo system will go into *oscillation*. This is because when the motor tries to go to the desired position, it will always overshoot slightly. When it tries to drive back to the proper position, it will overshoot again and the cycle will repeat indefinitely. The key to stability is to have *a gain of less than unity, around the loop, at the frequency where the phase shift around the loop is 180°* (because here, negative feedback becomes positive feedback). As we go around the loop in Figure 13-1, we see a *gain* due to the amplifier and various *losses* in the motor, mechanism, position sensor, and summing network. The net gain around this loop is called the *loop gain*. There are also various electrical and mechanical time constants, or *poles,* around the loop. Each time constant (pole) has a *corner frequency* (1/time constant) above which a phase shift approaching 90° is produced and gain falls off in direct proportion to frequency (6 dB/octave).

Bode plots of loop gain are often used to examine stability problems. Since phase shift is related to slope of the amplitude response, stability simply requires a slope of less than 12 db/octave (180° phase shift) at the point where loop gain is unity (zero db).

The loop gain (Fig. 13-1b) of a position servo starts out with a slope of 6 db per octave (a 90° phase shift) because position is the *integral* of motor velocity and therefore the integral of the amplifier output voltage. Because of this initial 6 db per octave slope, instability can be caused by the additional 6 db per octave slope from the slowest mechanical time constant in the system.

13.3 VELOCITY SERVOS

Figure 13.2 shows another type of servo called a velocity servo. In this case, *velocity* of the mechanism is sensed and fed back. The command voltage input thus specifies a velocity instead of a position. As Fig. 13.3 shows, there is no integration in a velocity servo so instability cannot occur until after the *second slowest* time constant in the system. Since *closed loop* gain (Fig. 13.2b) falls off when loop gain falls below unity this means that the frequency response of the system can extend way out past the slowest mechanical time constant. It is therefore possible, for example, to space paper in 17 msec with a motor having a mechanical time constant of 50 msec. Actually, bandwidth and loop gain of both position and velocity servos can be increased considerably by adding R-C lag or lead networks to the feedback path. These networks add pole-zero pairs which increase the slope of the loop gain, then decrease it again in the vicinity of unity gain.

Another advantage of a velocity servo is that acceleration forces can be controlled by changing the velocity command voltage with a gentle ramp, as shown in Fig. 13.4. Since position is the integral of velocity, velocity servos will make

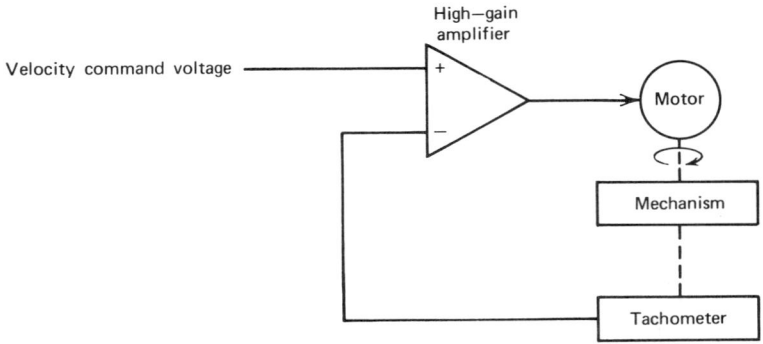

Fig. 13-2. Velocity servo.

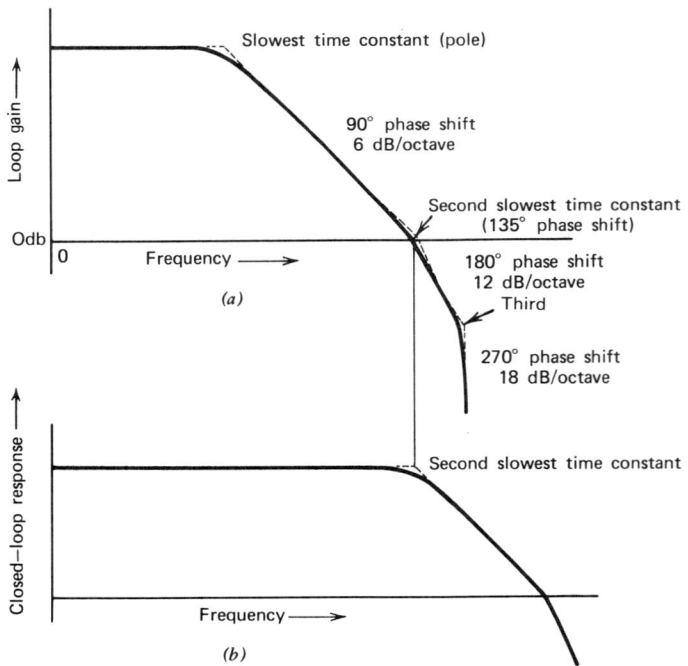

Fig. 13-3. (a) Loop gain. (b) Closed-loop response.

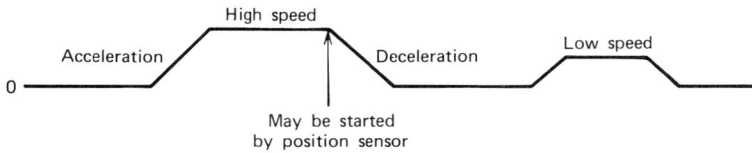

Fig. 13-4.   Velocity profile command voltage.

accurate *position changes* proportional to duration of the velocity command signal. If position feedback (e.g. from a photocell) is used to terminate a velocity command, accurate *absolute* positioning can be obtained with a velocity servo.

The velocity indication may come from a DC tachometer, which is simply a small electrical generator, or it can be generated electronically from a train of pulses as shown in Figure 13-5. A "one-shot" circuit (e.g., 74121) produces a pulse of a definite width and amplitude for each pulse from the photocell. The more rapidly the pulses come, the higher the duty cycle of the one-shot output and the higher the DC voltage out of the integrator. The linearity and accuracy of a circuit such as this are excellent, and no brushes are required.

For low-performance systems, *the back electromotive force (EMF) of the motor itself can be used as a tachometer* signal. A permanent-magnet motor generates a voltage directly proportional to rotational speed. The only reason this voltage can differ from the voltage across the terminals of the motor is that there is a drop across the armature resistance. By sensing the current through the motor, we can tell what the drop across that resistance is. The velocity signal is thus the voltage across the motor terminals, minus the voltage across the armature resistance (which is current sensed multiplied by armature resistance). The circuit of Figure 13-6 drives the motor at a velocity specified by the control input independent of load. AC motors are sometimes similarly controlled by comparing the back EMF to a control signal at the start of each half-cycle. If the back EMF indicates that the motor speed is too slow, *that half-cycle* is turned on by a silicon controlled rectifier (SCR) or Triac switch.

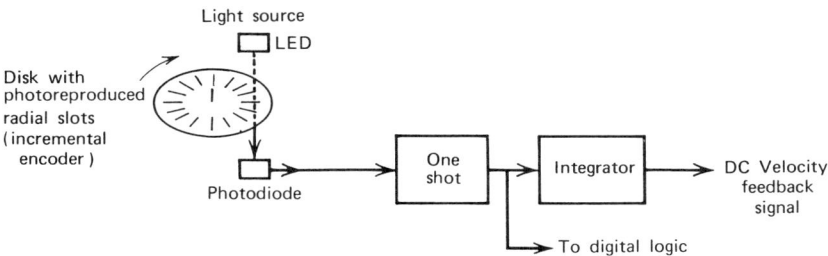

Fig. 13-5.   Generating a tachometer signal from pulses.

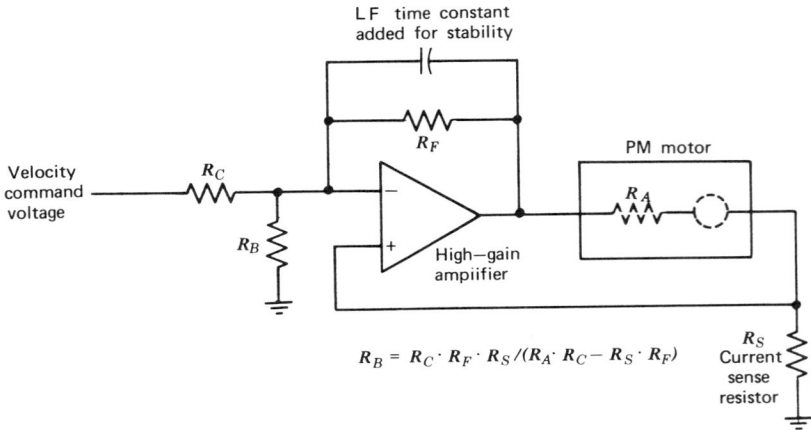

Fig. 13-6.   Using the motor itself as a tachometer.

13.4   DIGITAL OPERATION

Although the feedback systems we have been discussing work by summing analog voltages, the same principles can be applied digitally. Analog operation is generally used because *operational amplifier* ICs are very inexpensive, fast, and accurate. The subtraction of feedback voltage from command voltage (Figs. 13-1 and 13-3) normally requires nothing more than two resistors connected to the operational amplifier input.

Where fast response is not needed, negative feedback can be accomplished by digital logic. For example, long-term inaccuracies can be digitally subtracted out in response to a calibration input. A system that must make very accurate voltage measurements, for example, can use inexpensive analog circuits if it periodically measures a zero-voltage input. *The reading obtained at zero volts can then be digitally subtracted from all readings made.* By reading a calibration voltage periodically, a *scale factor* can also be computed. This scale factor can be multiplied by the zero-corrected readings for even greater accuracy. All of this digital computation can, of course, be done by *programmed* logic. It thus represents a direct tradeoff between more instructions in the program and simpler external hardware. We can go even further and "program out" *transducer nonlinearities,* mathematically or by table lookup.

Mechanical systems can also be made accurate by programmed correction. A simple microswitch or photocell can often give an accurate reference point for the thing that is actually being positioned. The program can then digitally add the correction to all outputs to put them in proper adjustment. It is thus possible to effectively "adjust out" wear and thermal expansion automatically every few seconds.

By programmed recalibration and linearization it is often possible to use amazingly crude transducers for precision jobs. Digital inputs can be directly generated by building an oscillator with a circuit element that varies with the parameter to be measured (e.g., pressure, temperature, position). The state of a counter clocked by that oscillator can then be periodically used to compute a corrected measurement. Signals of this type are called *pulse-rate* signals because the value is represented by the pulse rate. Preprocessing of pulse-rate signals is quite simple because *multiplication is easily performed* by special rate multiplier ICs (e.g., three—type 7496—six-bit multipliers will multiply by an *18-bit* binary number). A counter essentially acts as an integrator by accumulating pulses. By reading out the state of the counter periodically, a division is essentially performed, giving an *average* as a result.

Sometimes very accurate pulse-rate outputs can be obtained easily. For example, a simple *voltage-controlled oscillator* can provide a frequency that linearly tracks the voltage applied within 0.05%.

A quartz crystal is usually cut so that the frequency at which it oscillates is independent of temperature. However, specially cut crystals will oscillate at a frequency that varies linearly with temperature with an accuracy of 0.1% from 0 to 100°C. Such oscillators can resolve differences of temperature of 0.0001°C over a range of −80 to +250°C.

13.5 INCREMENTAL OPERATION

Another modern technique for providing accurate mechanical motion is *incremental operation*. All motions are done as a series of small, *discrete steps*. Since the number of steps is a *digital* quantity, the control interface can use digital ICs. Sometimes a velocity servo (see Fig. 13-5) is used with an *incremental encoder* to make an *incremental motion servo*. The velocity command voltage smoothly accelerates the mechanism and keeps it running at a constant velocity until the specified number of pulses (sometimes only one) have been produced by the incremental encoder. At this point the *deceleration* begins, causing a smooth stop.

An incremental encoder can also be used to provide *absolute* positioning feedback by letting it operate a *digital up/down counter* as shown in Figure 13-7. Each state of the counter thus represents a discrete mechanical position. A digital command to go to a new position causes a *velocity* signal in the required direction (as determined by a digital comparison) until the state of the counter corresponds to the desired position. When the system is first turned

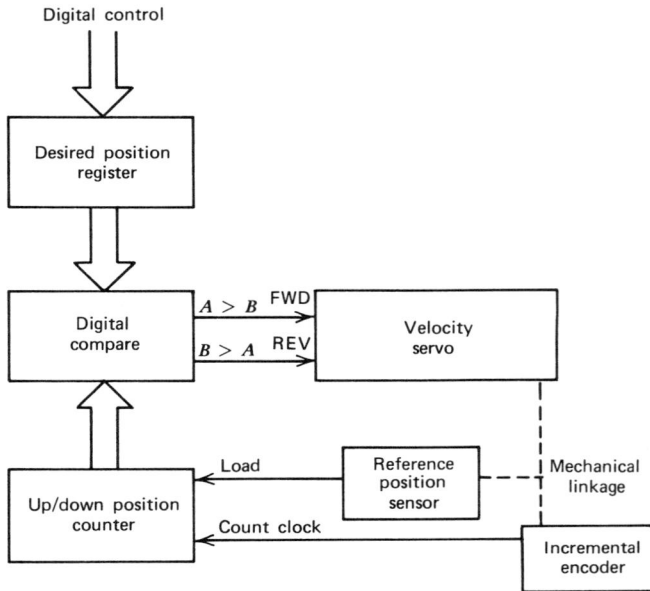

Fig. 13-7.   Digital positioning using velocity servo.

on, the position counter must be initialized to a correct *absolute* value. This is usually done by going through an *initialization sequence* to find a reference. A microswitch or optical mark and sensor on the mechanism indicates when the reference position is reached, causing the position counter to be set to the state corresponding to that position. As a *self-clearing* provision (Section 6.6) the counter is also set to that state if the reference position is detected during normal operation.

A similar system can be built using a *stepping motor* instead of a velocity servo. The stepping motor has the big reliability advantage of having *no brushes*. Also, the stepping motor will go *exactly one increment* for each change of state of the drive signals, so no feedback is needed. Figure 13-8 shows a typical system using a stepping motor. Note that although no incremental encoder is required for feedback, a reference position sensor is still needed to set up the absolute calibration. The stepping motor drivers are shown connected to the position counter because the drive signal required is a digital signal that follows a four-state *moebius* count sequence (Fig. 5-7e). It is therefore possible to generate the drive signals directly from the two least significant bits of the position counter.

As Figure 13-9 shows, a rotating force vector is produced by the moebius

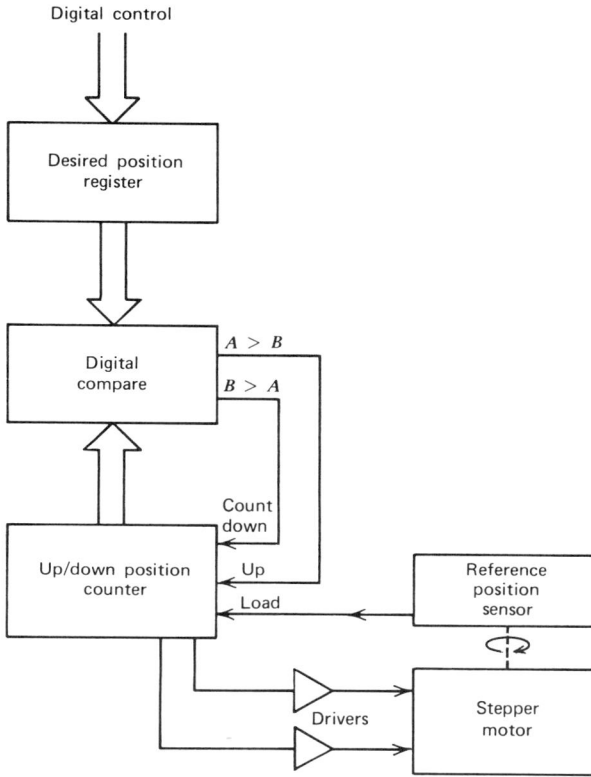

Fig. 13-8.   Digital positioning using stepping motor.

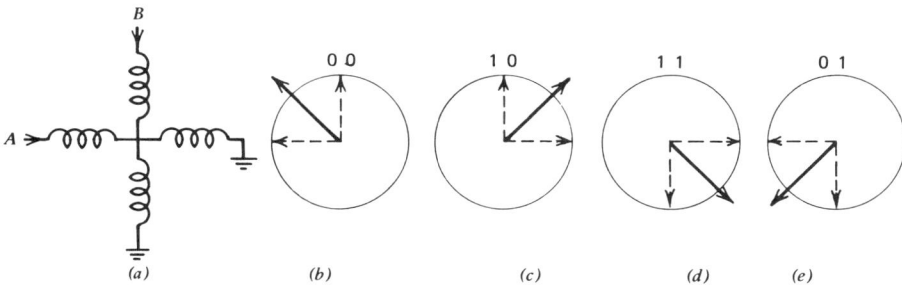

Fig. 13-9.   Driving a stepper motor from a Moebius counter: (a) arrow is direction of force for "1"; (b) first step; (c) second step; (d) third step; (e) fourth step.

count sequence. Going through the sequence in reverse order causes rotation in the opposite direction. The drivers simply drive the motor windings with (+) or (−) for 1 or 0. Bifilar-wound (center-tapped) motors are available that provide two windings for each phase. This makes it possible to use four single-polarity, IC drivers to drive one of two, oppositely wound windings for each phase. Actually each step is usually only $1/100$ or $1/200$ of a revolution because the motor is made with 100 or 200 poles. Positioning can thus be very accurate, yet speed can be as much as 1000 steps per second (600 rpm). Extremely accurate linear motion can be obtained by putting a lead screw on the shaft of a stepping motor as shown in Figure 13-10. Since there is *no* cumulative error, movements of tens of thousands of steps can be made within an accuracy of one step ($1/10,000 = 0.001\%$ accuracy). Since the increments are purely digital quantities, digital storage can be used, and digital logic can be used to mathematically interpolate between points, generate curves, and so on. In fact, the position counter, register, and comparison shown in Figures 13-7 and 13-8 can actually be just part of a microprocessor *program*.

Simple incremental encoders generate the same pulses for either direction of motion. However in some systems the actual direction of movement may not always agree with the polarity of the drive signal. In this case an *incremental encoder* can be used with *two outputs* generated by sensors spaced so as to produce outputs 90° out of phase (see Fig. 13-11*a*). As the encoding track moves to the right, the *A* and *B* outputs generate the familiar two-bit *moebius* counter sequence as shown in Figure 13-11*b*. Moving in reverse direction makes the sequence go in reverse. It is thus possible to *logically* sense direction of movement. Figure 13-11*c* shows a timing diagram of the outputs with a movement to the left, then stop, then movement to the right. A similar principle is used to sense clockwise and counterclockwise rotation with rotary incremental encoders.

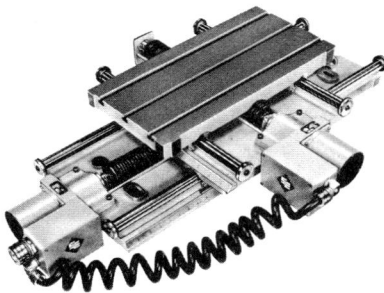

SLO-SYN NUMERICAL
POSITIONING TABLE

Fig. 13-10. Linear positioning with a lead screw.

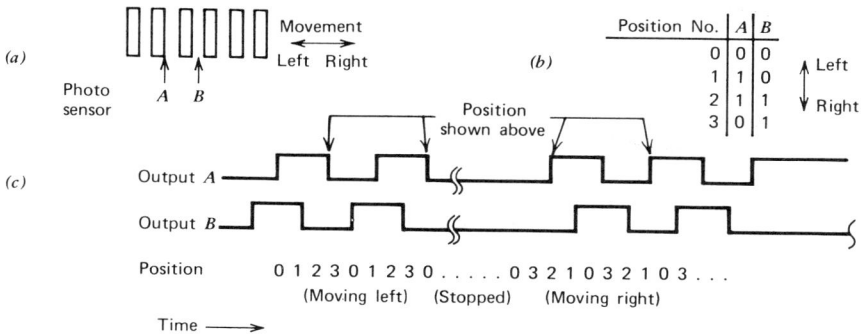

Fig. 13-11. Incremental encoder with direction sensing: (a) encoding track; (b) sensor output states; (c) outputs with motion reversal.

Now that we have covered some of the basic principles used in modern I/O equipment, we will look at a few successful examples of their use.

13.6   1600 BPI TAPE UNITS

Magnetic tape units have always been a supreme example of *time sharing* since a single nine-track head and read/write circuitry are used to read millions, or billions, of bits. This fantastic amount of time sharing makes the cost per bit of storage incredibly low, but at a great sacrifice of access time (since it may take minutes to find a particular record).

Early tape units used a solenoid-operated pinch roller to engage the tape against a *capstan* (a steel shaft about 0.25 in. in diameter) turning continuously at a constant speed. To go in both directions, two capstans and pinch rollers were required—one in each direction. When a pinch roller pulled in, it quickly accelerated the tape to the speed of the capstan. The pinching and uncontrolled acceleration, however, often damaged the tape.

Modern 1600-bit-per-inch (BPI) tape units use a *single capstan* to drive the tape. This capstan is about 2 in. in diameter and is directly on the shaft of a low-inductance servomotor (see Fig. 13-12). A tachometer on the back of the motor shaft gives negative feedback to a *velocity servo*. The tape can thus be smoothly accelerated to any of several precise speeds by simply generating analog command voltages as shown in Figure 13-4. A negative command voltage smoothly drives the tape in reverse using the same capstan.

The two tape reels are mounted directly on the shafts of two other DC motors. Each reel motor uses negative feedback from a photocell sensor to keep a loop of tape of the proper size between the reel and the capstan. When the

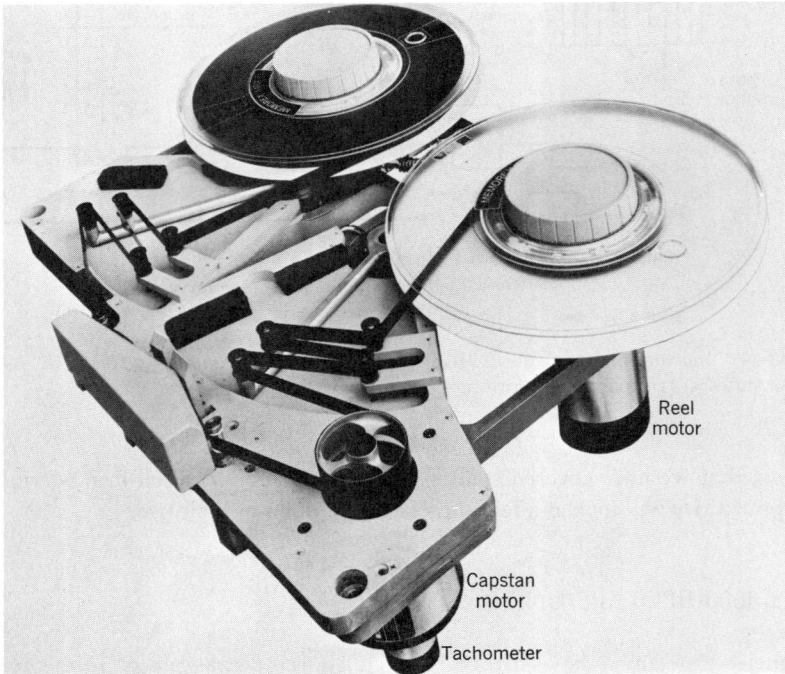

Fig. 13-12.   Single capstan tape unit (Precision Instruments).

tape starts moving, it can thus simply accelerate a small loop of tape instead of the whole reel. As the loop of tape grows too large on the takeup side, the photocell turns on the reel motor to make a correction. On the other side, the loop shrinks and causes the motor to turn on in the direction to feed tape. We thus have three separate servo systems operating, all as a result of a velocity command to the capstan servo. The tape loop sensors on some units simply turn the reel motors on or off. This type of servo is often called a "bang bang servo."

On high-performance tape units a fourth servo is sometimes used in decoding the data. This servo involves no mechanism but is used to electronically deal with tape speed variations beyond the control of the mechanical capstan servo. As data are read off the tape, a self-clocking signal, similar to Figure 12-17a, is produced. If a fixed timing circuit were used to decode this signal, speed errors would cause a degradation in error rate. For this reason a *phase-locked loop* is used to generate a clock that is in phase with the *average* signal received. The phase-locked loop is a servo system that uses negative feedback to set, not a position or velocity, but an *oscillator phase.* As Figure 13-13

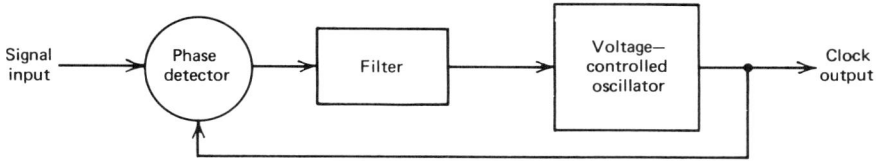

Fig. 13-13.   Phase-locked loop.

shows, a phase detector compares the phase of the incoming signal to an oscillator output. A DC voltage indicating the average phase error is used to control a *voltage-controlled oscillator* such that frequency increases if the oscillator phase is behind that of the input signal. Thus, the oscillator soon locks into phase with the incoming signal. This is such a generally useful circuit that complete IC phase-locked loops are available (e.g., NE562). Since it takes some time for the oscillator to "lock on" to the read signal, a *preamble* of 40 0's and one 1 precedes and follows (for reverse reading) each data record.

Another mechanical problem in traditional tape units, which is taken care of electronically in 1600-BPI tape units, is *skew*. Older tape unit types used **NRZ** coding, which requires all tracks to be clocked by the same clock. They therefore went to great lengths to eliminate the variation in timing in the read signals between one track and another. Though the tracks were all written in synchronism, variations in the relative timing of the tracks occurred due to uneven stretching and head and guide variations. Although previous tape units attempted to clock all tracks at the same time (or with small fixed corrections), the 1600-BPI unit uses a *self-clocking* signal on each channel. This makes it possible to have a variable-length digital buffer on each channel. Whenever all nine channels have a new bit ready, a character is output. Each channel, however, can get one or two bits ahead of the others if necessary.

13.7   CASSETTE TAPE UNITS

Cassette tape units are an attempt to provide a minimum cost magnetic tape unit that can sell for the price of a paper tape unit. By reading and writing on *one track* at a time (often switchable), an even greater degree of time sharing is obtained, at an even greater speed disadvantage. A small plastic tape cassette, which was originally designed to be used at 1 7/8 in./sec for audio, is used as an economical storage medium. Many disastrous attempts were made in 1971 and 1972 to make a reliable digital cassette transport. The tiny capstan and pinch roller used at 1 7/8 in./sec was inadequate for higher speeds and smooth starts and stops.

The newer cassette designs use no capstan at all and simply drive from the reel hubs. This is simple and reliable, but since the reel diameter varies from 0.8 in. when empty to 2 in. when full, tape speed may vary by a factor of 2.5 from one end of the tape to the other. We will discuss two electronic approaches to this problem.

The first approach just lets the tape speed vary and uses a recording format that is self-clocking yet speed-insensitive over a fourfold range. Figure 13-14 shows the recording scheme that has a positive "clocking transition" at the beginning of each bit cell, and a transition at the end of the first one-third of the cell for a "one," or two-thirds of the way through the cell for "zero." Decoding is done digitally with a six-bit up/down counter. At the beginning of the bit cell (the positive transition) the counter starts counting up; when the negative transition comes, the counter starts counting down. If the counter reaches zero before the next positive transition, a 1 is indicated (counter reversed *before* middle of cell); otherwise a 0 is indicated.

Although this scheme reads data accurately in spite of speed variation, it is inefficient for two reasons. First, bit density varies with tape speed, so with maximum density at the slow speed less than maximum density is obtained at high speed. Second, the encoding scheme in Figure 13-14 records one bit in the space of three transitions versus two for the phase-encoding scheme shown in Figure 12-17a.

By maintaining a constant speed and using phase encoding, *twice* as much data can be stored on the tape. Constant speed can be maintained, in spite of the reel diameter variation, by using *negative feedback*. One method of obtaining a tape velocity signal is to prerecord a special *clock track* on the tape. A tachometer voltage is generated from this track and used as negative feedback. This technique is also inefficient because the clock track, read amplifier, and so on, represent wasted storage capability. Also, special cassettes must be used with the clock track written on them.

Tape velocity can be maintained within $\pm$ 5% without a clock track by using the motors that drive the two reels of tape as tachometers. The voltage generated by the motor pulled by the tape (the *drag* motor) is proportional to its rotational speed S_2. A voltage proportional to the speed of the *drive* motor (S_1) can be obtained by simply subtracting the drop across its armature resistance (as sensed by a current sense resistor as in Fig. 13-6) from its ter-

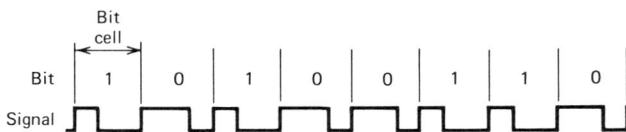

Fig. 13-14.   Pulse width encoding.

minal voltage. The speeds of the two hub motors can be related to tape velocity and reel diameters as follows:

$$V = \frac{S_1 D_1}{2} = \frac{S_2 D_2}{2} \tag{13-1}$$

The diameters of the reels are related to the hub diameter d, tape length L, and thickness t as follows:

$$\frac{\pi}{4}(D_1{}^2 - d^2) + \frac{\pi}{4}(D_2{}^2 - d^2) = Lt \tag{13-2}$$

or

$$D_1{}^2 + D_2{}^2 = \frac{4Lt}{\pi} + 2d^2 = K^2 \tag{13-3}$$

where K is a constant for any given hub size, tape length, and tape thickness. Substituting equation 13-1 into equation 13-3 and solving for velocity, we get

$$V = \frac{K S_1 S_2}{(S_1{}^2 - S_2{}^2)^{1/2}} \tag{13-4}$$

This function can easily be approximated with a diode function generator as two straight-line segments. The resulting velocity servo (Fig. 13-15) controls tape speed within $+ 5\%$ at 20 in./sec and attains stabilized speed within 35 msec. It is a good example of how we can electronically make use of signals

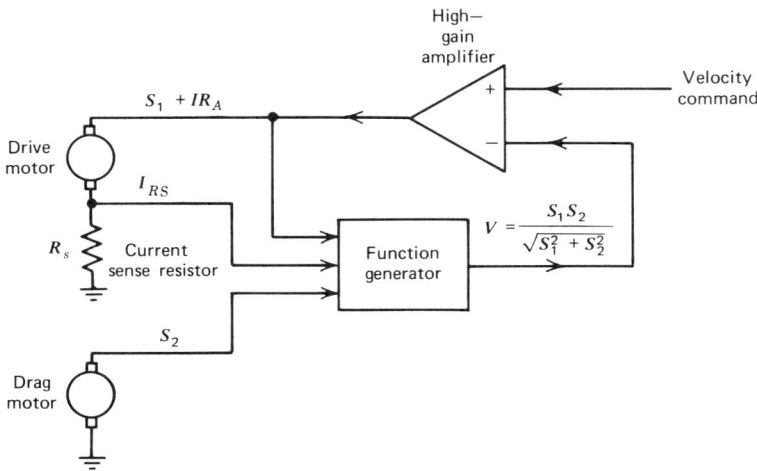

Fig. 13-15.   Cassette hub drive tape velocity servo.

that seem inappropriate at a much lower cost than that of adding more electromechanical components. We simply adapt to what is available—a process
not too different from adapting a logic design to a seemingly inappropriate
bargain component.

13.8   FLOPPY DISK DRIVES

Another peripheral device used both for auxiliary system storage and as a removable I/O medium is the *disk drive*. Large disk drives (e.g., the IBM 3331)
have removable *disk packs* that can store *billions* of bits at a cost of less than
1/200 of a cent per bit. The *floppy disk drive* (Fig. 13-16) is a minimum-cost
device that can store 3 million bits on a piece of magnetic tape material the
size of a 45-rpm phonograph record. The disk is always kept in a small plastic
envelope for protection and is loaded into the drive with the envelope. A *single
head* is moved over the surface of the disk by a *stepping motor*-driven worm
screw to read any one of 77 data tracks. A photocell indicates when the head is
on track zero, so the logic must go through an initialization sequence when
power is first turned on by stepping toward track zero until the photocell indicates the proper position of the head. The digital logic then outputs the correct
number of stepping signals to go to any track from any other track with circuitry (or program) as shown in Figure 13-8.

The Dynastore floppy disk eliminates the need for precision in its head posi-
tioner by using *negative feedback from the recorded tracks* to position its head.

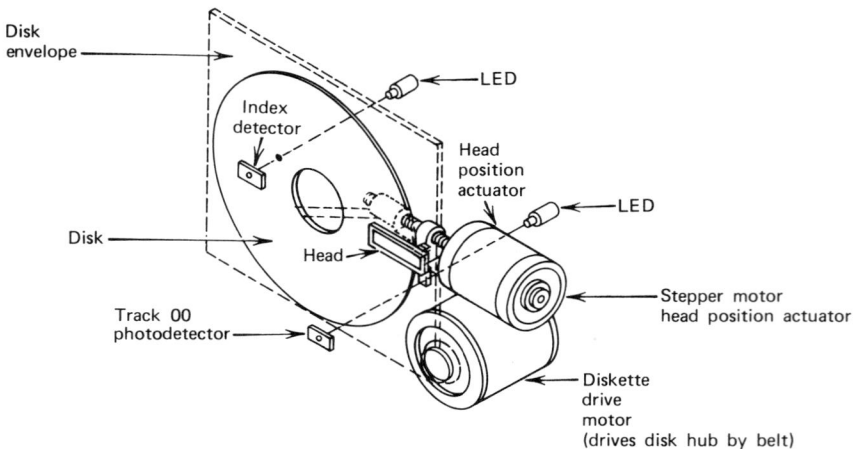

Fig. 13-16.   Standard floppy disk drive.

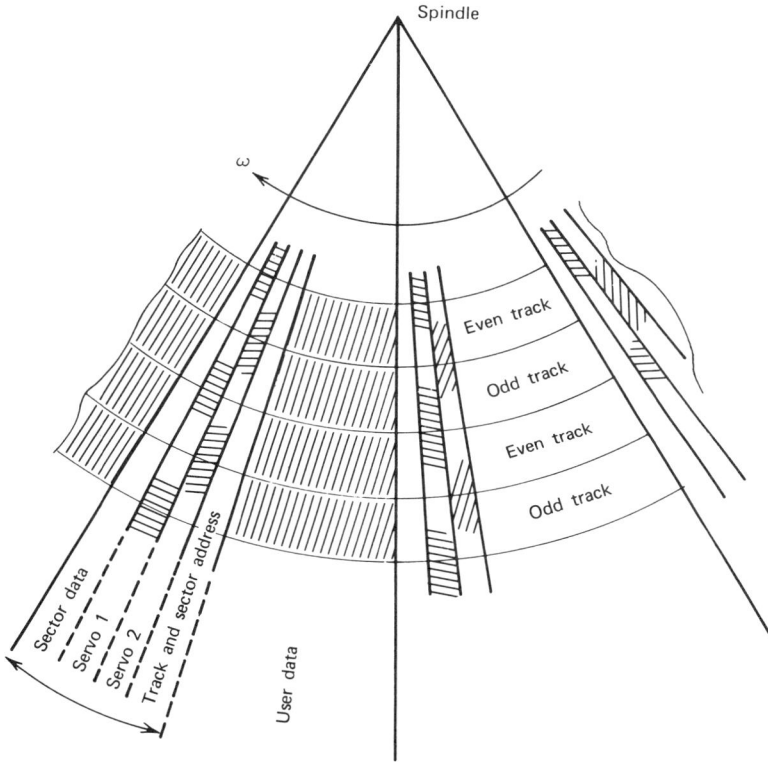

Fig. 13-17.   Dynastore floppy disk format.

The disk is prerecorded at the factory with sector marks for each track as shown in Figure 13-17; each track is divided into 32 sectors. Each sector begins with factory-prerecorded positioning information (SERVOS 1 and 2) and the track and sector address. All data are recorded in the space *between* the sector marks, leaving the prerecorded sector information intact.

The prerecorded SERVO 1 and SERVO 2 information is recorded in a checkerboard pattern as shown in Figure 13-17. (The crosshatched areas have a recorded signal, while the blank area has nothing recorded.) If the head is centered on the track, the amplitude of the signals during SERVOS 1 and 2 will be equal, because each is read by one-half of the head. If the head is off center toward the spindle, the signal will be stronger during SERVO 2 than during SERVO 1 beause more of the SERVO 2 recording will be under the head. If the head is off center toward the outside of the disk, SERVO 1 will be stronger. We can thus generate a head position feedback signal by simply storing the signal amplitude with a *sample and hold* circuit during **SERVO 1** and **SERVO 2** times and subtracting the results:

$$\text{POSITION ERROR} = \text{SERVO 1} - \text{SERVO 2} \qquad (13\text{-}5)$$

This signal will be zero when centered on the track, and plus when off center toward the spindle. This signal is used in a position servo for *final* positioning of the head.

When the head is *moved* to a new track position, the position servo is disabled until the head is within one-half track width of the null of the **POSITION ERROR** voltage for the correct track. Before this point is reached, control is *digital* with the head driven at a controlled velocity towards the correct track. By counting the number of polarity reversals of the **POSITION ERROR** signal (Fig. 13-18), the digital logic can maintain an up/down *track counter*. Velocity feedback is obtained from the *rate* of these polarity reversals. The velocity *command* voltage is generated by a two-bit up/down counter such that the head is gradually accelerated and decelerated during the first and last two tracks of the move. When the POSITION ER-ROR signal approaches zero for the desired track, the *position servo* mode of operation is enabled to provide final positioning.

It is thus possible to electronically provide extremely accurate centering of the head on the track without holding *any* particularly close tolerances. A simple DC motor, wire, and pulley drive can be used for positioning, with the accuracy provided *electronically* from the normal read signal. A similar head positioning system is used on the IBM 3331 (6 billion bit) disk drive. In that

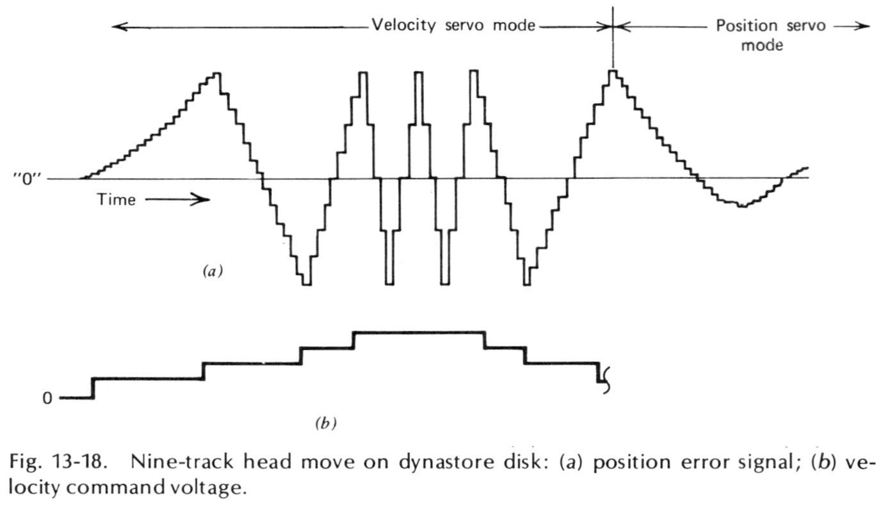

Fig. 13-18. Nine-track head move on dynastore disk: (a) position error signal; (b) velocity command voltage.

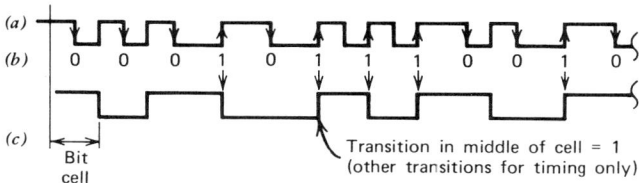

Fig. 13-19. (a) simple phase modulation; (b) data; (c) modified frequency modulation (MFM).

system one of the 24 disk surfaces is completely covered with the "checker-board" position servo information, so the servo information is *continuously* available. Due to the improved head positioning accuracy, *twice as many tracks* can be put on each disk surface as was possible with earlier units (e.g., 2311) that used feedback from the head actuator position.

Both the IBM 3331 and the Dynastore disk drives also use another interesting technique for *trading electronic complexity for mechanical* hardware. Both units use a *modified frequency modulation* (MFM) code for recording the data, which makes it possible to store *twice as many bits* on each track. The additional electronic complexity thus pays off handsomely by making each disk surface do the job of two surfaces. Figure 13-19 shows the MFM and simple phase modulation codes for comparison. Both schemes are self-clocking, but while the clock can be generated by just a one-shot circuit (as shown in Fig. 12-17) in simple phase modulation, a more complex decoder is required for the MFM system. The key to the waste in simple phase modulation is that a timing transition is sent for *every* bit, so *up to two transitions must be sent for each bit*. With MFM, timing transitions are sent *only when doing so will not cause transitions that are less than one bit time apart*. Each bit cell thus has a transition in the middle of it to indicate a 1, or no transition to indicate 0. Extra timing transitions are added at the bit cell boundaries between pairs of consecutive zeros. Transitions are thus always spaced by 1, 1.5, or 2 bit times. Clock is recovered with a phase-locked loop (Fig. 13-13), which generates a continuous clock, locked in phase with the received transitions. Decoding each bit consists of simply seeing if a transition is detected near the middle of the bit cell.

13.9 DOT MATRIX PRINTERS

A most successful example of the use of modern negative feedback and time-sharing techniques is the Centronics dot matrix printer. Instead of using a separate solenoid for each of the 132 characters across the line, seven very

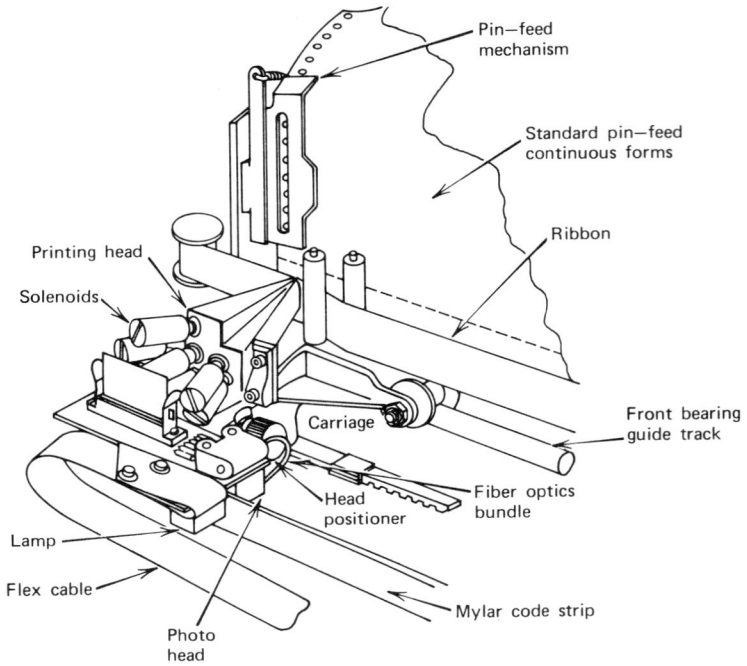

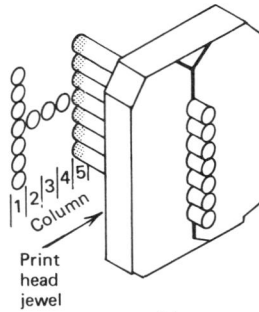

Fig. 13-20. Centronics printer: (a) carriage assembly; (b) print head detail.

high-speed solenoids are time shared by moving them across the line on a carriage as shown in Figure 13-20. The solenoids drive stiff wires that are brought together in a vertical column the height of one letter. As the carriage moves across each letter, the seven solenoids are pulsed or not pulsed during each of five positions to form a dot matrix representation of the character. A total of 64 characters can be printed by this method, as shown in Figure 13-21.

A simple mechanism moves the carriage across the paper. Horizontal spacing between columns is maintained exactly by optically sensing the *actual* position of the print head. A photocell on the print head simply senses light transmitted through a Mylar strip printed with alternate light and dark stripes. Perfect characters can thus be printed even while the print head is coming up to speed at the beginning of the line and decelerating at the end of the line.

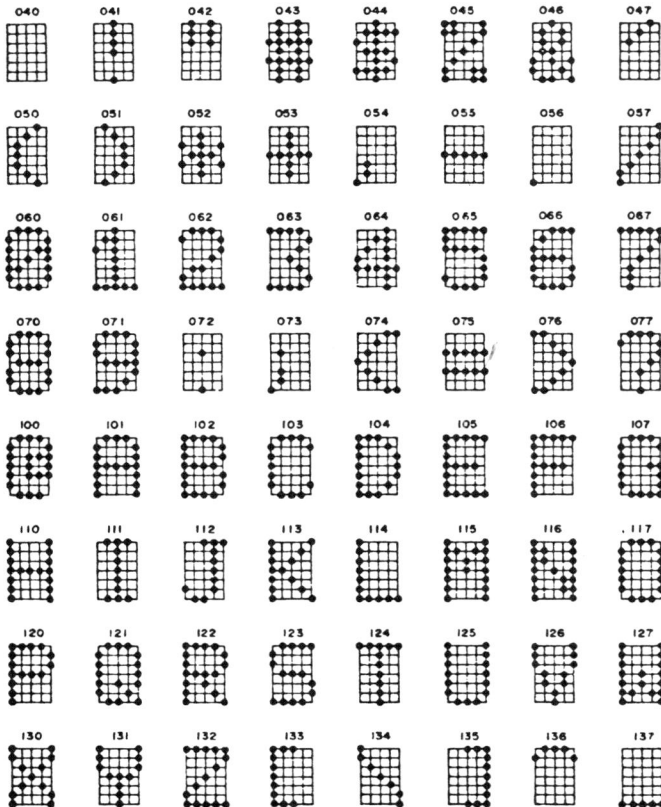

Fig. 13-21.   Standard 5 × 7 64-character set.

As with any time sharing, there is a speed tradeoff. The seven solenoids must fire up to $5 \times 132 = 600$ times for each line, instead of once for each line as required when there is one solenoid per character. Since there are only seven solenoids, an all-out effort was made to make them fast. On the Centronics printer the solenoids take only 1 msec to print a dot and recover, so characters can be printed at 165 characters per second. For an even faster print rate, another model printer uses *two* print heads moving together, each printing half the line. The print rate is thus doubled, giving 330 characters per second, or 125 full lines per minute. The time delay for carriage return is eliminated on the model *by printing alternate lines in reverse*. The digital logic easily takes care of reversing the letters by simply using up/down counters to count columns and character positions.

Since the shape of the characters is stored in a ROM, special character sets can easily be produced by simply changing the ROM bit pattern. Boldface characters are printed by just ignoring every other pulse from the carriage position coding strip. The important point is that we can electronically generate just about any signal we care to define to simplify the *mechanical* part of the system. Transforming the character codes into sequences of drive signals to the seven solenoids costs very little compared to the cost of a more complex mechanism.

13.10   DISPLAYS AND KEYBOARDS

A TV set is an example of the ultimate application of time sharing in that a picture containing about 250,000 elements is produced by modulating a *single* electron beam. Electronically generating a video signal for displaying 5×7 dot matrix characters simply requires reading the bits out of the ROM in a different order. Since a TV set is an extreme example of a bargain component, CRT displays have become the standard "people interface" output device where no *hard-copy* printout is required.

A CRT display is so flexible that it is usually unnecessary to use any other display devices in a system that includes one. For example, indicator lamps can be essentially mechanized by programming the appropriate message to appear on a particular part of the screen. The entire layout of a control panel can essentially be programmed as a CRT display format. System changes and various operating modes can thus be accomplished without changing any hardware.

A CRT *terminal* (see Fig. 13-22) combines a typewriter-like keyboard with a CRT display. Again, programming can make the standard keyboard do the job of *any* special purpose control panel. Special functions are simply programmed as keyboard sequences. For example, all the functions which

Fig. 13-22. CRT terminal.

used to be done on a computer control console can be done on a terminal. Each special button is simply programmed as a sequence on the keyboard. For example, pressing the *L* key on the keyboard can do the same thing as pressing the LOAD key on the special console. A special console does have the advantage of being somewhat self-explanatory, but it is often practical to display all the *meaningful* keys that can be pressed at any given time on the CRT. In this way the system is self-explanatory, and only pertinent functions are displayed at any given time.

Some CRT terminals include a programmed microprocessor for performing a whole range of logical functions *within the terminal*. These terminals are called *intelligent terminals*. They typically allow text to be prepared and altered by erasing, inserting, deleting, moving, and so on, directly on the CRT screen. This makes it possible, for example, to read a file (e.g., a customer name and address), change it, then send it back to the main computer system. As microprocessors become more powerful, intelligent CRT terminals will more fully resemble entire computer systems. Systems are already available with interfaces for disk memories, cassette tapes, and printers.

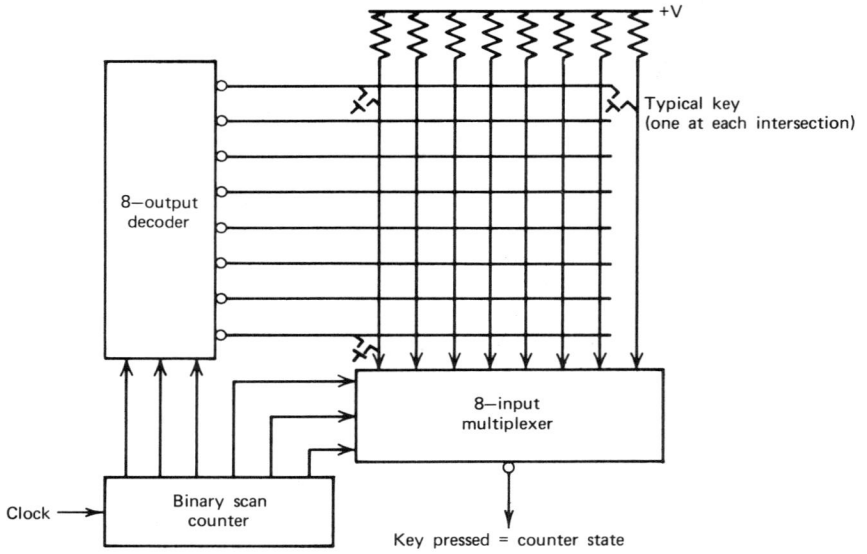

Fig. 13-23. Keyboard scanner.

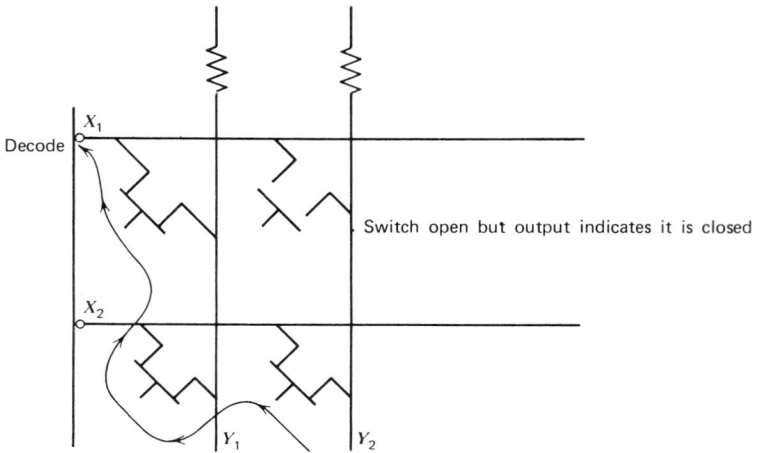

Fig. 13-24. Sneak path in switch matrix.

Special purpose keyboards and illuminated displays are still desirable for certain applications. Single-chip LSI circuits are available for interfacing keyboards to serial or parallel binary data inputs. Virtually all these circuits work by scanning the keys one at a time as shown in Figure 13-23. An X–Y matrix is formed of the key switches. For each of the 64 counter states a different key is addressed by a unique combination of decoder output and multiplexer input. The output of the multiplexer will thus go high if any of the keys is pressed, when *that key* is addressed by the counter. Since this output essentially consists of a time slot for each key, time-shared logic can be used for contact-bounce elimination and handling "rollover" (pressing of a key before the previously pressed key is released).

If *three or more* contacts are closed at once on an X–Y switch matrix, the results will be incorrect unless a diode is placed in series with each switch. Figure 13-24 shows an example of a sneak path when switches at X_1, Y_1 and X_2, Y_1 and X_2, Y_2 are closed. It will *appear* that the switch at X_1, Y_2 is closed,

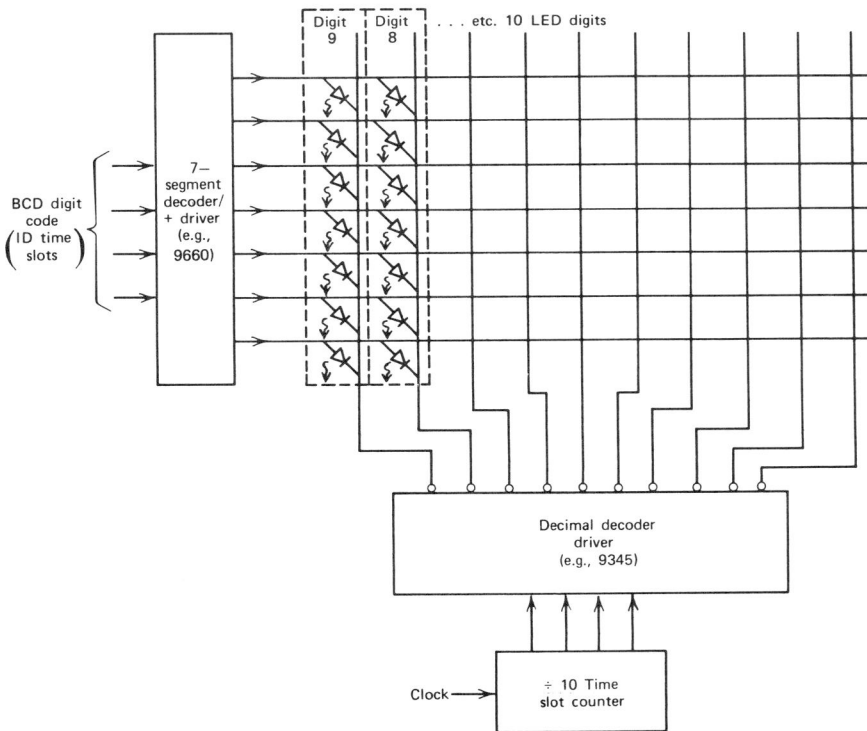

Fig. 13-25. LED 10-digit display with time-shared driver.

since multiplexer input Y_2 will be pulled down by X_1 via the *sneak path* through the other three switches. By putting a diode in series with each switch, the sneak path is blocked by the diode on the switch at X_2, Y_1.

Since LEDs pass current in one direction only, they can be driven in an X–Y matrix with no sneak path problems. In addition, LEDs actually *look brighter,* with the same average current, if they are *pulsed* instead of being driven continuously. It is thus possible to drive only one row of an X–Y matrix of LEDs at a time and thereby time share the drivers. A matrix of 80 LEDs, for example, can be driven by eight positive current drivers and a 10-output decoder to select the rows in turn. The display thus has 10 *time slots,* each sharing the same positive driver *and* any logic associated with it.

Seven-segment numeric displays can similarly use a single decoder driver to drive many digits, as shown in Figure 13-25. The nice thing about time sharing any I/O device is that *the time sharing can extend as far into the control logic as desirable.* Since each digit occupies a separate time slot, any logical or mathematical operation done on the digits can essentially time share the same gating between the digits with the *observer's eye* effectively demultiplexing the result.

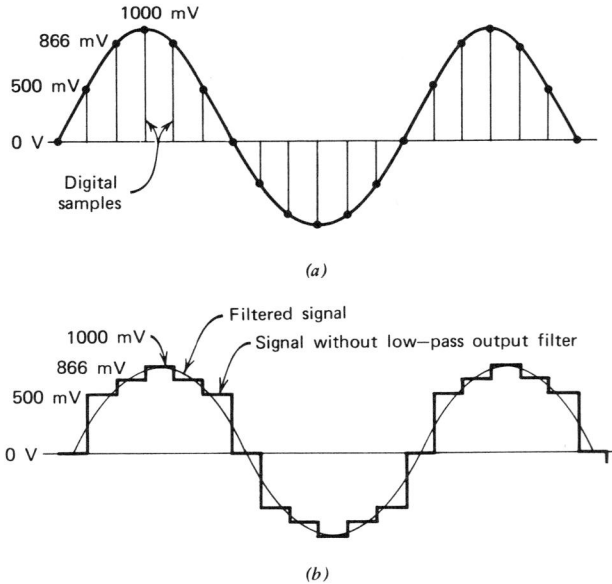

(a)

(b)

Fig. 13-26. Sampling and reproducing a continuous signal: (a) analog input signal; (b) reconstructed signal.

13.11   DIGITAL REPRESENTATION OF ANALOG SIGNALS

Analog voltages can be represented in a digital system just as numbers are represented. For example, a signal voltage range of ± 2.048 V can be represented, to a resolution of 1 mV, by a 12-bit binary number. The binary number thus represents polarity and the number of millivolts. The process of converting analog voltages to digital numbers is called *analog-to-digital (A/D) conversion*. Converting a digital number representing a signal voltage to an analog voltage is called *digital-to-analog (D/A) conversion*.

Continuous signal waveforms can be represented by periodically sampling the signal voltage. The digital representation of the signal is thus *a sequence of binary numbers*. The signal can be reconstructed by converting the sequence of binary numbers back into a sequence of voltage levels and filtering to smooth out the discontinuities in the signal (see Fig. 13-26). The rate of sampling must be *at least twice the highest frequency component in the signal.*

If higher frequency components are present in the input signal, they will cause distortion since they will be converted to lower frequency "difference signals" when the analog signal is reconstructed. For example, Figure 13-27 shows how a 6-kHz signal, sampled at 8 kHz, will be reconstructed as a 6 − 8/2 = *2 kHz* signal. To prevent this kind of distortion, *a low-pass* filter must be used to remove all signal components with frequencies higher than half the sampling rate before sampling. With an identical filter on the output signal, the reconstructed and filtered signal can be a perfect reproduction of the input signal within the frequency range reproduced. Of course the limited resolution of the digital representation will produce some *quantization noise,* and filter defects will allow some undesired frequencies through. But the sampling process itself is theoretically capable of *perfectly* reproducing the signal.

A general D/A system thus includes a low-pass filter and A/D converter on the input and a D/A converter and low-pass filter on the output, as shown in Figure 13-28. Sometimes actual low-pass filters are not required if the normal slow response of the transducers excludes frequencies greater than half of the sampling frequency.

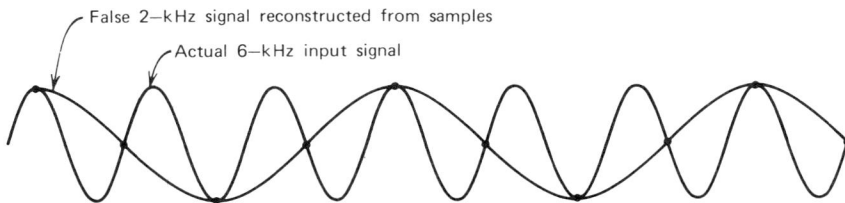

Fig. 13-27.   Incorrect reconstruction of signals > ½ sampling rate.

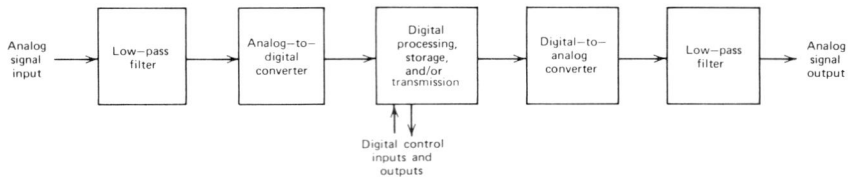

Fig. 13-28.   General digital-analog system.

For several reasons it is often advantageous to go to the trouble of digitizing analog signals. The most important reason is probably *accuracy.* When analog techniques are used, costs rise rapidly for accuracies of greater than about 1% for each operation. Since errors tend to be additive, complex operations quickly get out of hand. With digital operation accuracy is doubled each time word length is increased by one bit. With a 16-bit word, for example, an accuracy of 0.0015% is obtained with ordinary, mass-produced digital circuits. Digital data can be magnetically *stored* and retrieved perfectly. Error check codes and rewriting make *perfect* results possible, even with imperfections on the recording medium. Signals can be *transmitted* over long distances and regenerated with simple digital circuits for essentially perfect results. Although practical problems tend to limit the performance attainable with analog filters, there is essentially no limit to the filtering that can be done digitally. For example, a digital filter (where each output sample is a weighted sum of the 128 previous input samples) can produce a sharp cutoff to about *100 dB* below the passband response (see Ref. 2).

Another big advantage of digital operation is the *flexibility offered by digital programming.* An industrial process control system, for example, can be modified to improve the process by simply loading new programs. Programs can even be written to automatically optimize the process by adjusting themselves according to actual results obtained. Essentially any nonlinear function can easily be generated by "table lookup." Data gathered during experiments in space, on oil fields, and so on, can be later worked over creatively on a digital computer if held digitally on magnetic tape. Numerical control systems make it possible to digitally define machine operations using a special "language" and automatically do machining and drilling operations.

Another advantage is a result of the speed possible with digital operation. This speed capability makes it possible to *time share* one digital circuit among a large number of converted analog signals. Voice telephone signals, for example, can be represented digitally by 8000 seven-bit samples per second, or 56,000 bits per second. Since ordinary digital circuits can operate at hundreds of times this rate, a single digital multiplexer can switch *hundreds* of telephone conversations—each during a different *time slot.* Digital RAMs can be used to interchange time slots, so seven, high-speed, 1024-bit RAMs can do the job of a 1024×1024 crosspoint switch.

13.12 ANALOG-TO-DIGITAL CONVERSION

Now let us look at some actual conversion techniques. Digital-to-analog converters are available as self-contained, monolithic ICs (e.g., MC1406). Basically they work by having each input bit switch a current source with the proper binary weight. These currents are summed and converted to a proportional output voltage.

Analog-to-digital conversion is a more difficult problem. The most popular high-speed conversion technique requires logic as shown in Figure 13-29. *Successive approximations* to the input voltage are made by a D/A converter whose output is compared to the input signal by a *comparator* circuit. Whenever the input signal voltage is higher than the D/A converter output, the comparator output will be true. Starting with the most significant bit, a 1 is *tried* in each bit position, and *the 1 is kept if the signal is greater than the resulting D/A converter output.* After one comparison for each bit, the binary code remaining in the addressable latch represents the analog input voltage digitally. (Note that the D flip-flop is required because the latch is not a true clocked device, yet its output affects its input.)

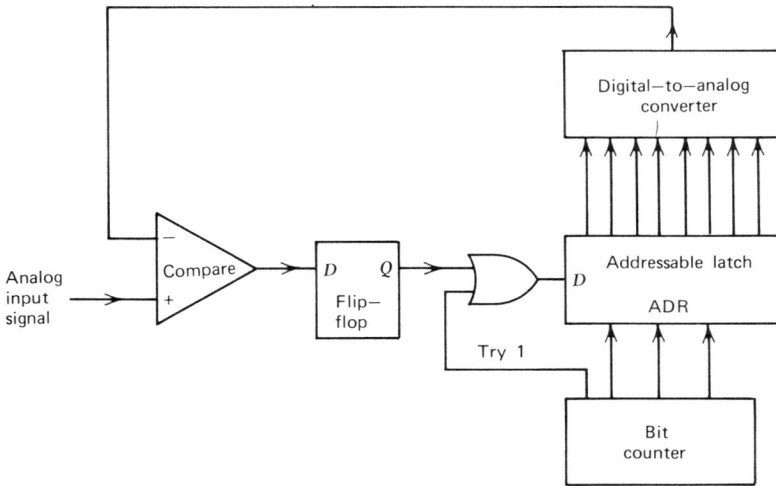

Fig. 13-29. Analog-to-digital conversion by successive approximation.

TABLE 13-1   SUCCESSIVE APPROXIMATION EXAMPLE: INPUT = 82 MV

Bit = 1	Latch Contents	D/A Output (mV)	<82 mV ? (store as this bit)
7 (sign)	10000000	0	1
6 (msb)	11000000	+64	1
5	11100000	+96	0
4	11010000	+80	1
3	11011000	+88	0
2	11010100	+84	0
1	11010010	+82	1
0	11010011	+83	0
(RESULT)	11010010		

As an example of an analog A/D conversion by succesive approximation, assume that we define a ±127 mV signal range with an eight-bit digital number. The most significant bit will thus represent the sign, and the remaining bits will represent 64, 32, 16, 8, 4, 2, and 1 mV, respectively. An input signal of 82 mV, for example, will result in the successive approximation sequence shown in Table 13-1.

The technique thus finds the proper binary code by essentially dividing the remaining range of possibilities in half with trial until the desired accuracy is obtained.

Of course the logic for making the successive approximations can be *programmed* if programmed logic is available. In that case an output D/A converter and a comparator input to the programmed processor are required. By using a *sample-and-hold* analog circuit, the same D/A converter can be used for many outputs as well as for comparison to inputs. A sample-and-hold circuit stores an analog voltage on a capacitor during a brief *sample* pulse. Between sample pulses a continuous output voltage is available, which is equal to the input voltage at the end of the sample period (typically a few microseconds). Many sample-and-hold circuits can thus be connected to a single D/A converter, which can produce the required analog outputs one at a time.

Sample-and-hold circuits are also used on fast-changing input signals to prevent them from changing during the A/D conversion process. An input voltage change (corresponding to a carry propagation) during a successive approximation A/D conversion could produce false results. When many analog inputs are present, the D/A converter, comparator, and sample-and-hold circuit are often *time shared* among many inputs. An *analog multiplexer* (e.g., CMOS CD4051) is used to connect one of the analog input voltages to the sample capacitor during the sample pulse (Fig. 13-30). The A/D conversion is then performed, then the next input channel in turn is sampled. Note that a *separate* low-pass filter is required for each input signal.

Since CMOS analog multiplexers look essentially like small resistors when turned on, they can also be used as demultiplexers. A single D/A converter output can thus drive eight HOLD capacitors for eight output channels. An amplifier with high input impedance and a low-pass filter is required for each output channel.

Much simpler A/D converters are possible if conversion speed is unimportant. Digital volt meters (DVMs), for example, digitize voltages for visual display or printing, so the signal has to be converted only 60 times/sec. The dual-slope integrating technique used in these meters is illustrated in Figure 13-31. An integrating capacitor is allowed to charge at a constant rate proportional to the input voltage, for exactly 1000 clock pulses. The capacitor is then discharged back to the same point by a calibrated reference current. The number of clock pulses required directly indicates the input voltage digitally (i.e., 1000 counts indicates the input rate was equal to the reference rate, 500 means it was one-half etc.). The dual-slope method can be very accurate with crude circuits, since clock rate, capacitor value, threshold voltage, and nonlinearities all cancel out. The disadvantage, of course, is that conversion is slow. Several LSI DVMs have been made using this technique to provide a resolution of 0.005%.

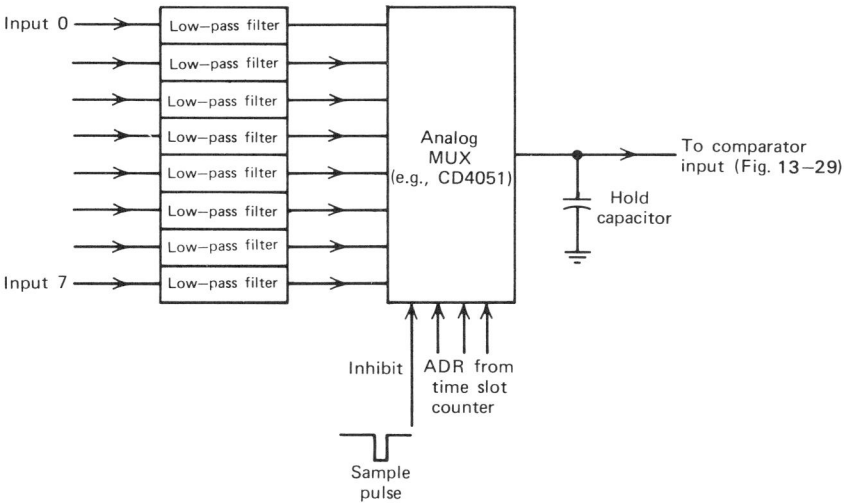

Fig. 13-30.   Multiplexing analog inputs.

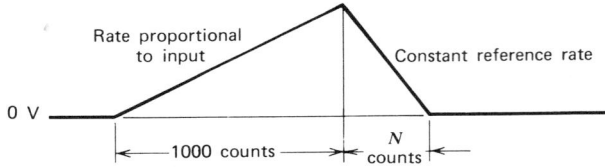

Fig. 13-31.   Dual-slope integrator A/D conversion.

Slow D/A conversion is likewise very simple. By controlling the *duty cycle* of a digital signal, which is then heavily filtered, accurate analog voltages can be produced. For example, a 5-V logic signal that is true for one out of 100 clocks has an average value of $5/100 = 50$ mV. Heavy filtering produces a 50-mV DC output. Simple digital counters can thus be used to produce pulse-width-modulated (PWM) signals which are then integrated to a DC level. Switching-mode DC power supplies effectively regulate output voltage with very little dissipation by controlling digital pulse width.

When a large range of signal amplitudes must be handled *linear* digital encoding of the signal can be inefficient. In PCM telephone systems, for example, a wide range of voice levels must be handled. To keep distortion at a very low level, a voice signal should be represented by at least about 64 quantum levels. However, telephone signal amplitudes can vary over a $128:1$ range; a linear coding would thus require $64 \times 128 = 8000$ steps, or 13 binary bits. This would be quite wasteful, as the highest level signal would be quantized too finely, into 8000 levels.

The solution is to use *nonlinear* coding steps that grow larger at larger amplitudes. An eight-bit code is used with three of the bits indicating one of eight *step sizes* (1, 2, 4, 8, 16, 32, 64, or 128), and the other five indicating polarity and one of 16 levels. Each binary code thus represents a unique level, but *the spacing between successive levels gets wider* the further the signal is from 0 V. If our smallest step size is 1 mV, then we can go up to $\pm$ 16 mV with 1-mV steps, the next 16 steps will be 2 mV each, then 16 steps of 4 mV, 16 steps of 8 mV, 16 steps of 16 mV, 16 steps of 32 mV, 16 steps of 64 mV, and 16 steps of 128 mV each. Our total voltage range with a 1-mV minimum step is thus

$$+16\,(1 + 2 + 4 + 8 + 16 + 32 + 64 + 128) =$$
$$\pm 16 \times 255 = \pm 4080\,\text{mV} \quad (13\text{-}6)$$

A D/A converter for such a compressed code can be made out of a five-bit D/A converter and an eight-bit *multiplying* D/A converter with each bit driven by a digital *decoder* output. Analog-to-digital conversion can be done by successive approximation (Fig. 13-29) using the nonlinear D/A converter.

TABLE 13-2   13-BIT LINEAR REPRESENTATION OF EIGHT-BIT
PCM CODE

													(S = Sign)									Step Size
		8-Bit Code									Equivalent 13-Bit Binary Value													(mV)
S	0	0	0	W	X	Y	Z		S	0	0	0	0	0	0	0	1	W	X	Y	Z			1
S	0	0	1	W	X	Y	Z		S	0	0	0	0	0	0	1	W	X	Y	Z	0			2
S	0	1	0	W	X	Y	Z		S	0	0	0	0	0	1	W	X	Y	Z	0	1			4
S	0	1	1	W	X	Y	Z		S	0	0	0	0	1	W	X	Y	Z	0	1	1			8
S	1	0	0	W	X	Y	Z		S	0	0	0	1	W	X	Y	Z	0	1	1	1			16
S	1	0	1	W	X	Y	Z		S	0	0	1	W	X	Y	Z	0	1	1	1	1			32
S	1	1	0	W	X	Y	Z		S	0	1	W	X	Y	Z	0	1	1	1	1	1			64
S	1	1	1	W	X	Y	Z		S	1	W	X	Y	Z	0	1	1	1	1	1	1			128

When mathematical operations are to be made on this type of nonlinear coding, conversion to a 13-bit linear code can be made easily as shown in Table 13-2.

13.13 OPERATIONAL AMPLIFIERS

Whenever digital techniques are used to handle analog signals, we are faced with the basic systems decision of where to draw the line between the digital and the analog representation of signals. Although digital techniques have many advantages in some cases, *analog techniques are often much simpler and more economical.* The digital designer must therefore have a basic understanding of what can be done easily with analog circuits.

Often it is better not to convert all analog signals to digital immediately. Analog techniques can often be used to process the signals much more economically before going digital. The trick, as always, is to use the technique that best fits the job. The servo system in Figure 13-1 is a good example. Although it is possible to convert the analog position sensor signal to a digital signal, subtract it from the digital command signal, and convert the result to an analog motor drive voltage, it is *much easier* simply to do an analog subtraction at the amplifier input.

Another example is the velocity command voltage in Figure 13-4. We could produce the ramps by counting up and down digitally, then converting the digital result to an analog signal. However, it is much easier simply to generate the two analog levels with two different resistors connected to a digital gate output and generate the ramps with a simple analog integrator circuit.

Very few analog ICs have emerged that are useful in a wide range of applications and have therefore developed the volume market necessary to make them bargain components. The big exception is *operational amplifiers,* which have become *the* standard linear component and are available, four to a package, for less than $1.00.

Most analog signal processing is done by various connections of *operational amplifiers.* As Figure 13-32a shows, an ideal operational amplifier has an infinite voltage gain (A_v) in response to the *difference* in input voltage at the two input terminals. Even inexpensive real circuits have typical voltage gains of 200,000 and minimum gains of 20,000.

Since the basic operational amplifier circuit has uncontrolled gain, it is not useful for analog signal processing without the addition of *negative feedback.* The circuit in Figure 13-32b *inverts* the signal and amplifies it, with gain precisely controlled by the ratio of two resistor values. We can see why this is true if we observe that any plus input voltage produces a current of $I_{in} = V_{in}/R1$ into the minus input of the amplifier. This tends to drive the amplifier output minus until an equal but opposite current is produced through $R2$ when

$$V_{out} = - I_{in} \cdot R2 \tag{13-7}$$

Substituting, we get

$$V_{out} = \frac{- V_{in}}{R1} \cdot R2 \tag{13-8}$$

Rearranging, we find that

$$V_{out} = \frac{- R2}{R1} \cdot V_{in} \tag{13-9}$$

The amplifier output thus produces just enough voltage to keep the voltage at its minus input terminal at 0 V. The input impedance is exactly equal to $R1$ since the minus input of the amplifier looks essentially like ground.

A *noninverting amplifier* can be made, as shown in Figure 13-32c, by dividing the output voltage with two resistors and feeding it back to the inverting $(-)$ amplifier input. Again, the amplifier will produce the output voltage necessary to make the voltage difference between the $(+)$ and $(-)$ input terminals zero:

$$(+)\, input = (-)\, input \tag{13-10}$$

(a)

(b)

(c)

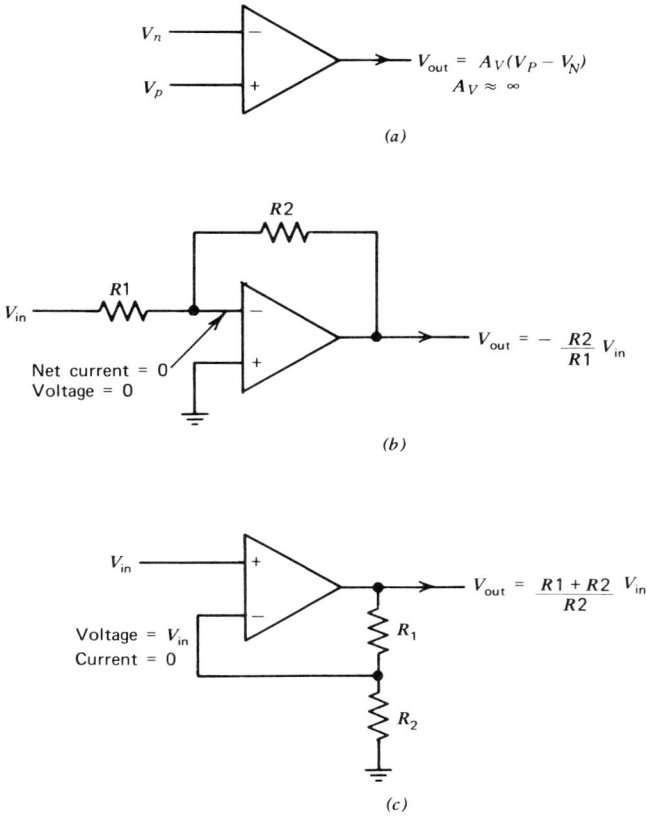

Fig. 13-32. Operational amplifiers: (a) basic circuit characteristic; (b) inverting amplifier with gain controlled by feedback; (c) noninverting amplifier with gain controlled by feedback.

Substituting, we get

$$V_{in} = V_{out} \cdot \frac{R2}{R1 + R2} \tag{13-11}$$

$$V_{out} = V_{in} \cdot \frac{R1 + R2}{R2} \tag{13-12}$$

Input impedance in this case will be essentially infinite.

Unless expensive resistors are used, gain accuracy will be limited to $\pm 1\%$ (for 1% resistors), which is equivalent to a seven-bit digital representation. If this accuracy is adequate, we can do many complex operations on analog signals with very little hardware. Figure 13-33 shows a few examples.

Figure 13-33a shows how many signals can be added together at the amplifier input, since the currents from the many input resistors are additive. Also, since different resistor values give a different scale factor for the input current produced for a given voltage, each input can be *weighted* as described. The result is equivalent to performing a *multiplication* (by the weighting factor) on each input voltage, then *adding* the results. Doing this *digitally* requires quite a bit of hardware, yet the analog approach requires only a *few resistors*.

Most servo systems use a circuit such as this to sum the *command voltage* and the *feedback* from the velocity or position transducer. Since the scale factors are determined by resistor values, adjustable resistors (trimpots) can be used to adjust out variations in tachometer (or other transducer) outputs from unit to unit. Note that the command voltage need not exist as such but can be produced directly from digital inputs. By connecting a digital signal through a resistor to the *current-summing point* (in Fig. 13-33), no current will flow when the digital signal is false (O V); and a current determined by the resistor value will flow when the signal is true (+). A fast and a slow velocity command thus requires simply a smaller and a larger resistor to the summing point. Many D/A converters work in exactly this manner using a binary sequence of resistor values (e.g., 1000, 2000, 4000, 8000, 16,000 ohms, etc.).

Figure 13-33b shows how *integration* can be performed with an operational amplifier. Since the current through the capacitor is

$$I = -C \frac{dv}{dt} \tag{13-13}$$

the *rate of change* (slope) of the output voltage will be the rate required to produce a feedback current that will cancel the input current. A constant slope proportional to the input voltage will thus be produced. Since the voltage on the minus input (summing point) is still zero, many inputs can be used as in Figure 13-33a, and the resulting output signal slope will equal the *weighted sum* of the inputs. Doing this digitally would require not only the multiplications and additions for the weighted sum, but the accumulation of a sum of the results. The dual-slope integration A/D conversion technique illustrated in Figure 13-31 is normally done with an operational amplifier integrator. The error-canceling properties of this technique illustrate how good accuracy can often be obtained by clever, error canceling configurations instead of precision components.

Figure 13-33c shows how capacitive input produces a *differentiator*. Since the input current is proportional to signal *slope,* the output voltage required to cancel it will be proportional to input signal slope. Digital differentiation requires storing the signal value of the previous sample and subtracting.

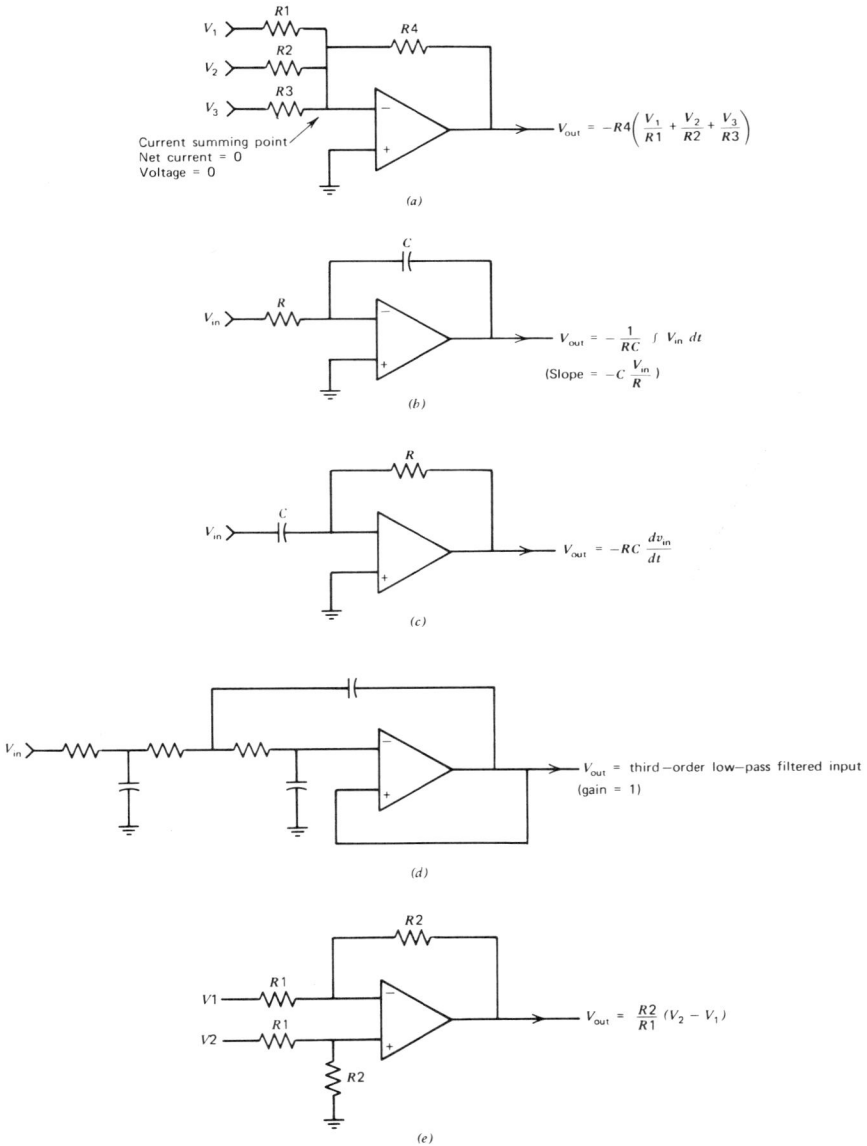

$$V_{out} = -R4\left(\frac{V_1}{R1} + \frac{V_2}{R2} + \frac{V_3}{R3}\right)$$

Current summing point
Net current = 0
Voltage = 0

(a)

$$V_{out} = -\frac{1}{RC} \int V_{in}\, dt$$

$$\left(\text{Slope} = -C\,\frac{V_{in}}{R}\right)$$

(b)

$$V_{out} = -RC\,\frac{dv_{in}}{dt}$$

(c)

V_{out} = third–order low–pass filtered input
(gain = 1)

(d)

$$V_{out} = \frac{R2}{R1}\,(V_2 - V_1)$$

(e)

Fig. 13-33. Other operational amplifier circuits: (a) weighted addition of several inputs; (b) integrator; (c) differentiator; (d) active low-pass filter (third order); (e) differential amplifier.

Figure 13-33*d* shows how an operational amplifier can be used to make a third-order low-pass filter with no inductors. By cascading two such circuits, a sixth-order filter can be obtained. By using capacitors where resistors are shown and vice versa, a high-pass filter can be made. Although filtering can be done digitally, it is a complicated process. For example, to match the performance of a sixth-order analog filter, about nine prior digital samples must be retained and each *multiplied* by a weighting coefficient and *summed* to obtain each digital filtered output sample. Analog filtering is thus much more economical than digital filtering unless extremely sharp cutoffs with high accuracy are needed.

Figure 13-33*e* shows how differential inputs can be amplified. Common-mode analog noise is canceled since it appears equally on both amplifier inputs. If the resistor ratios are inaccurate, however, they can return common-mode signals into differential mode.

EXERCISES

1. For a velocity servo system with a mechanical time constant of 50 msec and electrical time constants at 2 and 1 msec, what is the maximum allowable DC loop gain without compensating networks?

2. In exercise 1, what can be done to make a DC loop gain of 100 possible?

3. What is the effect of increasing DC loop gain on final positioning accuracy with a fixed torque load on the mechanism?

4. How quickly would you expect the position servo system of exercise 1 to reach its final velocity (approximately)?

5. Would a velocity servo be effective in preventing speed variations due to load fluctuations that are higher in frequency than the point where loop gain is unity?

6. Work out a digital drive sequence, similar to that shown in Figure 13-9, for a stepping motor with three windings.

7. Design clocked logic, for use with an incremental encoder as shown in Figure 13-11, that will produce a CHANGE pulse with a duration of one clock time each time the encoder output changes. The circuit should also produce a RIGHT pulse, with each CHANGE pulse produced by movement to the right.

8. Design a phase detector (Fig. 13-13), using digital logic, that will produce a duty cycle of greater or less than 50% depending on the relative phase of the 50% duty cycle CLOCK OUTPUT and the 50% duty cycle, logic level, SIGNAL INPUT.

9. Design encoding and decoding logic for the pulse width encoding signal illustrated in Figure 13-14.

10. Draw a block diagram of the head-positioning servo system of Figures 13-17 and 13-18. Include two sample-and-hold circuits and all logic necessary to generate the velocity command voltage and switch-to-position servo mode.

11. Design logic for producing a modified frequency modulation write signal (Fig. 13-19).

12. Design modified frequency modulation read decoding logic. Assume that a "peak detector" produces an output pulse for a signal transition in *either* direction. Assume that the first transition, after a pause of three bit times or more, is always a "1" transition.

13. Design decoding logic as in exercise 12 but make it capable of identifying the 1's transitions without any pause in the signal. Utilize the fact that 1, 0, 1 will never be decoded if the clock transitions are interpreted as 1's transitions.

14. Compare the relative performance of simple and modified frequency modulation (Fig. 13-19) where there are speed variations with a period of less than a few bit times.

15. Draw a block diagram of the logic necessary to produce the column drive signals for the printer illustrated in Figure 13-20. Indicate the ROM size required.

16. Draw a block diagram of logic to generate the video and vertical and horizontal sync signals for a CRT display using the 5 × 7 matrix characters shown in Figure 13-21. Assume that the data for 16 32-character lines are stored in an RAM and that the picture is scanned as 256 lines 60 times a second with no interlace. (Interlace is the offsetting of timing of horizontal sync pulses every other frame to give a picture with twice as many lines. It is used in TV broadcasting but is unnecessary here.)

17. Design logic that will eliminate bounce problems, in the keyboard scanner of Figure 13-23, by stopping scanning for 20 msec, or until the key is released, whenever a key is pressed. A strobe one clock period long should be produced for each key pressed.

18. Assuming Figure 13-24 shows part of an 8 × 8 matrix, what other keys will appear to be pressed as a result of the sneak path shown?

19. What will be the result if a matrix as shown in Figure 13-25 is used to drive light-bulbs?

20. What will be the percentage distortion in a reconstructed signal where the signal energy above half the sampling frequency equals the energy below?

21. Draw the noise signal waveform for the *unfiltered* output in Figure 13-26b. (Note that the noise signal is the difference between the actual signal and the correct signal.)

22. Make a table similar to Table 13-1 for an input voltage of 5 mV.

23. Combine Figures 13-29 and 13-30 and add design details to make a system that continuously samples eight analog inputs in turn and produces a digital strobe whenever the TIME SLOT COUNTER and addressable latch contain input number and digital value of an input signal.

24. (a) Design an eight-bit D/A converter using digital counters to produce a duty cycle output, then integrating to make a 0- 5V analog output.
    (b) What must the integrating rate of the integrator be to produce a maximum ripple equal to one-half the resolution of the converter?
    (c) How long will it take for the output to make a full-scale voltage change?

25. (a) Design an operational inverting amplifier with an input impedance of 2000 Ω and a gain of 50.

    (b) What is the *loop gain* if the operational amplifier has a gain of 100,000?

26. Design a noninverting amplifier with a gain of 1 without using any resistors.

27. The operational amplifiers shown in Figures 13-32 and 13-33 all have internal frequency compensation for stability that rolls off the gain to 1 at 1 mHz.

    (a) Where must this rolloff start if the DC gain of the amplifier is 100,000?

    (b) In the amplifier of exercise 25, how far could the beginning of this rolloff be moved over without sacrificing stability?

28. Design a six-bit D/A converter using a circuit as shown in Figure 13-33a with binary weighted resistors.

29. On the amplifier in exercise 25, what value of resistor, in series with the (+) input ground connection, will eliminate the effect of any equal *bias currents* flowing out of the amplifier input terminals?

30. (a) Design a dual-slope integrator A/D converter (Fig. 13-31) using a CD4066 CMOS quad analog switch to select current inputs. (This device is like four logic controlled analog "relays" with about 100-Ω "contact resistance.") Assume that a precision voltage standard of 3.6 V is available and the input signal range to be converted is 0 to 5 V.

    (b) Design a similar converter for converting seven different inputs in turn, using a CD4051 eight-input analog multiplexer to select input and reference current. A digital *strobe* signal should be produced whenever the counter has finished digitizing an input.

REFERENCES

1. T. K. Hemingway, *Electronic Designers Handbook,* Business Publications, London, 1965.

2. Lawrence Rabiner et al., "An Approach to the Approximation Problem for Nonrecursive Digital Filters," *IEEE Transactions on Audio and Electroacoustics,* June 1970, pp. 83–106.

BIBLIOGRAPHY

Feedback Control

Fitzgerald, Tom, "Poles and Zeroes in Peripherals," *Computer Design,* December 1968, pp. 36–43.

Hemingway, T. K., *Electronic Designers Handbook,* Business Publications, London. (Includes graphs showing response of 11 different phase-correcting networks.)

Van Allen, Leland, "D.C. Motor Control Circuit Cancels Armature Resistance," *Electronics,* April 12, 1973, pp. 104. (Further description of Figure 13-6.)

Peripheral Devices

Barton, David, "Why Multiplex LED's?" *IEEE Spectrum,* November 1972, pp. 30–32.

Flores, Ivan, *Peripheral Devices,* Prentice-Hall, Englewood Cliffs, N.J., 1973.

Kaye, David, "Pulse Width and Phase Encoding Compete to be Cassette Standard," *Electronic Design,* November 9, 1972, p. 32.

Ohm, William, "Reel-to-Reel Drive Design for a Cassette Recorder," *Computer Design,* August 1973, pp. 67–70 (describes velocity servo using motors).

Peatman, John, *The Design of Digital Systems,* McGraw-Hill, New York, 1972, Ch. 7.

Riley, Wallace, "The Technology Gap Starts to Close for Peripherals," *Electronics,* July 3, 1972, pp. 59–74.

Sidhu, Pawitter S., "Group-Coded Recording Reliability Doubles Diskette Capacity," *Computer Design,* December 1976, pp. 84–88.

Operational Amplifiers

Burr Brown Staff, *Operational Amplifiers: Design and Applications,* McGraw-Hill, New York, 1971.

National Semiconductor Staff, *Linear Applications,* National Semiconductor, Santa Clara, Calif., 1973.

Smith, John I., *Modern Operational Circuit Design,* Wiley, New York, 1971.

Brokaw, Paul, "An IC Amplifier Users Guide to Decoupling and Grounding," *Electronic Products,* December 1977, pp. 45–53.

Analog/Digital

Mayo, J. S., "Experimental 244 MC/S PCM Terminals," *Bell System Technical Journal,* November 1965.

Strong, Norman, "LSI Converts an Old Technique into Low-Cost-A-D Conversion," *Electronics,* September 11, 1972, pp. 102–105.

Filters

Al-Nasser, Farouk, "Tables Shorten Design Time for Active Filters," *Electronics,* October 23, 1972, pp. 113–118.

Brokaw, Paul, "Simplify 3 Pole Active Filter Design," *EDN,* December 15, 1970, pp. 23–28.

Rabiner, Lawrence, et al., "An Approach to the Approximation Problem for Nonrecursive Digital Filters," *IEEE Transactions on Audio and Electroacoustics,* June 1970, pp. 83–106.

Tufts, Donald W. and D. W. Rorabacher, "Designing Simple, Effective Digital Filters," *IEEE Transactions on Audio and Electroacoustics,* June 1970, pp. 142–158.

CHAPTER FOURTEEN

USE OF STATISTICS
In Digital Design

14.1 SYSTEM RELIABILITY

Techniques for using statistics to predict the reliability of digital systems have advanced in recent years to the point where equipment and components can often be purchased with the **mean time between failures (MTBF)** guaranteed to some specification. It is therefore possible to mathematically predict equipment failure rate with fair accuracy. It is important for the design engineer to have at least a basic knowledge of the mathematics of reliability, because reliability is always an important design goal. The cost of repairing failures has been rising almost as quickly as IC costs have dropped, so reliability may well be the most important design goal—even in inexpensive, consumer devices.

Unlike mechanical components, modern electronic components virtually never wear out. Failures can be predicted only on a statistical basis. Actually, on new system designs the failure rate is usually dominated by *early failures,* which can be avoided on future units by changing manufacturing procedures on the basis of experience. Typical early failure causes are design errors, manufacturing errors, transportation and handling stress, testing and debug errors, improper operating and maintenance procedures, and component deficiencies. As Figure 14-1 shows, system failure rate can be improved with each subsequent unit manufactured.

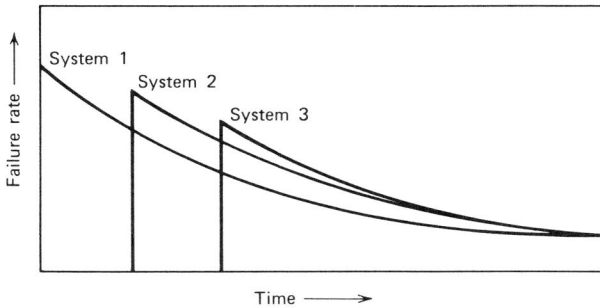

Fig. 14-1. Reliability improvement due to manufacturing experience.

As Figure 14-1 shows, all systems show a higher failure rate initially. Often power, temperature, or vibration stresses are intentionally accelerated during *burn-in* to encourage failure of weak components. This technique, and indeed all statistical predictions of failures, is based on *the assumption that all failures are a result of stresses exceeding the capabilities of some component.* If stresses fluctuate randomly with time, as shown in Figure 14-2, the *average* time before the stress causes a failure (MTBF) will depend on the stress applied to the part. If, for example, the stress is vibration and the failure is breakage of an electrical bond in an IC, intentional shock tests may uncover the problem in a very short time. If the stress is a voltage or temperature, testing under increased voltage and/or temperature should make the problem appear sooner.

Although the failure model shown in Figure 14-2 cannot be taken too literally, mathematical results obtained from its assumption seem to agree quite well with actual experimental results. For example, experiments have shown the following approximate relationships between applied stress and failure rate (1/MTBF):

1. The failure rates for deposited-carbon resistors double for every 20% increase in power dissipation or 38°C increase in ambient temperature.

2. The failure rates for solid-tantalum capacitors double for every 15°C temperature increase and for every 15% increase in applied voltage.

3. The failure rates for power transistors double for every 10°C rise in junction temperature and for every 7.5-V rise in applied voltage.

The designer can therefore design in improved reliability by operating all components well below their maximum ratings. Since temperature affects the reliability of all components, providing forced-air cooling can significantly improve reliability. Ideally the system should be capable of operating without the fans, so the failure rate of fans does not offset their reliability advantages.

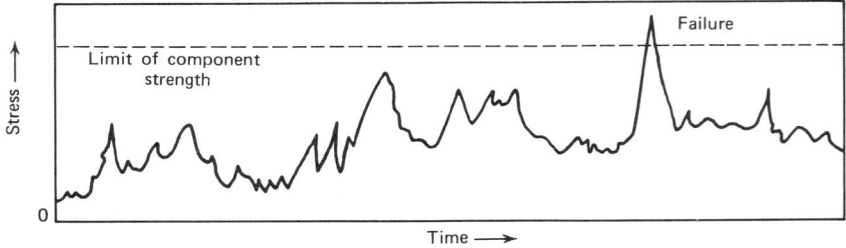

Fig. 14-2.   Theoretical component failure model.

14.2   CALCULATING MEAN TIME BETWEEN FAILURES

Calculation of MTBF for a system requires knowledge of the failure rate (1/MTBF) of each component in the system. If we assume that the system will fail if any of its components fails, *the failure rate of the system will equal the sum of the failure rate of its components,* or

$$\lambda_{\text{system}} = \lambda_1 + \lambda_2 + \lambda_3 + \cdots + \lambda_n \qquad (14\text{-}1)$$

where

$$\lambda = 1/\text{MTBF} \qquad (14\text{-}2)$$

The failure rate is usually represented as *percent per 1000 hours* or as failures per million hours. The most difficult thing about calculating system failure rate is determining the actual failure rate of the parts used; for example, IC failure rates, can vary from 0.00019%/1000 hr for high-reliability devices to 0.01%/1000 hr for standard, plastic-encapsulated ICs. New IC processes can have a higher failure rate due to unknown processing problems.

Because of the very low failure rates of modern electronic components, it is difficult to establish the failure rate with any confidence. As Figure 14-3 shows, MTBF calculations based on testing that produces only a few failures can give incorrect results. For example, if 100 IC's are operated for 1000 hr with only two failures, the observed MTBF is $100 \times 1000/2 = 50,000$ hr. Referring to Figure 14-3, however, we see that at the 60% confidence level, the actual MTBF could be anywhere between $2.5 \times 50,000 = 125,000$ hr and $0.7 \times 50,000 = 35,000$ hr. If we wait till more failures accumulate, we can get a more accurate idea of what MTBF actually is. Even after 10 failures, the 90% confidence level covers a range of $+180\%$ to -38% of the observed value.

With highly reliable ICs it becomes completely impractical to test enough circuits for a long enough time to get the 10 or 20 failures needed for a reasonably accurate determination of failure rate. For example, if circuits have a failure rate of 0.001%/1000 hr, 10^9 device hr would be required to get 10 failures. Even if 1000 devices were tested simultaneously, it would take 10^6 hr or *114* years of testing to get 10 failures!

To get usable reliability data, it has become necessary to use *accelerated testing*. By testing at accelerated temperatures, all thermally activated reactions that can cause failure are speeded up, and the failure rate is increased by an amount that can be calculated. By measuring the failure rate at elevated temperatures, then applying a correction factor, the failure rate at normal temperatures can be estimated. Testing at 125°C, for example, accelerates the failure rate by a factor of 100 when compared to operation at 55°C. The 117 years of testing can thus be accomplished in about 1 year. Although most process problems encountered so far in IC manufacture have been of the type

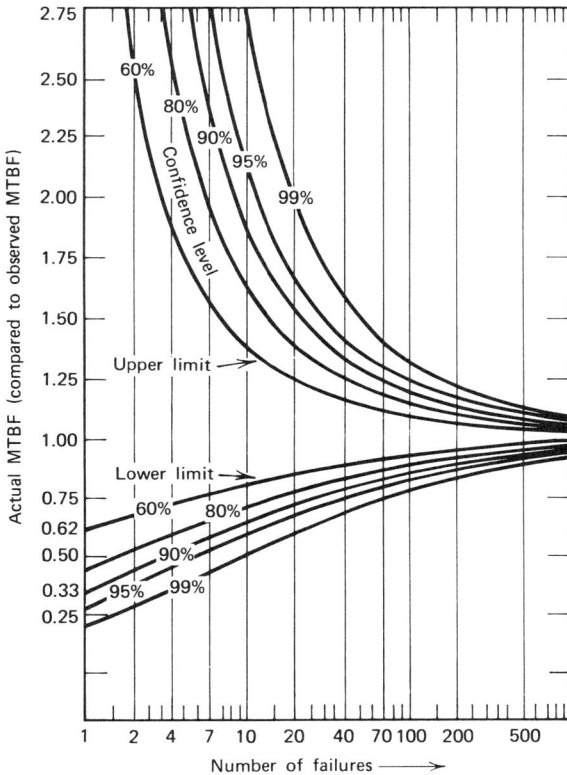

Fig. 14-3 MTBF uncertainty versus number of failures observed.

that are accelerated by temperature, there is a slight risk in any accelerated test data. As time goes by, we will gradually improve the accuracy of our reliability figures.

Tables of failure rates for various components are available from various sources. The most detailed source is probably MIL-HDBK-217B. It includes information for including the effects of temperature, operating environment, and device quality. Table 14-1 shows some typical failure rates for high-quality commercial components operated in an office environment with a stress of no more than 50% of rated current, power, etc.

Though earlier reliability figures for ICs in MIL-HDBK-217B indicated that the IC failure rate would increase with greater complexity, the newest version of that handbook omits the complexity factor. Although the possibility of failures due to misapplication decreases with LSI, the percentage of failures

TABLE 14-1 COMPONENT RANDOM FAILURE RATES

Component	(%/1000 hr) Failure Rate
Integrated circuits SSI, MSI, and LSI	0.01
High-reliability integrated circuits	0.001
Transistor (small signal)	0.005
Power transistor	0.05
Diode (small signal)	0.002
Power diode	0.02
Resistor (carbon comp)	0.001
Power resistor (wire-wound)	0.05
Variable resistor (cermet)	0.2
Capacitor (polystyrene)	0.01
Capacitor (Mylar)	0.05
Capacitor (solid tantalum)	0.02
Capacitor (aluminum electrolytic)	0.2
Quartz crystal	0.05
Connector contact	0.005
Printed-circuit connector contact	0.01
Switch contact (low level)	0.02
Relay (hermetically sealed, low level)	0.05
Fan (Rotron type)	2.0
Transformer (power)	0.5
Coil	0.03
Wire-wrapped joint	0.00001
Dip-soldered joint	0.0001

due to chip defects does rise. With SSI circuits, diodes, and transistors, *a fairly high percentage of field failures has been due to misapplication and design errors.* The usage of LSI circuits is much less subject to error, so this important source of failure is greatly reduced. Extensive testing on 1024-bit shift registers and RAMs has shown that a failure rate of 0.01% is achieved. Our design goal, in preceding chapters, of minimizing package count by using the highest level of integration possible will thus give maximum reliability.

In some applications, where extreme reliability is needed, all or part of the hardware must be duplicated completely. Often, for example, two central processors are used with means provided for automatic testing and switchover if one machine fails. Since the system will fail in this case only if the second processor fails *during repair* of the first processor, almost perfect reliability can theoretically be attained. If we assume, for example, that the *mean time to repair* (MTTR) of a processor is 1 hr and both processors have an MTBF of 1000 hr, then each processor will be unavailable 0.1% of the time. The probability of both processors being unavailable is thus $0.1\% \times 0.1\% = 10^{-6}$. This is equivalent to an MTBF of *both* processors of 10^6 hr, or *114 years*!

14.3 THE RELIABILITY FUNCTION

The *Poisson distribution* is useful where a large number of independent events occur at some constant average rate m. We can calculate the probability (P_n) of exactly n occurrences as follows:

$$P_n = \frac{m^n e^{-m}}{n!} \tag{14-3}$$

If we assume that the stress level in Figure 14-2 has a Poisson distribution of peaks causing failure, the average number of failures during some interval of time T will be $m = T/\text{MTBF}$. We can then use equation 14-3 to find the probability of *no failures* ($n = 0$) over time interval T as follows:

$$P_0 = \frac{(T/\text{MTBF})^0 \, e^{-T/\text{MTBF}}}{0!}$$

$$P_0 = e^{-T/\text{MTBF}}$$

This is sometimes called the reliability function R:

$$R = e^{-T/\text{MTBF}} \tag{14-4}$$

Application of this equation gives somewhat surprising results. For example, though MTBF is the average time between failures, the probability of no failures over an interval of time *equal to the MTBF* is only

$$R = e^{-1} = 0.367 \qquad (14\text{-}5)$$

The probability of no failures over half the MTBF is

$$R = e^{-1/2} = 0.607 \qquad (14\text{-}6)$$

We can turn equation 14-4 around as follows:

$$T/\text{MTBF} = -\ln(R) \qquad (14\text{-}7)$$

Using this equation, we find that to have 90% confidence of success ($R = 0.9$), we can only depend on operation without failure for 10.5% of the MTBF ($-\ln(0.9) = 0.105$). The exponential nature of the reliability function makes the high confidence required for space missions really hard to achieve. For 99% confidence, for example, the MTBF must be *100 times* the length of the mission ($-\ln 0.99 = 0.01$)!

If more than one system must work for the mission to be a success, the resulting reliability is the product of the reliability of each of the systems. If two systems, each with 90% reliability, must both work, the overall reliability is thus $0.9 \times 0.9 = 81\%$. Sometimes duplicate systems are provided to improve reliability. In this case system reliability is given by

$$R_{\text{system}} = R1 + R2 - (R1R2) \qquad (14\text{-}8)$$

Note that this equation assumes that repair is *not* possible during the mission.

Statistical theory can also be used to reduce the cost of receiving inspection of parts. It is naive to assume that a manufacturer will always deliver parts that meet all specifications just because the purchase contract says he will. One approach to keeping the suppliers honest is to test *all* parts received. This approach, however, represents a gross duplication of effort, no *statistical sampling* procedures are usually used. The Poisson distribution can be used to design sampling plans to allow testing only enough devices to establish a reasonable confidence level that the desired acceptance quality level (AQL) is met.

When parts are received from a new vendor, testing must be more cautious than when they have established their reliability. One hundred percent testing, of course, gives 100% confidence that the *lot tolerance percent defective* (LTPD) is not greater than the measured percent defective. We can often test only part of a large lot and attain some reasonable confidence level C. For any number of failures n the confidence level is $1 - P_n$ using equation 14-1. The ratio of percent defective (PD) to AQL is m in equation 14-1:

$$1 - C_n = P_n = \frac{(\text{PD}/\text{AQL})^n \times e^{-\text{PD}/\text{AQL}}}{n!} \qquad (14\text{-}9)$$

14.4 COMPONENT TESTING AND BURN-IN

Rising labor costs and falling component costs have brought us to the point where manufacturing checkout cost and service cost during the first year of service on a system can exceed the IC cost (see Table 2-1). One solution to this problem is to use components that have been put through a special, high-reliability testing and burn-in program. As Figure 14-4 shows, these programs can reduce the percentage of bad ICs assembled into equipment by an order of magnitude and reduce failure rate during the first year by a factor of 4.

Low-priced commercial-grade ICs are normally shipped with an *acceptable quality level*, (AQL) of 1%. This means that up to 1% of the parts received may be bad. If 100 such ICs are assembled together on a large PC board, *the chances of at least one IC on the board being bad is 63%!* If the percentage of defective ICs can be reduced to 0.1%, the chances of a bad IC on the board are reduced to 10%. A special testing and burn-in program may thus reduce rework costs by a factor of 6. Figure 14-5 shows how the percentage of boards reworked is related to board size and quality level.

Most manufacturers offer special, high-reliability ICs for a small additional charge. The savings in checkout and rework costs can often more than cover this additional charge. Improved reliability and lower field service costs during the first year then represent an additional savings.

For the desired level of confidence, the PD/AQL ratio for each n can be found, and a *sampling plan* can be made out for the desired AQL. For a 90%

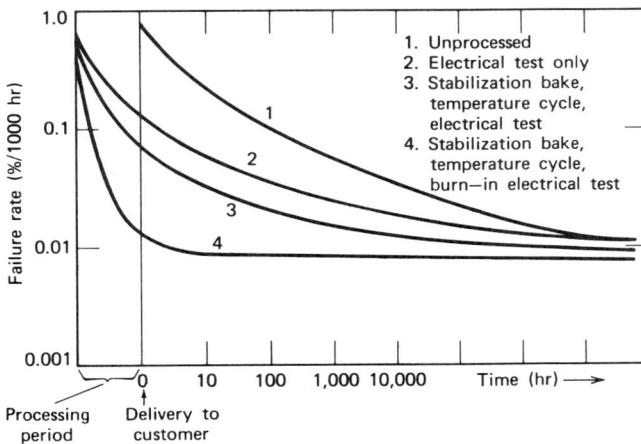

Fig. 14-4.   Improvement in early failures of high-reliability ICs. (Reliability Inc. Houston, Texas) data based on input from six customers on 1 million ICs.

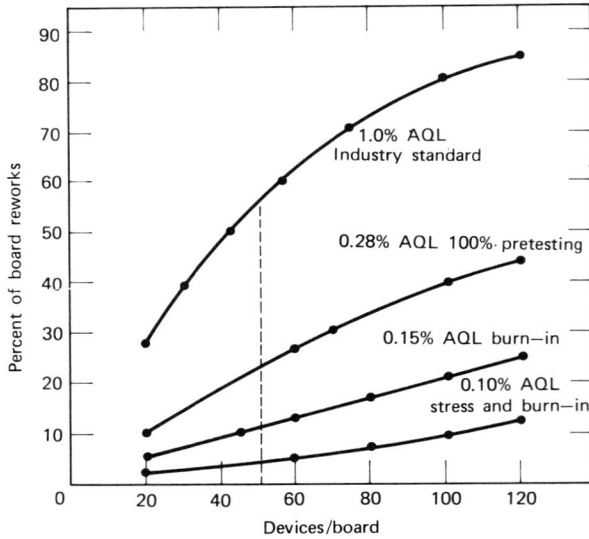

Fig. 14-5.   Reworks versus PC board size and IC quality.

confidence level an 1% AQL, for example, the plan would be as shown in
Table 14-2. If 231 units are tested without a failure, the entire lot, regardless
of size, can be accepted with 90% confidence that it meets the 1% AQL specification. If, however, a bad part is found before the 231st part, testing continues
until the 390th part if there are no more failures. For each sample size number
on the chart we thus accept the whole lot if the number of bad parts at that
point is not greater than the acceptance number.

To test for a 0.01% AQL the sample size numbers on Table 14-2 would be
10 times greater, so at least 2310 parts would have to be tested to accept the
lot. Lowering the confidence level reduces the required sample size, but since
0.1% represents one bad part in 1000, we would approach 1000 as a minimum
sample size.

TABLE 14-2   SAMPLING PLAN (1% AQL, 90% confidence level)

Acceptance No.	0	1	2	3	4	5	6	7 etc.
Minimum sample size	231	390	533	668	798	927	1054	1178

14.5   SHARED FACILITIES

The Poisson distribution is useful to digital designers for more than just reliability calculations. It can also be used to save hardware when a system must
serve a large number of independent, relatively inactive "customers." A
telephone exchange is the classic example of this type of problem, but it also
occurs in many time-shared systems where one central processor serves a large
number of *terminals* used for information retrieval, data entry, and so on. Although all terminals could theoretically be used simultaneously average usage
is much less. By using statistics, system facilities can be shared among large
numbers of customers, yet the possibility of system overload can be kept at an
acceptable level.

In a telephone exchange, for example, the number of *trunks* (voice paths)
between two cities is calculated to give a probability of *blocking* (no path
available) during peak periods of about $1/100$ or better. If the possibility of
blocking was not allowed, one trunk would be required for each telephone to
provide for the theoretical possibility of everybody in one city simultaneously
calling somebody in the other city. The cost of this kind of service would be out
of the question as thousands of trunks would have to be provided in place of each
trunk now provided.

In a typical time-shared computer system we have a similar problem. If, for
example, a customer credit-checking system must handle requests from 100
terminals scattered throughout a large department store, we could provide 100
input buffers in memory to accumulate 100 simultaneous 10-digit (40-bit)
requests. The central processor in this case would have to be capable of
processing requests at the maximum rate from all channels simultaneously.

A much more reasonable approach is to provide buffers and processing capability for a rate of requests that will occur only very occasionally. When this
rate is exceeded, a busy signal or even no response is acceptable, as long as it
happens infrequently. The problem, then, is to determine how much capacity
to provide for a given probability of getting a busy signal or no response.

Equation 14-3 can be applied to this type of problem as well as to reliability
problems. If m is the *average* number of terminals busy, then P_n is the
probability of exactly n terminals being busy at any given time:

$$P_n = \frac{m^n e^{-m}}{n!} \tag{14-3}$$

If we provide facilities for handling k terminals at a time, *the system will be
busy whenever $n > k$.* The probability of being busy is thus

$$P = \sum_{n=k+1}^{\infty} P_n = \sum_{n=k+1}^{\infty} \frac{m^n e^{-m}}{n!} \tag{14-10}$$

This is called *Poisson's exponential summation*. Since the probability usually falls off quickly as n increases, it is usually adequate just to compute P_{k+1} and possibly P_{k+2}. However, an exact solution can be obtained by computing the probability of *not* being busy and subtracting from unity:

$$P = 1 - \sum_{n=0}^{k} \frac{m^n e^{-m}}{n!} \qquad (14\text{-}11)$$

Figure 14-6 shows the relationship between the average number of circuits required and facilities required for several different probabilities of being busy.

For the credit-checking system mentioned previously, there may be an average of two requests at any given time from the 100 terminals. If we provide facilities (buffers and processor thruput) for handling seven simultaneous calls, the chart tells us that a busy signal will occur only 0.1% of the time, or once every 1000 requests. The system can thus have $7/100$ = $1/14$th of the *worst-case* capability yet provide excellent service.

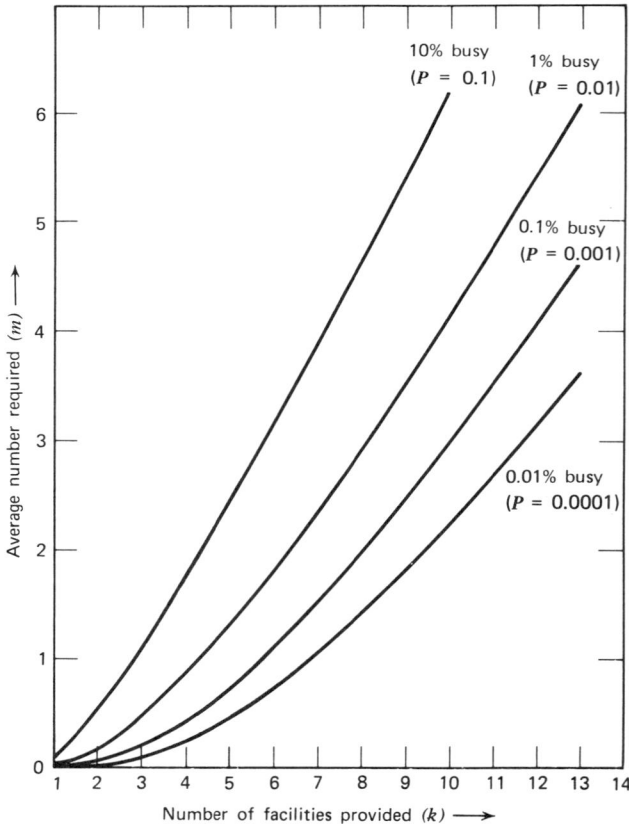

Fig. 14-6. Probability of exceeding facilities (busy).

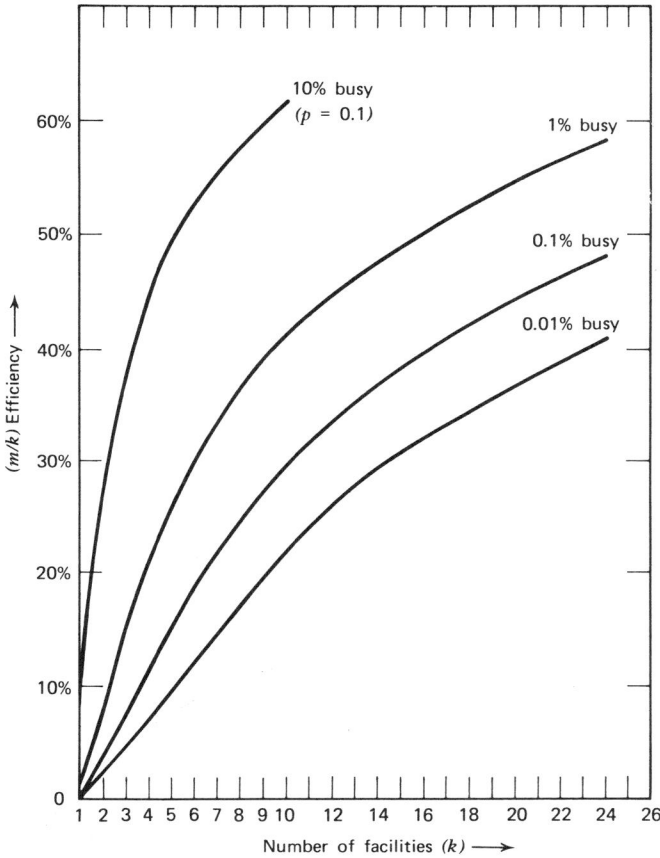

Fig. 14-7.   Efficiency of use of shared facilities (trunking efficiency).

The curvature of the lines on Figure 14-6 represents the inefficiency of the statistical approach when the number of facilities is small. This is analogous to the problem of determining failure rate accurately from a small number of failures. Figure 14-7 plots this efficiency as a function of the number of facilities for various busy probabilities. All of the curves approach 100% for very large numbers, which simply means that the random fluctuations average out so well that the load essentially looks continuous. An efficiency of 10%, for example, simply means that each facility is used an average of only 10% of the time. Note that the curves in Figure 14-7 have a "knee" for each busy probability below which efficiency falls off quickly as the number of facilities provided is reduced.

The telephone industry has developed extensive tables (see Bibliography) for *blocking probability (grade of service)* as a function of number of trunks and load in *Erlangs* (average number of calls). Slightly different results are obtained by assuming that lost calls are *held* (Poisson equation), *cleared* (Erlang *B* equation), or *delayed* (Erlang *C* equation).

All of the equations above (and Fig. 14-6) are accurate *only if the number of customers is much larger than the number of facilities provided.* If, for example, we try to apply the Poisson equation where there are only two customers who are each active 50% of the time ($M = 1$), Figure 14-6 would tell us that with four facilities we would still have about 1% blocking. Since two facilities can obviously serve two customers with no blocking at all, it is clear that the Poisson equation is not suitable. *Since the Poisson and Erlang equations assume infinite sources, they are not suitable for estimating blocking in many small systems.*

Whenever the number of customers is *not* much greater than the number of facilities, the *binomial* equation must be used. In this case the probability of blocking is

$$P = 1 - \sum_{i=0}^{i=k} \frac{n!}{i!(n-i)!} p^i \cdot (1-p)^{n-i} \tag{14-12}$$

where p = the probability of each customer being active
n = the total number of customers
k = the number of facilities provided

While this equation cannot be reduced to a simple table, it is not as difficult to compute as it looks. The quantity $n!/i!(n-i)!$ can be found on "binomial coefficient" tables, and some calculators compute the entire function automatically.

EXERCISES

1. What is the MTBF of a system containing the following parts: 100 ICs (standard), 20 resistors, 15 mylar capacitors, and two tantalum capacitors?

2. What is the effect on the MTBF of the system of exercise 1 if a fan is added, assuming the fan lowers IC junction temperature from 70 to 55°C and ambient temperatures from 50 to 30°C?

3. Assume a system is operated for 3 years with eight failures:
   (a) What is the observed MTBF?
   (b) What is the minimum MTBF at 60% confidence?
   (c) What is the minimum MTBF at 80% confidence?

4. Assuming that we must prove with 60% confidence that a system has an MTBF of 300 hr, how long must it run before the third failure occurs?

5. What is the MTBF of a duplicated system if each half of the system has an MTBF of 400 hr and an MTTR of 2 hr?

6. What is the probability of an LSI fuel injection controller in an automobile lasting

the life of the car? Assume that the life of the car is 10 years and the MTBF of the IC is 200,000 hr.

7. What is the probability of no IC failures during the life of the automobile in exercise 6 if it also has an IC voltage regulator with an MTBF of 300,000 hr and an IC ignition system with a 150,000-failure rate?

8. If the average cost of repair and rework on a PC board with 50 ICs is $100, how much of a premium could be paid per IC to get a 0.15% AQL instead of 1%?

9. (a) Assuming a system containing 5000 ICs is shipped after 10 hr of system burn-in, approximately how many failures during the first year in the field could be avoided by using burned-in ICs instead of ICs with electrical test only? Use a linear approximation to Figure 14-4 to estimate.
   (b) How much of a premium per IC could be paid to break even if each service call costs $300?

10. Make a table, similar to Table 14-2, for 60% confidence level, 1% AQL, $n = 0, 1, 2, 3,$ and 4.

11. (a) How many units must be tested without failure to establish an AQL of 0.15% to 90% confidence?
    (b) To 60% confidence?

12. Input words are received from many sources at an average rate of 10 words per second. They are stored in a buffer that is emptied every 0.5 sec. How large a buffer is required to keep the probability of overflow less than 0.01%?

13. How many trunks are required over a particular route in a telephone system if there are 10,000 lines, each active 5% of the time with 5% of the calls going over the route in question? Assume that 0.1% blocking is acceptable.

14. At the close of the stock market, a stock price information retrieval system receives an average of five requests per second. If the disk file can handle 11 requests per second, what is the probability of having to ignore a request due to overload during this time?

BIBLIOGRAPHY

Reliability

Bell Telephone Labs, *Physical Design of Electronic Systems, Vol. IV, Design Process,* Prentice-Hall, Englewood Cliffs, New Jersey, 1972. (Chapter 3, "Statistics"; Chapter 7, "Reliability"; Chapter 8, "Reliability of Electronic Parts.")

Brauger, Joseph, "A Logical Approach to Testing IC's," *Electronic Products,* August 1969, pp. 140–149. (Cost benefits of burn-in and testing.)

Grant and Leavenworth, *Statistical Quality Control,* McGraw-Hill, New York, 1972.

Green and Bourne, *Reliability Technology,* Wiley, New York, 1972. (Includes complete statistical and component reliability tables.)

ITT Staff, *Reference Data for Radio Engineers,* Howard W. Sams, Indianapolis, 1972, Chapter 39, "Probability and Statistics"; Chapter 40, "Reliability and Life Testing."

MIL-HDBK-217B, *Reliability Stress and Failure Rate Data for Electronic Equipment.*

MIL-STD-414, *Sampling Procedures and Tables for Inspection by Variables for Percent Defective.*

MIL-STD-883, *Test Methods and Procedures for Microelectronics.*

RADC Reliability Handbook, Rome Air Development Center (Def. Doc. Cen. ASTIA Doc. AD-821640), September 1967.

Reliability Analysis Center, *Digital Failure Rate Data,* Rome Air Development Center, Griffiss AFB, New York 13441, order #MDR-8 (1978), $50.00, phone (315)330-4151.

Taylor, Gordon M., "Forced Air Cooling in High Density Systems," *Electronics,* January 24, 1974, pp. 87–89. (Examines system reliability aspects.)

Vyenielo, Martin, "Component quality assurance: Which plan is better?" *Electronics,* September 30, 1976, pp. 97–99.

"Mathematics of Reliability," *Electronics,* November 30, 1962, pp. 54–75.

"Special Issue on Reliability of Semiconductor Devices," *Proceedings of IEEE,* February 1974.

Shared Facilities

A. T. & T, *Switching Systems,* American Telephone and Telegraph Co., New York. 1961, (Includes graphs for delays and blocking.)

ITT Staff, *Reference Data for Radio Engineers,* Howard W. Sams, Indianapolis, 1972, Chapter 32, "Traffic Concepts," including trunking tables.

Panice, Joseph A., *Queing Theory,* Prentice-Hall, New York, 1969.

CHAPTER FIFTEEN

THE SOCIAL CONSEQUENCES of Engineering

15.1 INTRODUCTION

Although this chapter may seem off the subject of this book, it is, in a sense, the most important. If the reader forgets some of the ideas in the previous chapters, the worst that will happen is that a few IC packages may be wasted. Ignoring this chapter, however, can cause serious consequences.

Traditionally, the engineer has tended to stick with his formulas and designs and leave the philosophizing to others. His value system has been limited to "good design" in the technical sense. Under this system a well-designed machine gun is as much a source of satisfaction to the engineer as a device that helps blind people read books. Under this value system the decision to string power lines between poles or towers above ground, or to bury them underground is made strictly on the basis of the relative installation costs.

We are at the point today where we desperately need a new kind of engineer who will use technology to improve the *quality* of life for everyone. The possibilities opened up by integrated circuits are fantastic, but without a basic change of awareness in the engineering profession, ICs may only complicate our lives with more useless gadgets.

15.2 THE "GOOD OLD DAYS"

In the "good old days" most people worked from sunrise to sunset just to stay alive. Instead of having a furnace controlled by an automatic thermostat, wood had to be chopped, hauled, and split, and the fire had to be fed by hand all winter. To cook a meal, first a fire had to be built in the stove. Before taking a bath, the water had to be pumped, hauled, and heated on the stove. To keep food cold in summer, blocks of ice cut in winter had to be cut, stored, and hauled daily to the ice box. Now we earn the money to pay for heat, hot water, and refrigeration each day in a few minutes of work.

Technology, it would seem, has freed us from our slavery by making it possible to do a lot with a small amount of work. Today's coal miner presses a button and starts a machine that digs as much coal in an hour as 50 miners with picks used to dig in a day. A modern tractor ploughs a field in an hour—work that used to take days of backbreaking toil. Instead of hunching over a desk all day adding figures, a modern clerk simply presses a button on a computer and does the same work in a minute.

The really interesting question is why, if things are so much easier, do we call those days of backbreaking work the "good old days"? The fact is that technology has been a mixed blessing: with advertising and just plain love of gimmicks, we have created "needs" that people never knew existed. To satisfy these "needs," we work much harder than we have to.

We have misused our technology to make most of our cities a nightmarish world of polluted air, traffic, noise, and giant rotating neon signs. We have used our new strength to carve graceful mountains and hills into staircases strewn with identical tract homes and forests of power poles and TV antennas. Even, when camping in the forests, we are assaulted by the squawk of transistor radios, the roar of motorcycle engines, and campgrounds that look like parking lots. Our computer systems have become enemies that deny our identity, invade our privacy, then send us irrelevant form letters in reply to our passionate complaints.

15.3 THE RAT RACE

Technology provides so well for our real and imagined needs that it even provides solutions to the problems it creates—and solutions to the problems created by the solutions. We can buy a power lawnmower, for example, to eliminate the work of pushing the mower. When our health suffers due to lack of exercise, we can buy a mechanical exerciser to *create work* to replace that eliminated by the power mower. We thus end up doing the same amount of exercise, but instead of working outdoors and having a sense of accomplishment, we can stare at the clock indoors while we pump the exerciser.

The power mower and exercise machine are thus a matched set in that they nullify each other except for one problem: we must not only work to pay for both of them but also maintain them in operating condition and suffer the noise and smell of the mower's exhaust. If we are truly successful, we can have the ultimate in both—a mower with a seat and a canopy ($500) so we need not walk or have the sun shine on us, and a Dynavit exerciser that has an electrical generator on foot pedals and a variable-load resistor connected to a microcomputer which computes the calories consumed ($1745). We can also buy a sun lamp to offset the lack of sunshine due to the canopy on the mower!

Unfortunately, such combinations are not usually sold as matched pairs, because nobody would buy them if they saw the irony. Instead, they are promoted separately without either manufacturer really considering his product as part of a subtle trap.

This, in fact, is the real problem. We have a process at work that is caused by a way of thinking rather than a conspiracy. Each person simply goes about his job mindlessly: the engineer designs the products, the advertiser sells them, and the manager optimizes profits. The advertiser does not care whether he is selling cigarettes that cause lung cancer or ineffective medicine—he just does his job. The engineer likewise just does his job and probably even tries to do a good job, but he does not question *the effects of the product on a society.*

If we are to break these destructive patterns, each engineer must take a personal responsibility for his actions and their effect on society. When a patient asks a doctor for a drug that is harmful, it is quite natural for the doctor to refuse since he can personally see the harmful effects on the patient if he does otherwise. The engineering profession, however, has no such face-to-face contact with its clients. *The engineer's client is the entire society,* rather than an individual. It therefore takes a much greater awareness on his part to see the ultimate results of his actions. The engineer can, in effect, produce heroin by the carload without ever looking into the addict's face.

15.4 THE ENGINEER AS A DOPE PUSHER

The IC revolution has put us in a position of tremendous power. Suddenly we can do very complicated things for very little money. The backlog of jobs to be done is fantastic, and the order of priority for doing them will be determined partly by engineers. Already much of what has been done amounts to nothing more than complicated gimmicks that complicate life rather than improve its quality.

One of the first consumer applications of LSI, for example, is a "digital electric range" (see Fig. 15-1), which allows serially punching in on a keyboard the desired temperature. The number entered is automatically checked for reasonableness, and an error message is produced if it is out of range. A digital display indicates the temperature setting as well as time. Digital times can also be entered to program the oven on and off at various times. Our exciting new world of electronics has thus produced an $800 toy for adults with essentially *no real advantages* over the simple mechanical approach used previously.

Where the power of ICs could have been used to help the blind to see or relieve the tedium of routine jobs, it has been prostituted to provide gadgets to keep gadget addicts working 80 hours a week to support their habit. The simple reason people today never seem to have enough money, even though they can earn enough to satisfy their basic needs with 2 hours of labor a day, is that they are working to pay for such gadgets. Just as with heroin addiction, the need is never satisfied but only grows the more it is fed. The engineer, with his love of gadgets, has been the all-too-willing creator of such traps.

Fig. 15-1. Frigidaire Touch Control Range.

Gordon Moore has extrapolated the present growth rate in IC circuit functions (gates, bits, etc.) produced per year and has found that *by 1985 we will be producing a quarter of a million circuit functions annually for every person on earth.* These new circuits can be a blessing or a curse depending on what they are used for. Since almost any job *can be done* automatically there is a tendency to automate things that can just as easily be done manually.

A perfect example of automation for its own sake is the Heathkit GR-2001 preprogrammed TV option. It uses a 60-chip bipolar processor to allow the user to enter in advance programmed channel selection for two days of TV viewing. Since remote control has already eliminated the need to get up and change channels, this option is apparently designed for people whose fingers get tired from pushing buttons!

As the IC revolution progresses it becomes more and more important that engineers ask themselves more than "can it be done?" The important question in a world where essentially anything *can* be done is *"Is it useful?"*

It is interesting to note that, while we have put so much effort into developing LSI circuits to turn our ovens off and on, virtually 100% of our telephone calls are switched by *relays*. Likewise, the telephone itself is basically unchanged since the turn of the century. The same carbon microphone used by Alexander Graham Bell in 1877 is packaged with the same dial used by Strowger in 1895. The only real change is a sleek, molded plastic case. Who knows what kind of rate reductions would be possible if IC technology were applied to our telephone system!

15.5   THE NEW BREED OF ENGINEER

If technology is to liberate man, instead of enslaving him, we need a new breed of engineer who is more than just a narrow technologist. He must be aware of the social consequences of his actions and must consider the aesthetics as well as the economics of a problem. If we are affluent enough to have digital ovens, surely we can afford the extra 10% it costs to make a computer system respond to people's *names* instead of numbers, or squeeze them into one of 100 categories instead of four. Computer systems can be dehumanizing or not, depending on how they are programmed. They can send inappropriate form letters that make us feel like numbers, or with a little more effort they can efficiently take care of *routine* correspondence, yet refer the interesting problems to *people* for solution.

A perfect example of the result of applying pure economics to a problem is the decision to put power lines above ground (Fig. 15-2). A pure economic analysis shows that it is cheaper to string the lines above ground. However,

Fig. 15-2.   Telephone Poles, Richmond, California (photo by Ron Partridge)

what is the hidden cost of the ugliness of the power lines, and the deaths due to electrocution or crashing cars into the poles. In recent years many communities finally passed laws prohibiting aboveground utilities. Once the power companies were forced to develop the techniques for installing underground cables, equipment was developed that actually made it *cheaper* than going above ground.

Early digital systems often tried to fit the people to the machine rather than vice versa. Now that digital logic complexity is so inexpensive, we can begin to make devices that do exactly what we want them to do. As storage costs come down, for example, the additional cost of using someone's name instead of a number becomes insignificant. When machines used to do their work like obnoxious, inflexible clerks, they can now be programmed to do routine work like a splendid, efficient employee with a perfect memory.

Machine-made parts have always meant boring standardization, but now we can begin to make machines that can be programmed to make an endless variety of parts with the same cost advantage formerly available only if all parts produced were identical. The individuality of furniture, trim, and so on, of the good old days, when everything was made by hand, can now be produced by programmable machines.

In the early days of automation there was much fear that it would produce widespread unemployment. Time has proved, however, that this is not so. New industries created by automation have provided as many new jobs as have been eliminated. Though we can certainly work fewer hours to produce the same things, the work taken over by machines is the backbreaking, boring, and routine part of the job. The portion left for people is the interesting, creative work. Just how successfully we make technology our servant instead of our master will depend on the awareness of our new breed of engineers.

BIBLIOGRAPHY

Balabanian, Norman, "Technology and Values" (letter), *IEEE Spectrum,* February 1974, pp. 26–27.

Blake, Peter, *God's Own Junkyard,* Holt, Rinehart & Winston, New York, 1964.

Braden, William, *The Age of Aquarius,* Quadrangle, Chicago, 1970.

Christiansen, Don, "The New Professionalism," *IEEE Spectrum,* June 1972, p. 17.

Florman, Samuel C., *The Existential Pleasures of Engineering,* Marin's Press, New York, 1976.

Fromm, Eric, *The Revolution of Hope,* Harper & Row, New York, 1968.

Grosch, Herb, "Problems and Priorities," *Datamation,* March 1972, p. 48.

I.E.E.E., Committee on Social Implications of Technology Newsletter (a free newsletter to members of IEEE).

Kemeny, *Man and the Computer,* Scribner's, New York, 1972.

Lindgren, Nilo, "Semiconductors in the 80's," *IEEE Spectrum,* October 1977, pp. 42–47.

Toffler, Alvin, *Future Shock,* Bantam Books, New York, 1970.

Ullrich, Manfred and Hegendorfer, Max, "TV receiver puts two pictures on screen at same time," *Electronics,* September 1, 1977, p. 102.

Weizenbaum, Joseph, *Computer Power and Human Reason: From Judgement to Calculation,* W. H. Freeman, San Francisco, Calif., 1977.

"Bipolar processor in TV set leads to preprogrammed channel selection," *Electronic Design* 3, February 1, 1977.

"Electric range holds cooking programs," *Electronics,* March 4, 1976, p. 33.

GLOSSARY

Analog—Indicates continuous, nondigital representation of quantities. An analog voltage, for example, can take any value.

ASCII—See USASCII.

Asynchronous—Not synchronized with a system clock signal.

Baud—Usually simply the number of bits per second sent serially, but some MODEMS can send more than one bit per band.

BCD—Binary coded decimal. A system of number representation that indicates each decimal digit of a number as a four-bit binary number. Binary values 10 to 15, therefore, never occur.

Benchmark—A sample application program which is used to complete microcomputers. By comparing program storage space and execution time when it is executed, a measure of each microcomputer's efficiency for the specific job can be obtained.

Binary—Though the term "digital" could mean any number system, it has come to mean binary almost exclusively. A binary digit (bit) can have only two values (1 or 0), just as a decimal digit can have 10 values. We can represent any number in the binary system by using many digits representing powers of 2, just as we represent any number in the decimal system using many digits representing powers of 10. (See Figure G-1.)

Biquinary—A method of representing decimal digits with four binary bits weighted 5, 4, 2, 1.

Bit—An acronym for *binary digit*. A bit is the smallest divisible piece of information and has only two values: 1 or 0. See also *binary*.

Bit Slice—A microprogrammed processor chip that handles (usually) four bits of logic and registers. Carry inputs and outputs allow any number of bit slices to be interconnected to make a very wide processor word.

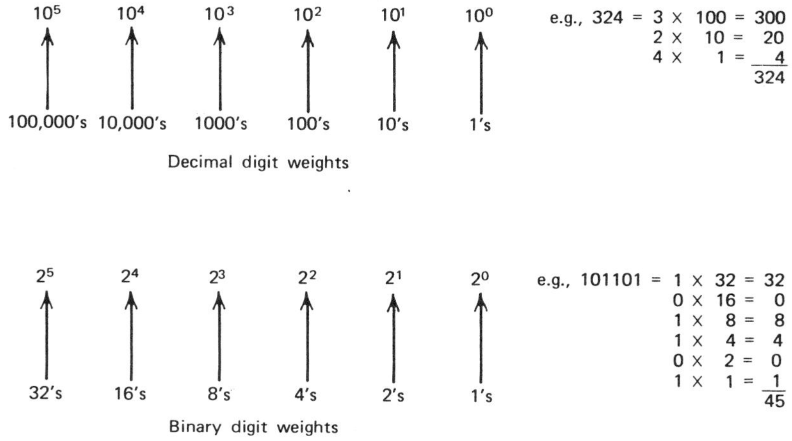

$$10^5 \quad 10^4 \quad 10^3 \quad 10^2 \quad 10^1 \quad 10^0$$

```
e.g., 324 = 3 × 100 = 300
            2 × 10  =  20
            4 × 1   =   4
                       324
```

100,000's 10,000's 1000's 100's 10's 1's

Decimal digit weights

$$2^5 \quad 2^4 \quad 2^3 \quad 2^2 \quad 2^1 \quad 2^0$$

```
e.g., 101101 = 1 × 32 = 32
               0 × 16 =  0
               1 ×  8 =  8
               1 ×  4 =  4
               0 ×  2 =  0
               1 ×  1 =  1
                         45
```

32's 16's 8's 4's 2's 1's

Binary digit weights

Fig. G-1

Buffer—(1) An amplifier used to increase drive capability. (2) Intermediate digital storage used to hold data temporarily.

BUS—A group of wires that carry related signals. Devices can be connected in parallel along the bus.

Byte—An eight-bit binary code.

CRT—Cathode-ray tube.

Chip—A tiny piece of the silicon wafer that, when packaged, will be an integrated circuit.

Clear—Usually unclocked reset. Sets the state to zero.

Clock—A signal that causes logic decisions to be made, one at a time, for each clock transition.

CMOS—Complementary metal oxide semiconductor. A logic family made by combining N-channel and P-channel MOS transistors. This gives fairly high-speed operation (65 nsec maximum), excellent noise rejection, low power consumption, and extremely large fan-out. Also called COS/MOS or McMOS. (See Figure G-2.)

CPU—Central processing unit. Another name for a computer.

Cross Compiler—A compiler that is written in ANSI Standard Fortran, so it can be run on virtually any mini or main-frame computer.

Decade Counter—A ÷ 10 counter.

Die—See *Chip*.

Dielectric Isolation—A method of isolating devices on an IC by etching away the silicon between them. Gives higher density and speed.

Differential Comparator—A circuit that compares two analog inputs and produces a 1 or 0 logic level output, depending on which voltage input is more positive.

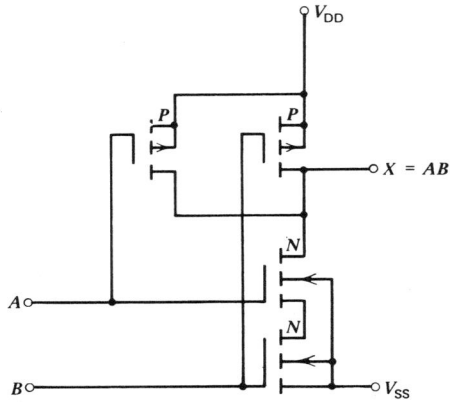

Fig. G-2.   CMOS NAND gate (1/4 CD4011).

Digital—Represented by discrete, as opposed to continuous (analog), values. Usually means binary.

DIP—Dual inline package. The industry standard IC package. Available in plastic or ceramic, usually with 14, 16, or sometimes 24 or more pins arranged in two rows. (See Figure G-3.)

Disassembler—A program that converts machine language programs into assembly language mnemonics.

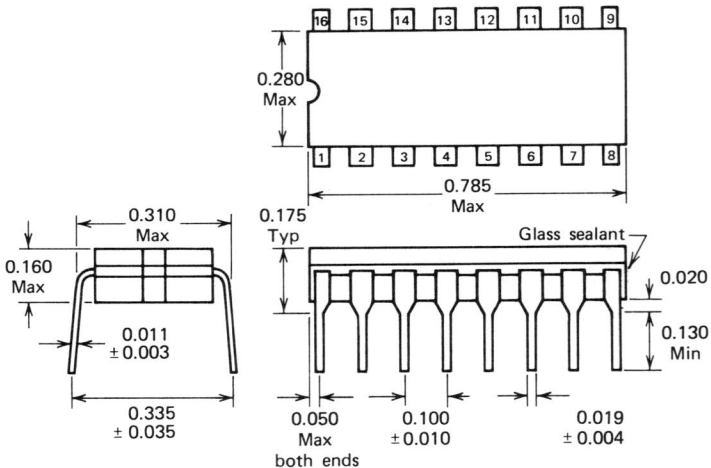

Fig. G-3.   Sixteen-pin dual inline package (DIP).

DMA—Direct memory access. Used to read or write data directly into a computer memory.

DTL—Diode transistor logic. The first successful logic family. Now superceded by TTL.

Dynamic—Dynamic memories and shift registers must be clocked at least once every millisecond or so, or they lose their data. Dynamic ROMs produce an output only just after the clock.

ECL—Emitter-coupled logic. The fastest logic family available today. MECL III has only 1-nsec delay per gate. The industry standard ECL 10,000 series has a maximum propagation delay of 3 nsec. Logic levels are −0.9 and −1.9 V. Terminating resistors are required for all but the shortest connections. Virtual OR connections are possible by connecting outputs in parallel. (See Figure G-4.)

EOF—End of file.

EPROM—A PROM that can be erased and reused indefinitely. Generally EPROMs are erased by shining ultraviolet light on the chip through a quartz window on the package.

Flip-flop—A digital circuit used to store one bit of information.

Fan-in—The number of inputs connected to a gate.

Fan-out—The number of outputs connected to a gate or the maximum number that can be connected without current overload.

FPLA—Field programmable logic array. A PLA that can be programmed by the customer.

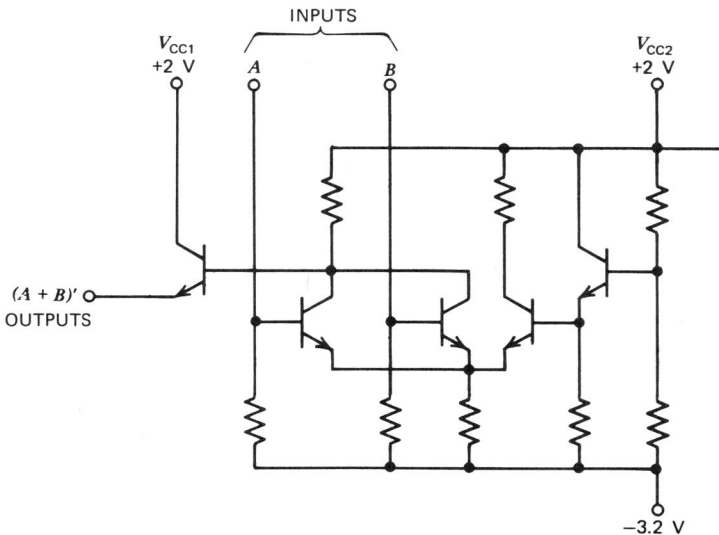

Fig. G-4.   ECL two-input NOR gate (1/4 10102).

Gate—A digital logic element where the binary value of the output depends on the values of the inputs according to some logic rule.

Gray Code—A binary code sequence where only one bit changes between successive values.

Half Duplex—A circuit that is capable of sending data or receiving data but not simultaneously.

Hamming Codes—Binary check codes that make it possible to both detect and correct errors.

Hexadecimal—A numbering system based on the scale of 16. Each digit represents a power of 16, just as each decimal digit represents a power of 10 (see *Binary*). Digits are represented as decimal numbers 0 through 9, and letters *A* through *F* are used to indicate 11 through 15. Long binary numbers can be thought of as one hexadecimal digit for every four bits; for example: 9F = 10011111. (See Table G-1.)

Inverter—A logic device that complements a logic variable.

IC—Integrated circuit. A complete circuit made on a "chip" of silicon.

I/O—Input/Output. A microprocessor's interface with the outside world.

LED—Light-emitting diode. A semiconductor device that emits light whenever current passes through it. Typically, a few milliamps at 1 V is enough to give visible light. Red, green, yellow, and infrared units are available.

Logic Design—Design using digital logic circuits.

LSI—Large-scale integration. An IC with complexity equivalent to over 100 logic gates.

Majority Logic—A combinational logic function that is true when more than half the inputs are true. A kind of threshold logic.

Mask—A glass photographic plate whose image is transferred to the wafer to define, for example, patterns of diffusion in IC production.

MOS—Metal oxide semiconductor. Circuits using metal-oxide field-effect transistors in which currect flow through a channel of *N*- or *P*-type semiconductor material is controlled by the electric field around a metal gate.

MODEM—*Mo*dulator/*dem*odulator. Used for sending data over telephone lines.

MSI—Medium-scale integration. Usually contains a minimum of 12 gates but less than 100 (LSI).

MTBF—Mean time between failures. Average hours between random failures after initial "burn-in" period.

Multilayer—Printed-circuit boards with many layers of conductors made by laminating double-sided printed-circuit boards together and plating through the holes.

Nanosecond—One billionth of a second, or one millimicrosecond. Abbreviated nsec.

Nixie Tube—A gas-discharge numeric display tube with each number formed of wire.

NMOS—MOS devices using *N*-type source and drain contacts (*N* channel on a *P*-type substrate). Faster than PMOS.

TABLE G-1 HEXADECIMAL-DECIMAL CONVERSION TABLE

	0	1	2	3	4	5	6	7	8	9	A	B	C	D	E	F
0X	000	001	002	003	004	005	006	007	008	009	010	011	012	013	014	015
1X	016	017	018	019	020	021	022	023	024	025	026	027	028	029	030	031
2X	032	033	034	035	036	037	038	039	040	041	042	043	044	045	046	047
3X	048	049	050	051	052	053	054	055	056	057	058	059	060	061	062	063
4X	064	065	066	067	068	069	070	071	072	073	074	075	076	077	078	079
5X	080	081	082	083	084	085	086	087	088	089	090	091	092	093	094	095
6X	096	097	098	099	100	101	102	103	104	105	106	107	108	109	110	111
7X	112	113	114	115	116	117	118	119	120	121	122	123	124	125	126	127
8X	128	129	130	131	132	133	134	135	136	137	138	139	140	141	142	143
9X	144	145	146	147	148	149	150	151	152	153	154	155	156	157	158	159
AX	160	161	162	163	164	165	166	167	168	169	170	171	172	173	174	175
BX	176	177	178	179	180	181	182	183	184	185	186	187	188	189	190	191
CX	192	193	194	195	196	197	198	199	200	201	202	203	204	205	206	207
DX	208	209	210	211	212	213	214	215	216	217	218	219	220	221	222	223
EX	224	225	226	227	228	229	230	231	232	233	234	235	236	237	238	239
FX	240	241	242	243	244	245	246	247	248	249	250	251	252	253	254	255

Noise Margin—The noise voltage required to make logic circuits malfunction. The difference between output voltage and input threshold voltage.

NRZ—Nonreturn to zero. Direct 1 or 0 representation of a signal without extra transitions for clocking.

Object Program—The actual program to be used. Generated by an assembler or compiler from a source program.

Octal—A number system based on the scale of eight. Each digit represents a power of eight and can have a value of 0 through 7. Long binary numbers can be thought of as one octal digit for every three bits; for example, octal 37 = 011111.

Optical Isolator—A semiconductor device consisting of an LED and a photodiode or phototransistor in close proximity. Current through the LED causes internal light emission, which causes current to flow in the phototransistor. Since the two devices are electrically separated, voltage differences of thousands of volts between input and output have no effect.

Parity Check—A one-bit check code where the total number of 1's in the word, including the parity bit, is odd (odd parity) or even (even parity). Odd parity has the advantage of indicating error for all zeros. Double errors may go undetected with either scheme.

PC Board—Printed-circuit board. Made by silk screening a pattern of acid resist on a copper-plated plastic board and etching off the unprotected areas with acid. Plated-through holes are often used to connect one side to the other.

PCM—Pulse code modulation. Digital transmission of analog signals by sending periodic binary coded samples of the signal value.

PLA—Programmable logic array. A ROM with only certain specified words decoded.

PMOS—MOS devices using *P*-channel MOS transistors. Slower than NMOS.

Port—A place through which inputs or outputs can pass. Sometimes each input/output address in a microcomputer is referred to as a certain port number. A dual ported memory, on the other hand, is a memory that can connect to two separate memory buses.

Program Status Word—(PSW). A word that contains various machine states which must be saved when answering an interrupt. It includes such things as interrupt enables and condition codes.

PROM—Programmable read-only memory. An ROM that can be programmed by the customer.

Propagation Delay—The time difference between the change of an input signal, or clock, and the change of the output.

Rack—A steel cabinet used for enclosing electronic equipment with standard, drilled and tapped (RETMA) rails down each side for mounting panels, subracks, and so on, that are any multiple of 1 ¾ in. high.

RAM—Random-access memory. Has come to mean a memory in which data can be read or written to any location indicated by a binary address input. A write pulse causes input data to be written in the indicated address. Otherwise the content of the indicated address just appears on the data output.

Real Time—Indicates that data are processed as they happen, producing immediate output.

ROM—Read-only memory. A fixed pattern of data, permanently defined during manufacture, appears on the outputs for each binary (address) input code.

RTL—An obsolete logic system based on *R*esistors and *T*ransistors.

Schmitt Trigger—A circuit with "snap action" used to produce a sharp, single transition from slowly changing inputs. Positive feedback shifts the threshold as soon as it is first reached. If the signal goes back slightly due to noise, the output remains steady.

Schottky TTL—An improved version of TTL logic in which saturation of the transistors is prevented by Schottky diode clamping from collector to base, and the impedance of the (+) output circuit is reduced. The result is that 74S logic gates have a maximum delay of 7 versus 15 nsec for TTL without Schottky. Power dissipation, however, is about twice that of TTL. Low-power Schottky TTL (74LS series) has maximum delays of 28 nsec, with only 20% as much power required as for normal TTL.

Sense Amplifier—An amplifier used to sense the small output voltage of a core memory. It is like a **differential comparator** with a built-in offset (threshold) voltage.

Setup Time—The time, before the clock transition, that the data input to a flip-flop must be stabilized for proper operation.

Shift Register—A linear chain of storage elements that shifts the contents of the storage elements one position for each clock transition.

Silicon Gate—A type of MOS in which the gate is made of silicon instead of metal. Faster and denser.

Sink Current—The ability of a device to accept current from external loads. A saturated transistor "sinks" current from loads through its collector.

SOS—Silicon-on-sapphire. A faster MOS technology in which silicon is grown on a sapphire wafer only where needed. Each device is thus isolated by air from other devices.

Source Program—A program written in a language that must be translated by computer to an object program.

SSI—Small-scale integration. Integrated circuits containing fewer than 12 logic gates.

Static—A static memory or shift register does not need to be continually clocked to keep its data as a dynamic one does.

Statemenet—A line of source program.

Subrack—a sheet-metal assembly that holds many PC boards. Several subracks can be mounted in one rack.

Switch-Tail Ring Counter—Another name for a moebius counter, or twisted-tail ring counter.

Synchronous—Synchronized by a common system clock.

System(s) Design—Conceptual design of a system without worrying about small details.

TABLE G-2 ASCII CODE

Bits Shown in Transmission Sequence: 4, 3, 2, 1	Bits Shown in Transmission Sequence: 8,[a] 7, 6, 5							
	0000	0001	0010	0011	0100	0101	0110	0111
0000	NUL	DLE	SP	0	@	P		p
0001	SOH	DC1	!	1	A	Q	a	q
0010	STX	DC2	"	2	B	R	b	r
0011	ETX	DC3	#	3	C	S	c	s
0100	EOT	DC4	S	4	D	T	d	t
0101	ENQ	NAK	%	5	E	U	e	u
0110	ACK	SYN	&	6	F	V	f	v
0111	BEL	ETB	'	7	G	W	g	w
1000	BS	CAN	(	8	H	X	h	x
1001	HT	EM	)	9	I	Y	i	y
1010	LF	SUB	*	:	J	Z	j	z
1011	VT	ESC	+	;	K	[	k	{
1100	FF	FS	,	<	L	\	l	:
1101	CR	GS	–	=	M	]	m	}
1110	SO	RS	.	>	N	⌐	n	~
1111	SI	US	/	?	O	–	o	DEL

[a] Bit 8 is used as parity. Can be 1 or 0 depending on whether odd or even parity is used.

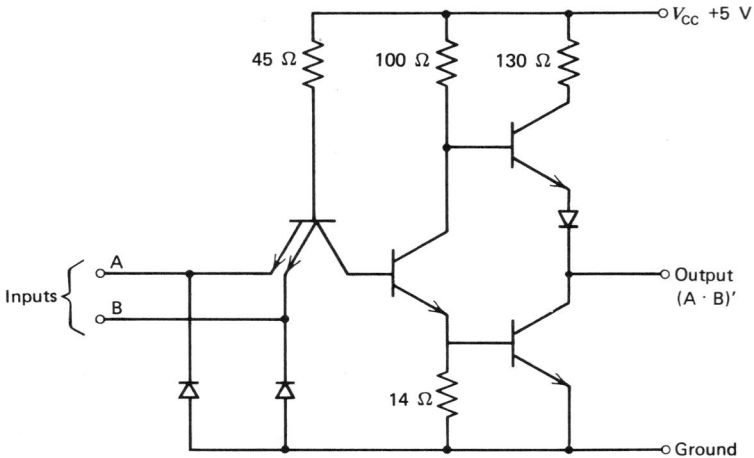

Fig. G-5. TTL two-input NAND gate (1/4 7400).

TDM—Time division multiplex. Sending several signals during different time slots over the same channel.

Threshold Logic—Combinational logic where the output goes true if some number of inputs are true.

Time Sharing—Using the same hardware for different jobs during different, brief, periods of time (time slots).

Tristate—A logic output that can be inactive, high, or low. This makes it possible to connect many outputs to a single bus and have only one active at a time.

Transducer—A device that converts energy from one form to another. For example, a motor converts electrical energy to mechanical energy.

Trimpot—A small, multiturn, variable resistor that is simply dip-soldered on PC boards.

TTL (T^2L)—Transistor transistor logic. A very popular logic family that uses multiple transistor emitters to perform logic functions. The gates have a maximum delay of 15 nsec and require +5 V power of about 2 mA per gate. Logic levels are normally +0.2 and +3.4 V, and input threshold is nominally 1.2 V. (See Figure G-5.)

Turnkey—A complete ready to use system where the customer has simply to "turn the key."

USASCII—United States of America Standard Code for Information Interchange. An eight-bit code used for sending information. More commonly called ASCII. (See Table G-2.)

Wafer—A thin slice of silicon on which a large number of ICs are simultaneously produced. It is later scribed and broken up into individual IC chips.

Wire Wrap—A very reliable solderless interconnection technique in which wire is simply wrapped tightly around a square gold-plated post. The stress concentrated at the corners of the post effectively produces a large number of tiny welds. Automatic programmed machines can wire-wrap large panels without human intervention.

Index